Who are the Native Americans? When and how did they colonize the New World? What proportion of the biological variation in contemporary Amerindian populations was 'made in America' and what was brought from Siberia? This book is a unique synthesis of the genetic, archeological and demographic evidence concerning the Native peoples of the Americas, using case studies from contemporary Amerindian and Siberian indigenous groups to unravel the mysteries. It culminates in an examination of the devastating collision between European and Native American cultures following Contact, and the legacy of increased incidence of chronic diseases that still accompanies the acculturation of Native peoples today. This compelling account will be required reading for all those interested in the anthropology of Native Americans, past, present and future.

The Origins of Native Americans

Evidence from anthropological genetics

The Origins of
Native Americans

Evidence from anthropological genetics

MICHAEL H. CRAWFORD

Department of Anthropology,
University of Kansas, U.S.A.

CAMBRIDGE
UNIVERSITY PRESS

CAMBRIDGE UNIVERSITY PRESS
Cambridge, New York, Melbourne, Madrid, Cape Town,
Singapore, São Paulo, Delhi, Tokyo, Mexico City

Cambridge University Press
32 Avenue of the Americas, New York NY 10013-2473, USA

www.cambridge.org
Information on this title: www.cambridge.org/9780521004107

First published in Spanish as Antropología Biología de los Indios Americanos,
Editorial MAPFRE, S.A., 1992; revised English version (by the author) 1998.

First paperback edition 2001

A catalogue record for this publication is available from the British Library.

ISBN 978-0-521-59280-2 Hardback
ISBN 978-0-521-00410-7 Paperback

to The native people who made the long journey from Siberia, *and to* their progeny who founded civilizations; *and to* Marshall T. Newman (1911–1994), a friend and colleague

Contents

Preface

Like the histories – both biological and cultural – of the native peoples to whom this book is dedicated, the making of this book is a story of expansion and confrontation, admixture and adaptation, though, fortunately, on a gentler plain: the plain of science. My interests in the human biology of New World populations were originally stimulated by my association with the late Marshall T. Newman at the University of Washington. Bud, as he was known to his friends, joined the anthropology faculty there in 1966, as I was completing my final year of graduate studies. Although I never received any formal instruction from him, I did sit in on a number of his seminars, and, over the course of that year, we discussed many aspects of Amerindian biology.

The idea for this particular volume came into existence more than twenty years ago, at a dinner at Bud's house in Seattle. Although we had shared many interests and ideas, Bud and I had never published anything together, and, that night, we decided that we should begin work on a volume concerning the biology of the native populations of the New World. Unfortunately, this project never came to fruition, pushed to the back burner by the protracted illness and eventual death of Bud's wife, Judy, and last year by Bud's death.

The next stage in the evolution of this volume occured six years ago, when I was approached by the MAPFRE Foundation to write a volume on the physical anthropology of American Indians. This impetus led to a volume entitled *Antropologia Biologica de los Indios Americanos*, which was released in October, 1992 at the World's Fair in Seville, Spain, as part of a book series commemorating Columbus' journeys to the Americas.

From 1992 this volume underwent a considerable update and metamorphosis as I refocused it from general biological anthropology of the native peoples of the Americas to their anthropological genetics. I decided to emphasize the subject matter that I knew best, and to personalize the text with examples primarily from my own field work. Thus, the Tlaxcaltecans of Mexico, the Eskimos of St. Lawrence Island and Alaska, the Black Caribs of St. Vincent and the Caribbean, and several founder populations from Siberia are given disproportionate coverage. On the other hand, the interrelationship of the questions that governed the original studies of these populations provides the 'glue' that holds this volume together; the use of case studies adds a degree of depth which would otherwise be lacking in a general survey of the literature.

Although there are many other fascinating areas of biological anthropology, such as osteology, nutrition, growth and development, and high-altitude adaptations, I have limited this book to the anthropological genetics of New World groups in the broadest sense. Chapter 1 focuses on the origins of New World populations and reviews the historical, archeological, morphological and genetic evidence for a Siberian migration. This chapter includes attempts to answer the following questions. Who were the founders? When did they come? How did they colonize two continents? The second chapter estimates the Amerindian population size at contact and examines the impact of disease on the subsequent depopulation. Examples are given from Tlaxcala, St. Lawrence Island and St. Vincent Island, each with its distinct pattern of depopulation and with differing severity of morbidity and mortality. Chapter 3 includes a discussion of the demography of the dead and the living populations. It contains a reconstruction of the demographic patterns prior to the arrival of Europeans into the New World and links these patterns to mortality, fertility and migration of living populations of Siberia, North, Middle, and South America. In Chapter 4 the breadth of the observed genetic variation and its evolutionary consequences are assessed by studying the blood-group, serum-protein, red-cell-protein and DNA polymorphisms. In the fifth chapter I apply some of the contemporary genetic models to the observed genetic variability in the New World. Population structure is examined based upon isolation by distance, spatial autocorrelation, synthetic gene maps and the fission–fusion model. The focus of the sixth chapter is the vast morphological diversity observed among the native peoples scattered from Alaska to the tip of South America. Their bodily morphology (anthropometrically assessed), skin color (based on reflectance), dermatoglyphics, and dental morphology are discussed. A seventh chapter, focuses upon the cost of survivorship, both medically and evolutionarily, to those who lived through the tragic depopulation of the Americas after contact. These effects include a series of chronic diseases, such as the so-called New World Syndrome, essential hypertension, coronary heart disease, and alcoholism, plus genetic hybridization on a massive scale.

This volume contains a greater emphasis on the Indians and Eskimos of North America, the indigenous groups of Central America and the founder Siberian groups. There is less emphasis on the South American populations because a current and well-written volume by Francisco Salzano and Sidia M. Callegari-Jacques is available to the reader. However, most of the publications on North American Indians are either strongly slanted towards osteology or are badly out of date.

Throughout this volume, I have referred to the native peoples of the Americas as Amerindians or American Indians. This terminology is not meant in disrespect nor does it carry any pejorative meaning. I have used Amerindian, Native American, and American Indian interchangeably as a means of identifying, for purposes of communication, a particular group, in much the same vein that I refer to Europeans, Africans, and Asians.

Despite the fact that I am used to referring to genes, genotypes, and phenotypes in the traditional ways, I have applied to this volume a standard notational system. An international system for gene nomenclature (ISGN, 1987) has been utilized throughout this book (Shows *et al.*, 1987). The exceptions occur in the discussions of the discoveries of some of these systems. The notational system that was appropriate for that particular time period is employed. The relatively new notational system minimizes some of the ambiguities associated with whether an author is referring to a genetic locus, genotype, phenotype or gene. The literature contains considerable notation sloppiness and the use of the new international standard notation should aid in distinguishing between the gene-product markers from the DNA haplotypes.

There are many people who have contributed to the successful completion of this volume. My wife Carolynn and son Kenneth have made this writing project possible by taking over many of my duties at home, thus freeing me to concentrate on the writing. During the past few years my research assistants have contributed to this volume by compiling bibliographic materials, dermatoglyphic counts and some analyses. I thank my graduate students, including Ravi Duggirala, Lisa Martin, Tony Comuzzie, Rector Arya and Joe McComb, for their assistance. Joe McComb scanned some of the figures into a computer and improved these rough-hewn illustrations. Several colleagues have kindly read various versions of this manuscript or have discussed and criticized some of my half-baked ideas with me. They include Tibor and Audry Koertvelyessy, Peter Nute and Moses Schanfield. I am particularly grateful to Peter Nute, who has spent enormous time trying to 'exorcise' from my prose Russian and German sentence structures. Although I thank my colleagues for their help, I alone am responsible for any oversights or inaccuracies in this volume.

M. H. Crawford

Acknowledgement

Supported by the MAPFRE America Foundation of Madrid, Spain, in commemoration of the five hundredth anniversary of Columbus' Contact with the New World.

1 Origins of New World populations

To the best of our knowledge, the aboriginal inhabitants of the Americas were initially contacted by Europeans in the eleventh century AD during Viking voyages of exploration. It is well documented historically that Norsemen had charted lands located to the west of Greenland, and that they referred to the landmass as Vinland. Apparently, the Vikings visited Vinland to collect lumber and established a settlement. Archeological evidence locates a Viking settlement at the L'Anse au Meadow site in Newfoundland. However, these first European settlers failed to permanently colonize this region of North America and either returned to Iceland or were massacred by hostile local inhabitants.

On 12 October 1492, Columbus, while searching for a route to the East Indies, set foot on the so-called New World (San Salvador, in what is now the Bahamas). He paraded the captive 'Indians' before the Spanish court and with the support of the Spanish Crown and Papal blessing brought Europeans to the New World in search of gold and spices. These European adventurers encountered a large native population that was distributed over a massive geographical expanse from the Arctic regions of North America through the Amazonian forests of Brazil to the bleak landscape of Tierra del Fuego, South America. These aboriginal populations practiced a wide variety of subsistence patterns, ranging from hunting and gathering, to fishing, to intensive agriculture. In some regions, such as the highlands of Mexico and Peru, their agricultural technology was highly sophisticated and sustained dense populations that developed civilizations rivaling those of Europe at that time. The origins of these 'mysterious' peoples of the New World perplexed the intelligentsia of Europe for many centuries.

1.2   EARLY THEORIES OF ORIGIN

Because Columbus failed to reach the East Indies, the natives that he encountered were obviously not East Indians. Who, then, were the inhabitants of the New World? Since the Bible 'explained' the creation and origin of all humanity, many of the scholars of the sixteenth and seventeenth centuries went scurrying to the Bible for an answer. For example, a Dutch theologian named Joannes Lumnius connected the ten Hebrew tribes with the New World inhabitants through the Bible (II Kings 17:6). The fourth book of Esdras described the meanderings of ten tribes of Israel who were driven from their homeland by the Assyrian King Shalmaneser in approximately 720 BC (Stewart, 1973). The Church of sixteenth century Europe accepted the New World inhabitants as the descendants of the lost tribes. In support of this biblically derived explanation, scholars and theologians of the time searched for cultural parallels between the contemporary Jews and the Amerindians (Stewart, 1973). This view of American Indians having descended from the so-called lost tribes, or 'remnant of the House of Israel,' reappeared in the nineteenth-century in the Book of Mormon and persists to this day in the Mormon religion.

The conquistadors of Mexico and their accompanying clergy returned to Europe with reports about the existence of massive pyramids in the Valley of Mexico at Tenochtitlan and Teotihuacan. The existence of such monumental stone constructions and the graphic descriptions of specific artistic motifs (such as the representation of the human body in profile) reminded the Europeans of the art and architecture of dynastic Egypt. Thus, during the sixteenth and seventeenth centuries, theories and speculations proliferated in Europe about possible migrations from Egypt to the Americas. One of the speculations fueling the Egyptian–Mexican connection concerned the founding of Atlantis (a massive continent in the Atlantic Ocean) by Phoenician seafarers. After the settlement of Atlantis, it was hypothesized that a volcanic catastrophe drove the inhabitants of the continent to the New World. Some kernels of this fanciful explanation could be traced to the descriptions by the Greek philosopher Plato, who wrote about the lost civilization of Atlantis. Vestiges of such a theory can be seen in Thor Heyerdahl's much-publicized voyages on replicas of ancient Egyptian papyrus reed boats to demonstrate the possibility of cultural contact between Egypt and the Americas. Jett (1978) summarized the evidence for pre-Columbian transoceanic contacts between the Old and New Worlds, and concluded: 'transoceanic influences must be seriously reckoned with in any consideration of these fundamental cultural developments' (p. 604). He further argued that this statement should not be misconstrued to imply that 'any New World culture "is" Egyptian, Chinese or Cambodian.' However, he added that the Old World may have contributed to the evolution of these New World cultures.

Since the eighteenth century, most scientists have been convinced that the first

Americans originated from Northeastern Asia (Stewart, 1973). George Louis Leclerc, Conte de Buffon, a noted naturalist, recognized the existence of a morphological resemblance between Amerindians and various Asian groups such as Tartars, Chinese and Japanese. He concluded in his writings of 1749 that these Asians and American Indians shared a common origin and that the New World was originally peopled by Asians (Buffon, 1749). Similarly, Johann F. Blumenbach, a naturalist of the eighteenth century, presented his earliest classification of human races (1775) in a volume entitled *De Genetica Humani Varietate Nativa*. He originally subdivided the human species into four varieties or races, Caucasoids, Mongoloids, American Indians and Africans, and concluded that the American Indians probably originated in Northeastern Asia as Mongoloids. He further hypothesized that several migrations occurred at different times and that the Eskimos more closely resembled the Asian Mongoloids because of their more recent migration into the New World.

During the 1940s and 1950s, it was suggested that the Amerindian founders were a mixture of two 'racial ingredients', namely Caucasoid and Mongoloid (Birdsell, 1951). Fresh from his highly acclaimed analysis of the origins of Australian Aborigines, Joseph Birdsell applied the same typological principles to the origins of American Indians. He argued that a proto-Caucasoid group, the Amurians (a people who originated from the Amur region of Siberia) gave rise to the contemporary Ainu of Hokkaido, Japan, and contributed to the Amerindian gene pool in the New World. The second racial 'ingredient' that contributed to the creation of the American Indians was the Mongoloid, who had evolved in response to cold stress in the subarctic regions of Siberia. However, this creative theory by Birdsell on the origins of Amerindian populations was 'slain by ugly fact'. Comparisons of traditional hematological and mitochondrial DNA markers and their frequencies among the Ainu, Europeans and other Asian populations reveal that genetically the Ainu most closely resemble the Japanese and fail to exhibit any of the European genetic markers (Omoto, 1972; Harihara *et al.*, 1988). Thus, despite their hirsute appearance, the Ainu do not appear to be descendants of some 'proto-Caucasoid' group but are Japanese. Most likely, they may be descendants of the early Jomon hunters and gatherers, who may have been the earliest inhabitants of Japan. This genetic evidence refutes Birdsell's theory on the multiple racial origins of the American Indian.

1.3 SCIENTIFIC EVIDENCE FOR AMERINDIAN ORIGINS

Since the initial European contact with the native peoples of the New World, a considerable body of scientific evidence has been compiled about the origins of these populations. This evidence indicates extremely strong biological and cultural affinities between New World and Asian populations and leaves no doubt that the first migrants into the Americas were Asians, possibly from Siberia. I will briefly review some of the relevant data that support the Asian origins of the Amerindians. This evidence does not preclude the possibility of some small-scale cultural contacts between specific Amerindian societies and Asian or Oceanic seafarers.

The evidence in support of an Asian origin of New World populations can be grouped roughly into four categories: (1) genetic similarities; (2) morphological resemblance in contemporary populations; (3) craniometric affinities; and (4) cultural similarities.

Genetic evidence

The Amerindians resemble Siberian and other Asian populations in the kinds and frequencies of various genetic markers of the blood. For a more complete description of genetic markers and their use in tracing populational affinities and origins, see Crawford (1973)[1]. Szathmary (1993) has recently summarized the genetic diversity of North American Indian populations based upon gene frequency distributions. Amerindian and northeastern Siberian populations have similar frequencies of many blood types, forms of serum proteins, red-cell enzymes, distributions of DNA variable numbers of tandem repeats (VNTRs), and haplotypes or haplogroups of mitochondrial DNA. When compared to other geographical populations of the world, on the basis of multivariate statistical analyses of gene frequencies, the Siberian or Asian populations tend to cluster together with those of the New World (see Fig. 1).

Cavalli-Sforza et al. (1988) used an average linkage analysis of Nei's genetic distances to construct a genetic tree based upon 120 alleles from 42 world populations. A bootstrap method (a resampling technique for obtaining standard errors) was utilized to test the reproducibility of the sequence of the splits in the phylogenetic tree (dendrogram). This tree shows two main branches, the African and non-African. The North Eurasian branch divides into Europeans (Caucasians) and Northeast Asians, including the Amerindians. Thus, this multivariate approach to population affinities reveals a close genetic relationship between Amerindian and Asian groups (see Fig. 1).

Some genetic markers occur only in New World and Asian populations. These include the following: the Diego allele, DI*A; gamma globulin allotypes, GM*A T; Factor 13B*3; transferrin, TF*C4; and complement, C6*B2 alleles. Szathmary (1993) adds SGOT*2 (glutamic oxaloacetic transaminase), TF*D, GC*TK1 (GC 1A9) and GC*N (GC 1A3), to a list of markers that indicate an Asian connection. Although the Diego DI*A gene is not always present in all Amerindian groups, when it is

---

[1] The term genetic marker will be restricted here to discrete, segregating, genetic traits that characterize populations by virtue of their presence, absence or high frequency in some populations in contrast to others (Crawford, 1973).

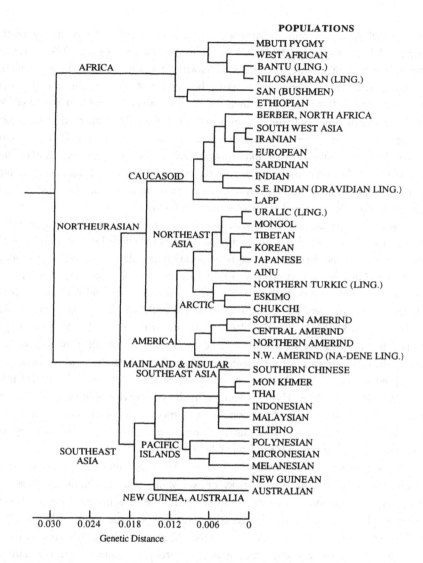

FIGURE 1 A genetic tree based on 120 alleles from 42 world populations. Average linkage analysis of Nei's genetic distance and a bootstrap resampling procedure was used by Cavalli-Sforza et al. (1988) to test the reproducibility of the sequence splits in this dendrogram.

detected DI*A occurs only in American Indians or Asians. The frequency of the immunoglobulin haplotype GM*A T in Asian populations reaches 50% in central Mongolia but is at a lower frequency in North American Indian groups. Similarly, GM*A G is found at frequencies varying between 86% in the Chukchi of Siberia (Schanfield et al., 1990) to 56% among the Ainu of Japan (Matsumoto and Miyazaki, 1972). In North American Amerindian populations, this GM marker varies from 98% among the Northern Cree to 47% in a mixed Alaskan group (Schanfield et al., 1990). Less is known about the geographic distribution of the complement B2 allele and Factor 13B*3; however, preliminary analysis suggests that these alleles occur at high frequencies in both the Amerindian and Asian groups (a more complete discussion of genetic markers is contained in chapter 4 of this volume).

In many of the other genetic systems, e.g. the human leukocyte antigen (HLA) system, the various blood groups, and even the mitochondrial DNA (mt-DNA) Asian haplotypes, most of the forms occur in some other populations of the world, but at different frequencies. Amerindians share the four major haplogroups (A–D) with Asian populations (Torroni and Wallace, 1995). In addition, Siberian and Amerindian populations share two identical mitochondrial DNA (mtDNA) haplotypes, namely S26 (AM43) and S13 (AM88). The S and AM designations represent the same haplotypes, defined by the presence or absence of the specific restriction sites, in Siberian and Amerindian populations. From these two haplotypes, Torroni et al. (1993a) have attempted to reconstruct the time of divergence of the Asian and New World mitochondrial DNA variation. These differences in the frequencies of some of the genetic markers are not surprising: the contemporary Amerindian populations are the result of small founding groups, unique historical events and possibly the action of natural selection over a span of 15 000 to possibly 40 000 years.

New World and Asian populations both exhibit a high incidence of dry or brittle earwax (cerumen) instead of the sticky or wet variety that is commonly present in most other populations of the world. Apparently, the presence or absence of a dry cerumen phenotype is under the genetic control of a single locus with two alleles (Matsunaga, 1962). This form of cerumen has been linked with an increased risk of otitis media (middle ear infection), which is particularly common throughout the indigenous populations of the Arctic on both sides of the Bering Strait (Pawson and Milan, 1974). McCullough and Giles (1970) demonstrated the existence of a statistical association between mean midwinter dew point temperature and the prevalence of wet-type cerumen. They have proposed a possible selective advantage for the high prevalence of the wet cerumen type in hot climates, and suggest that health complications are associated with the cerumen polymorphism.

During field investigations in Alaska and Siberia, I observed numerous cases of middle ear infections and was informed by the Public Health medical officers that otitis media was one of the leading diseases in that region. According to the Indian Health Services (IHS) report in 1989, the most common diagnosis for outpatient visits by males to IHS clinics or hospitals in 1988 was otitis media, with a total of 109 124 cases reported in the United States. Thus, it appears that both Amerindian and Siberian populations share a genetic predisposition to otitis media.

Morphological resemblances

Asians and Amerindians have a number of superficial morphological characteristics in common. These include: straight black hair; relatively glabrous, i.e. sparse, beards and bodily hair (with the exception of the Ainu of Hokkaido, who often exhibit beards); Mongoloid sacral spot; small brow ridges; and broad zygomatic arches, giving a high-cheek-bone appearance. The face is relatively flat, and some American Indians and most Asians exhibit an eye fold that covers at least the inner corner (canthus) and often the whole free margin of the upper eyelid (see Fig. 2). These epicanthic folds give the eye a distinctive appearance. There is considerable individual variation in all of these traits in both Amerindian and Siberian populations (see chapter 6 for a discussion of morphological variation).

The dental evidence strongly supports an Asian origin of the Amerindians and Eskimos. The shovel shape of the incisors of Northeast Asians and the New World populations occurs in 50–100% of subjects, in contrast to European and Asian groups where this dental trait is extremely rare (Turner, 1987). The incidence of shovel-shaped incisors varies between 0–10% in Africa and Europe and 20–40% in southern Asia. In addition, Turner has described the presence of three-rooted lower first molars in Asian, Eskimo and Amerindian groups at frequencies from 6% to 41% (Turner, 1971). The highest incidence of three-rooted molars is among Aleut–Eskimo samples (27–41%); the lowest is in some of the North and South Amerindians (6–11%). Thus, the dental evidence also strongly supports the hypothesis of an Asian origin of Amerindians (see Chapter 6).

Craniometric affinities

In multivariate comparisons of measurements of samples of crania collected from 26 or 28 diffferent geographic sites, Howells (1989) noted close genetic affinities between Siberian, Asian and New World populations. He made a total of 57 measurements from each of the cranial samples (males and females were analyzed separately) representing human variation in the world prior to the European expansion of 1492. The New World was represented by samples from the Arikara (South Dakota), Santa Cruz Island (California) and Yauos District (Peru). Siberia was represented by only the

(a)

(b)

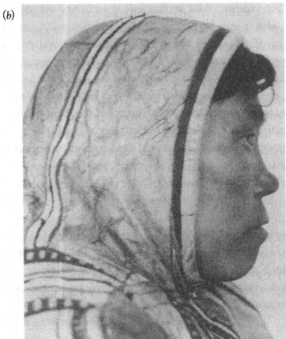

FIGURE 2 A male and female Nganasan, one shown in profile and the other full-face, to illustrate some of the Asian facial features that are also characteristic of New World populations (provided by Dr. R.I. Sukernik).

Buriats. Samples from Hokkaido, northern Kyushu, and Hainan Island provided Asian samples. The Buriat females clustered with the Arikara, and with other Asian cranial samples (see Fig. 3). These results suggest that some Amerindians craniometrically resemble Asian populations and support the Siberian origin of New World populations.

The morphological similarities between Siberian and Amerindian crania are not limited to craniometric studies. Recently, Hajime Ishida (1993) described populational affinities based upon the incidence of 22 non-metric cranial traits, e.g. supraorbital foramen, metopism, and ossicle at the lambda. Principal coordinate and cluster analyses revealed that the indigenous Amerindians and Arctic populations loosely cluster together. This study shows a morphological continuity between Asian and Amerindian populations, with Eskimos and Aleuts in an intermediate position.

Cultural similarities

The Siberians who crossed Beringia were most likely hunters and gatherers, who brought with them the rich cultural traditions of Asia. However, as the descendants of these founding groups spread geographically, their cultures and their biologies had to respond to the new environmental conditions encountered. Obviously, descendants of Arctic hunters do not continue to wear furs when their new place of residence is the tropical forest of Amazonia! Thus, during the 30 000 or more possible years in the New World each population had to adapt to specific environmental conditions but within the constraints of their biological and cultural attributes. A few cultural traditions or practices from Siberia persisted in Amerindian groups up to European contact times; however, most of the culture of the New World peoples bears a 'made in the Americas' label. The following similarities between the cultures of Siberian native populations and those of the New World may have been brought to the Americas and persisted until recently. I am not implying that cultural expressions cannot be reinvented *in situ*, nor that all cultural similarities require a diffusionist explanation.

The majority of both northeastern Siberian and New World societies exhibit the following cultural phenomena (Underhill, 1957). (1) There is a belief in spirits and a need to occasionally placate them with prescribed rituals. Spirits of animals, plants and natural phenomena were an integral part of the belief systems of these cultures. (2) Shamans or medicine men occured widely in the New World and in Siberia. These shamans obtained their visionary powers from direct contact with the supernatural and their chief function in these societies was curing patients of disease. (3) The Plains of the midwest United States and the Taiga of Siberia both exhibited dwellings that were of similar construction and materials. I was particularly struck by the similarities between the 'chuums' of the Evenk reindeer herders of Siberia and the 'tepees' of the Plains Indians (see Fig. 4). (4) Calendar sticks were utilized by peoples on both sides of the Bering Strait. According to Alexander Marshack (a noted archeologist/paleontologist) the calendar sticks of American Indians represent an ancient system of astronomical observation. He claims that the Amerindians brought this tradition from Asia and that it eventually led to the development of devices such as the solstice markers. It is believed that, like the Amerindians, the Neolithic tribes of Siberia tracked the lunar month and named these months after various activities.

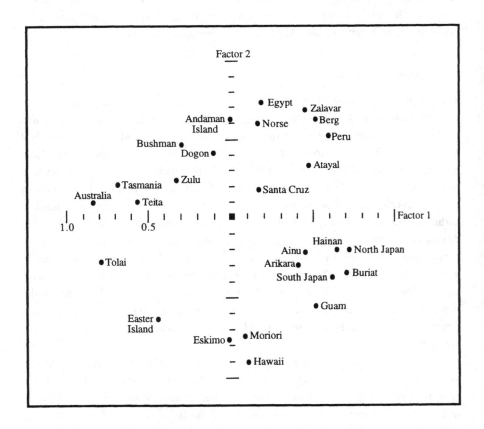

FIGURE 3 A multivariate comparison based on 57 measurements of female crania from 26 geographic sites (Howells, 1989).

This list of cultural similarities is not meant to be exhaustive, but provides a few examples of possible cultural connections spanning the Bering Strait. When all of the

FIGURE 4 An Evenki family posed in front of a summer 'chuum' located in the taiga of Central Siberia. Note the resemblance of this tent to an Amerindian 'tepee'.

cultural, genetic and morphological similarities between New World and Siberian peoples are considered, the only reasonable and the most parsimonious conclusion is that Amerindians and Eskimos are descendants of Asian hunters who crossed the Bering Strait. These first migrants came into the New World either by walking across the land bridge, or by paddling across on a boat or raft, or by trekking across the frozen Bering Sea. Since we now know 'who' the Amerindians are, and 'where' they came from, the next puzzles to be explored are 'when' and 'how many' of them came.

1.4   ECOLOGICAL BACKGROUND

The ecology of northeastern Siberia and northwestern Alaska during different periods of the Pleistocene provides clues as to the most likely windows of opportunity for the peopling of the New World. These so-called windows were defined by the major glaciations that interrupted the normal cycle of precipitation and return of water to the sea (see Fig. 5). During cycles of glacial advance, much of the Earth's water was

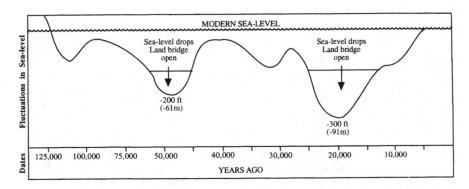

FIGURE 5 An illustration of sea-water fluctuations during the Wisconsin glacial activities (after Hopkins, 1967, 1979).

stored in ice, thus lowering the sea level by as much as 100 m (Hopkins, 1979). During the maxima of the last glaciation (the Wisconsin) which occurred from 70 000 to 12 000 years ago, a land bridge appeared across the Bering Strait, linking Siberia and Alaska (Hopkins, 1979). This land bridge, referred to as Beringia by geologists, almost 1500 kilometers wide was easily accessible to human migration from 70 000 to 45 000 years ago and again 25 000 to 14 000 years ago. These periods of opportunity do not imply that the New World was entirely sealed off from Siberia during the less frigid periods and that the movement of people was not possible. In fact the Bering Strait may not ever have been an effective barrier to human migration. During my field research in Wales, Alaska, it was obvious to me that on clear days Siberia was visible both from the Alaskan side and from the Diomede Islands (located in the Bering Strait equidistant between Siberia and Alaska). Caribou and fox move freely from Siberia and Alaska to the Diomedes. During particularly severe winters the sea between Alaska and Siberia freezes and the ice is sufficiently firm to sustain the weight of a human or a dog sled. Despite the most challenging political barriers and boundaries separating the USA and former USSR, there was evidence (in the form of Siberian brides on St. Lawrence Island) of mate exchanges between the Eskimos on both sides of the Bering Strait accomplished by contacts across the frozen sea.

Two giant ice sheets, acting in concert during periods of maximum glaciation, may have served as barriers to human migration (see Fig. 6). These two glaciers were: the western Cordilleran sheet, which was centered in the mountains of British Columbia (Canada) and stretched from the Columbia River Valley to the Aleutian Islands; and the Laurentian sheet, which covered approximately five million square miles (or thirteen million square kilometers) during glacial maxima, and extended northward from the Ohio River Valley to the Arctic Ocean, eastward to the Atlantic Ocean, and westward

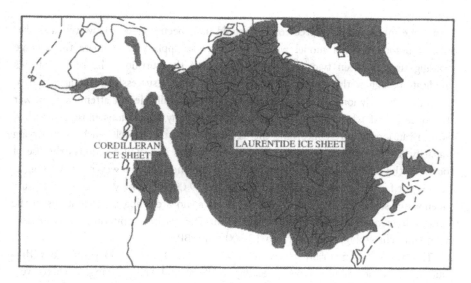

FIGURE 6 The extent of the two glaciers, Cordilleran and Laurentide, during the Wisconsin glacial maxima.

to the eastern foothills of the Rockies of central and northern Alberta. Both of these gigantic ice sheets measured approximately 2500 m in thickness and at their maxima may have fused, thus blocking an ice-free corridor that sometimes linked Alaska to the Central Plains. During these periods of fusion it would have been impossible for the first Americans to move from Alaska to the Central Plains through an interior route. Most likely the two glaciers were confluent during the maxima of the Wisconsin glaciation 18 000–15 000 years BP and 65 000–62 000 years BP. It is uncertain whether humans actually traversed the narrow ice-free strip between the glaciers. Fladmark (1979) suggests that this strip was of insufficient size to maintain the herds of game needed to sustain the Siberian hunters during their trek from Alaska to the central Plains. An alternate route along the coast of northwest North America would be a possiblity because of the presence of game in the coastal, ice-free refugia. Rogers *et al.* (1991, 1992) and Pielou (1991) have argued in favor of human dispersal via coastal routes. Aoki (1993) has attempted to determine, by using a deterministic reaction–diffusion model, whether the dispersal of Siberians through the ice-free corridor was possible. A diffusion coefficient of 3.01 square kilometers per year was utilized. The environment in the ice-free corridor was considered to be severe, but not to the degree envisaged by Pielou (1991: 157) who concluded that 'food, clothing and firewood would have been impossible to come by . . . ' in the corridor. On the basis of his deterministic model, Aoki (1993: 83) concluded that given the demographic param-

eters 'the inhospitable ice free corridor could have been crossed.' However, a proba-
bilistic branching-chain model with migration was applied to estimate the average
waiting time between two successful crossings of the corridor. The results of this
application suggest that the average waiting time until a successful crossing may have
been prohibitively long. Although these are mathematically elegant attempts to answer
the question of which route was the more likely, any conclusions must be reached
with a cautionary note concerning the assumptions of the models, such as a carrying
capacity of 500 persons per tribe. The coastal refugia theory also requires the use of
boats by the Siberian hunters for movement along the coast. However, such a suggestion is not far-fetched: the peopling of Australia, 39 000 or more years ago, did require
open-water sailing by the Asian settlers. To date, no archeological excavations in the
refugia have revealed early human habitation. The earliest habitation site in Alaska
with solid chronology dates back to 12 000 years BP.

The glaciers dramatically receded between 12 000 and 9 000 years ago with a
number of concomitant climatic, geological, floral and faunal changes. There were
shifts in the post-Wisconsin flora and fauna with a large number of species disappearing
from the fossil record. Approximately 90% of all large mammals (70 species) became
extinct. This prompted Paul Martin (1973) to postulate an 'overkill hypothesis,' arguing
that the Amerindians bearing fluted projectile points (termed Clovis after the site)
were in part responsible for the widespread extinction of large mammals. He posited
that the arrival of extremely efficient hunters from Siberia resulted in the overhunting
and extermination of Pleistocene mammals. Many archeologists and biologists criticize
this hypothesis because it ignores the major environmental modifications, such as the
warming trends associated with the termination of the Pleistocene. Such major changes
in the climate might have modified the flora, then the fauna, of the postglacial New
World and eventually caused the extinction of the large mammals.

1.5   SIBERIAN CHRONOLOGY AND NEW WORLD PEOPLING

Another important piece of the puzzle in the peopling of the New World is the
chronology of human entry into Siberia. Obviously, if Siberia was uninhabited until
recently then a chronologically early human expansion into the New World was not
possible. However, excavations by Mochanov (1978a), a Russian archeologist, suggest
that humans were present in northeastern Siberia by 35 000 years BP. He describes
a Siberian culture (the Dyuktai culture) with a subsistence pattern of hunting mam-
moths, woolly rhinoceros, bison and other Pleistocene fauna (Michael, 1984). To date,
Mochanov has reported a large number of these radiocarbon-dated Dyuktai sites
scattered widely throughout northeastern Siberia. He has also uncovered various stone

implements near the Lena River that are purported to be in a excess of 1 million years old. Such an early entry into Siberia, probably from China, requires that *Homo erectus* move into the Arctic at a very early time. In addition, if the mitochondrial DNA evidence is considered and 'Eve' accepted, then the earliest Siberians would have been replaced at a later date by modern *Homo sapiens*. Most scientists are extremely skeptical both of the dates and of assertions that these so-called tools were in fact manufactured by hominids (Morell, 1994). Pebbles that are fractured by natural geological processes sometimes appear to be anthropogenic. In addition, there is some question as to whether early people had the technology to successfully deal with the severity of an Arctic environment.

If humans were already present in Siberia from 35 000 to 40 000 years BP, then it was possible for them to demically expand into the New World earlier than 12 000 or 13 000 years ago. What form did this movement into the New World take? Several archeologists have suggested that small groups of Siberians followed herds of Pleistocene fauna into the New World and then (almost as an afterthought) remained there. Are there any clues from contemporary or historical colonizing populations to the dynamics of peopling an uninhabited continent?

One lesson from previous studies of hunting and gathering groups is that such societies usually have circumscribed territories of sufficient area to permit the harvesting of nutrients necessary for survival. Based upon the Australian Aborigine model, the size of the territory is a function of the abundance of flora and fauna (Birdsell, 1941). Similarly, the sizes of early Siberian Arctic hunters' territories were most likely a function of the availability of Pleistocene flora, fauna and the extractive efficiency of the group. Judging from the Siberian archeological evidence, there was a geographic expansion of human groups into northeastern Siberia approximately 40 000–35 000 years ago. Most likely, this expansion was made possible by technological improvements in hunting implements and by cultural advances in coping with the severe cold. With the presence of a land bridge 1500 km wide, it is likely that the Siberian hunting groups moved into the New World during one of the windows of opportunity, as defined by the lowering of the sea level and the possible existence of an interior ice-free corridor (see Fig. 6).

Demographic theory provides some suggestions about the dynamics of expanding or colonizing populations. Given a 'virgin' land expanse containing flora and fauna that had previously not been harvested, the food supply should have been effectively unlimited. This abundance of food within the territory of the population should have reduced the mortality due to periodic famines while supporting previous levels of fertility. When compared with sedentary agriculturalists, these groups should have had low mortality due to infectious diseases because the population's periodic relocations within their territories prevented extensive contamination of water or food supplies.

Under these conditions, colonizing populations can increase in size dramatically; as hunting units become too large for their territory, they undergo fission. The offshoot of the expanding population relocates itself within a new territory. Thus, these hunting groups expanded geographically into the New World, oblivious to the fact that they had crossed the Bering land bridge. A demographic model of population expansion into the New World, without the necessity of having to follow herds anywhere, is a more reasonable approximation of the process of human demic expansion. According to Cavalli-Sforza et al. (1993) the rate of expansion is a function of the growth and migration rates of the population. Radial rates of expansion of more than 1 km per year have been proposed (Cavalli-Sforza et al., 1993). Given such expansion rates, the Americas could have been peopled in a few thousand years.

Similar dynamics could be seen in Black Carib fishing–horticultural populations as they numerically and geographically expanded from a core of fewer than 2000 persons residing in a single village to approximately 100 000 people living in 54 villages, and all in fewer than 190 years (Crawford, 1983). When did the Siberian hunters expand into the New World? Did this expansion occur entirely between 13 000 and 12 000 years ago or could some of the groups have moved into the New World earlier, e.g. 30 000–40 000 years BP? What is the cultural and skeletal evidence for human habitation in the New World prior to 12 000 years BP?

1.6   ARCHEOLOGICAL EVIDENCE

There are several 'schools of archeological thought' as to when humans first entered the New World. One school, which at this time appears to be the vocal majority, interprets the archeological evidence to indicate a late peopling of the Americas, from approximately 11 000 to 14 000 years ago (Meltzer, 1989; Haynes, 1969). Other scientists argue that evidence exists for entry into the New World 30 000–40 000 years BP (Guidon and Delibrias, 1985; Bryan et al., 1978; Adovasio and Carlisle, 1986).

A few researchers have proposed a peopling of the Americas 50 000–100 000 years (Krieger, 1964; Dillehay and Collins, 1988; Leakey et al., 1968). Most of the evidence cited by these three groups is cultural and revolves around the existence of a pre-Clovis point culture (Krieger, 1964; MacNeish, 1976). This reliance on cultural artifacts and their dates is in part necessitated by the paucity of excavated skeletal materials dated prior to 12 000 years ago. Figure 7 locates many of the major archeological sites that have been uncovered in the New World.

Until the 1920s, most archeologists and physical anthropologists concluded that the material culture and skeletal evidence indicated that humans arrived in the New

FIGURE 7 The distribution of the major archeological sites of the New World (after Shutler, 1983).

World no earlier than 5 000 years BP. Most vocal among these scientists were Hrdlička (1917) and Holmes (1925). However, in the mid-1920s several sites in the southwestern region of the United States revealed an association between projectile points and the skeletal remains of extinct Pleistocene mammals. Most notable of these sites was at Folsom, New Mexico, where a Paleoindian fluted projectile point embedded between the ribs of an extinct Pleistocene bison was discovered in 1926. This Folsom discovery pushed the date of human arrival in the New World back to nearly 12 000 years BP (Hrdlička, 1937). The development of radiocarbon dating further confirmed the Pleistocene origins of early Americans and provided dates for the few Paleoindian remains at 10 000–11 000 years BP. The archeological evidence was used to reconstruct the big-game hunter subsistence pattern of the Paleoindians, who hunted Pleistocene fauna (such as giant long-horned bison and mammoth). Until recently no dated Paleoindian materials in Beringia had been found. Kunz and Reanier (1994) described the discovery of Paleoindian fluted points in a region adjacent to the Bering Strait (Mesa site) dating to 10 000–12 000 years BP. This research indicated that a 1300-year cold period in this region of Alaska (11 640 ± 250 and 12 940 ± 250 years BP) was coupled with a concomitant interruption in the Paleoindian habitation. Kunz and Reanier interpret these data to suggest that the region became uninhabitable because of the severe cold and that this climatic change stimulated the movement of human groups towards the south. This interpretation provides an explanation for the movement of Paleoindian populations to less severe climes.

A very early peopling of the New World, from 50 000 to possibly 150 000 years ago, has been proposed by a few paleontologists and archeologists. Such theories imply pre-*sapiens* hominids moving into the Americas. Although there is not a single shred of reliable evidence in support of this extremely early peopling, *Homo erectus* remains have been found as early as 500 000 years BP in the Choukoutien (Zhoukoutian) site of northern China. In addition, Mochanov (1978b) claims the existence of a site in northeastern Siberia of more than 1 million years' antiquity. The Calico Hills, California, site excavated by Simpson and Leakey in 1964 was judged to harbor stone artifacts dated to 200 000 years ago (Leakey *et al.*, 1972). Geologists and archeologists have discredited this site and concluded that the so-called simple choppers, found on an alluvial soil where the ground is unsettled, are probably 'naturefacts', i.e. stone fractured by geological processes. Another site that has been used to bolster the early peopling hypothesis is the Old Crow Basin site of the Yukon Territory. A few bone tools were recovered and were originally dated to 30 000 years of age. However, recent dates based upon an accelerated radiocarbon method suggest a new age of only 1300 years. Based upon amino-acid racemization, two California sites, Del Mar and Sunnyvale, were originally assigned dates of 48 000 and 70 000 years BP (Bada *et al.*, 1974). However, Bischoff and Rosenbauer (1981) have redated these sites by utilizing uranium series analyses and have concluded that the skeletal materials are of much lesser antiquity, 11 000 and 8300 years BP, respectively. MacNeish (1976) subdivided the probable existence of *Homo sapiens* in the New World into four stages. His earliest stage represents Siberians who crossed the land bridge 70 000–30 000 years ago and were possibly scavengers of big game rather than hunters. He provides meager evidence in support of this early migration, but argues that early Americans can be characterized by a tool tradition of hand-held choppers and unspecialized bifaces.

A few archeologists continue to adhere to the interpretation of the skeletal and cultural artifactual evidence that the New World was peopled no later than 12 000 years BP (Marshall, 1990). They base these conclusions on the fact that to date no pre-12 000 BP human skeletal remains have been discovered and that the studies of sites purported to be earlier contain flaws associated with either the excavation techniques or the dating methods. Roger Lewin (1987), in a review article in *Science*, posed the question 'what does it take to show that a site is really old?' The criteria attributed to archeologist Donald Grayson (1987) by Lewin include the following: (1) the site must yield artifacts or human bones that are not in dispute; (2) the human material must be found in undisturbed strata; (3) dates must be secure beyond a reasonable doubt; (4) the excavation and publication of the data found at the site should permit scholars to evaluate the evidence and to determine whether the criteria have been met. Given these criteria, Lewin argues, the Clovis sites qualify as old and the Meadowcroft Rock Shelter site comes close to qualifying. Of the plethora of sites discovered in North and South America during the past decade, do any of them fulfill the listed criteria and qualify as legitimate pre-12 000 BP sites?

Out of the more than 150 sites shown in Fig. 7, at least three and possibly four sites appear to be promising candidates for providing evidence of early (pre-12 000 BP) habitation by Amerindians in the New World. These sites are Meadowcroft Rock Shelter, Monte Verde Site, Boqueirao da Pedra Furada Site, and Caverna da Pedra Pintada.

Meadowcroft Rock Shelter

This site was excavated by James Adovasio and a large team of specialists during the 1970s (Adovasio and Carlisle, 1986). This campsite in Pennsylvania, located less than 50 km south of the Laurentide ice sheet, stands beneath a standstone overhang 50 ft (15 m) long and up to 20 ft (6 m) deep. The charcoal from the fire pits was dated at approximately 15 120 years BP. The oldest directly dated (^{14}C) artifact from this site, a remnant of a carbonized piece of basket or mat, appears to be 19 600 ± 2400 years. The deepest stratum containing indisputable traces of human occupation has been termed IIa. Some skeptics have argued that the charcoal may have been contaminated by groundwater; however, this is a well-excavated and well-described site with eight radiocarbon dates ranging between 19 600 and 12 800 years BP. In addition, Adovasio states that contamination is unlikely because it would have disrupted the near-perfect sequence of dates from older to younger in the strata from the lower to the upper (Lewin, 1987).

Monte Verde site, Chile

This site, once a peat bog, was excavated by T. Dillehay from 1976 to the present (Dillehay, 1987). He unearthed two open settlements along the tributaries of Rio Maullin: the first is on Chinchihuapi Creek, the second further downstream. The first site revealed the remains of wooden structures (12 dwellings joined together to form two rows), a human footprint, and some mastodon flesh, all in stratigraphic order and dated to about 13 000 years BP. The second site contained mastodon bones (from at least five or six individuals), fragments of animal skins, and what appear to be a foundation, stone tools, and worked wood. Dillehay interprets this site as a place where the inhabitants prepared meat and hides and manufactured stone tools. The radiocarbon dates for the charcoal from the three apparent hearths are close to 34 000 years BP. The evidence from the Monte Verde site is suggestive that humans entered the New World prior to the appearance of the big-game hunters with their fluted stone projectile points: i.e. more than 12 000 years ago. The preservation of the wooden artifacts, the remnants of wooden structures and the organic debris from a 13 000-year-old cultural period allows the reconstruction of the subsistence patterns of these early Americans. Theirs was not entirely a big-game-hunting existence, but one of mixed hunting and gathering.

Boqueirao da Pedra Furada Site

This Brazilian site, on the remote northeast plateau, was first excavated by a French archeologist, N. Guidon (Guidon and Delibrias, 1985). This prehistoric rock shelter consists of five distinctive sedimentary layers, each containing artifacts and presumed hearths. The upper layers have been dated at approximately 8000 years BP, whereras a hearth at the lowest stratum was dated at 32 000 years BP (Guidon and Delibrias, 1986). Guidon claims human habitation by people using quartz and quartzite pebble cores and flakes, with limited retouching. There has been some suspicion that these stones may have been chipped as a result of falling into the shelter, and do not represent human tool-making. This controversial site appears to contribute to the recently emerging pattern of the habitation of the earliest Amerindians in the New World by 40 000 years BP. This early date is based upon the premise that in order for the South American Indians to have reached Chile and Brazil by 33 000 years BP, the initial migration must have occurred 7000–8000 years earlier.

Caverna da Pedra Pintada

A Paleoindian campsite has recently been excavated in the Brazilian Amazon (Roosevelt et al., 1996). This painted, sandstone cave near Monte Alegre exhibited stratified

Paleoindian deposits that yielded 56 radiocarbon dates from carbonized plant remains ranging from 10 000 to 11 000 years BP. The more variable thermoluminescence and radioactivity dates from 13 lithics and sediment range from approximately 16 000 to 9500 years BP. The cave contained rock paintings of concentric circles, hand prints, and an inverted figure with rays emanating from the head (Roosevelt *et al.*, 1996). The discovery of triangular bifacial spear points and other stone implements demonstrated that this culture was distinct from the lithic culture observed in North American Paleoindian sites. What separates this site from the earlier excavations is that the Monte Alegre Paleoindians were river foragers instead of the specialized big-game hunters usually associated with North American Paleoindian cultures. Apparently these foragers of Caverna de Pedra Pintada had simpler tools, subsisted on plants, smaller game, and fish and began colonizing the Amazon forest during the Pleistocene. Their culture was apparently contemporaneous with the Clovis big-game hunters of the North American high plains.

There are a number of other sites that may eventually be shown to predate 12 000 years BP. However, various questions persist as to the chronology or the excavation techniques. There are three other candidate sites in South America that may qualify as pre-Clovis: Los Toldos Cave in Argentina, Tagua-Tagua in Chile, and Taima-Taima in Venezuela (Bryan *et al.*, 1978). In a possible fourth site, T.F. Lynch and Kennedy (1970) obtained radiocarbon dates (using accelerator mass spectrometry) of 12 560 years at Guitarrero Cave in Northern Peru. Such evidence of early human habitation in Peru is also suggestive of pre-Clovis populations in the New World. The dated stratum also contained a human mandible and teeth (Lynch and Kennedy, 1970).

Given all of these pre-Clovis archeological sites, spread geographically from Alaska to Argentina, why have no skeletal remains have been uncovered to date? One explanation that has been posited is that geologic conditions before 12 000 years BP were unfavorable for preserving human remains. However, osseous remains of other large mammalian species from the same time frame are plentiful. Undoubtedly, the population density of early Americans must have been low; however, tools (even wooden implements) have been located in many sites. Could the early settlers into the New World have practised above-ground burials, as is so often the case with Arctic populations?

1.7   THE NEW SYNTHESIS?

A few years ago, an apparent multidisciplinary consensus was reached on the chronology and the number of Siberian migrations entering the New World. In a series of articles by Greenberg, Zegura and Turner (Greenberg *et al.*, 1985; Turner, 1985; Zegura, 1985) it appeared that the genetic, dental and linguistic evidence were reconciled in favor of a late entry (no earlier than 12 000 years BP) by three population movements. According to Greenberg, the earliest Siberian migration is represented by the Amerind speakers; next came the group that gave rise to the Na-Dene speakers; and the final population movement resulted in the Aleut–Eskimos. The geographical distribution of the three language families is shown in Fig. 8. Greenberg's massive comparison of basic vocabulary and word similarities in American Indian groups (approximately 600 different native languages surviving to the present) and his linguistic classifications serve as the solder to this newly forged consensus. Greenberg (1987) concluded that American Indian languages can be classified into three linguistic groups and that each of these groups represents a separate migration into the Americas. These three linguistic groups are: (1) Amerind, which is the oldest, according to Greenberg, because it centers furthest south and shows greatest internal differentiation; (2) Na-Dene, a language spoken mainly in the northwest United States and in an enclave of the southwest; (3) Aleut–Eskimo, a linguistic group located in the northern peripheries of North America and most closely related to what Greenberg terms the Euroasiatic stock. Within this Euroasiatic stock, Greenberg includes the Indo-European, Uralic, Altaic, Ainu, Gilyak, Japanese and Korean languages. The dental evidence was offered as further support for the three-migration theory of the peopling of the New World. However, Turner fails to statistically test the relationship between a linguistic dendrogram based upon Greenberg's classification and his dental dendrogram. Instead he chooses simply to declare 'a remarkably good fit.' Turner's selective use of the dental evidence prompted Szathmary to remark:

> 'Turner's equating the label Na-Dene with the Greater Northwest
> Coast group suggests that he is not prepared to question, let alone
> reject, the three-migration hypothesis . . . Rather, he interprets his ana-
> lytical results in the light of a preexisting hypothesis that he simply
> assumes to be true.'
>
> (*Szathmary*, 1986a)

Zegura (1985) presents his genetically based interpretation in a more guarded fashion, stating that he views the genetic data as constituting secondary support for the primary inferences based upon linguistic and dental data. He also indicates that the genetic data can be and have been interpreted by various authors to be indicative of other patterns of migration and evidence for an earlier peopling of the New World.

At a conference in Colorado (reported by V. Morell, 1990) Greenberg's linguistic classification of the Americas received harsh criticism from other professional linguists in attendance. The primary bone of contention was the extreme lumping by Greenberg

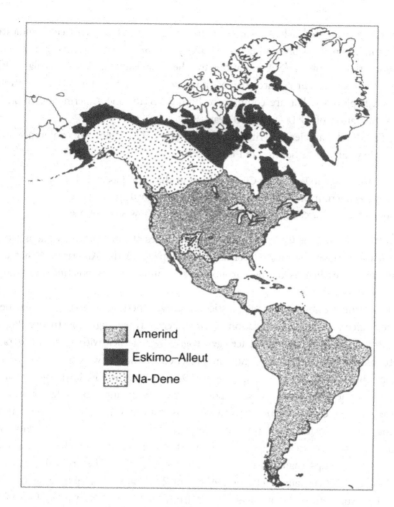

FIGURE 8 A map depicting the geographic distributions of the three
Amerindian language groups championed in the 'new synthesis'.

of the vast linguistic variation observed in American Indian languages into a single
Amerind language group. Most of the linguists concur with the classification of the
Na-Dene and Aleut–Eskimo families, but reject the lumping of 'everything left-over'
into a single language group. Without this tripartite linguistic subdivision, the so-called
consensus began to unravel. Recent mitochondrial DNA research by Pääbo and his
colleagues (1990) on Amerindians primarily from the Pacific Northwest have revealed
the existence of 30 major lineages, thus implying that additional migrations might be
necessary to explain the accumulated genetic variation. In this same conference,

Rebecca Cann presented a biological clock based upon mtDNA. This clock indicates an early migration, possibly dating to 40 000 years BP. There is some controversy surrounding the use of mtDNA mutations for the establishment of a molecular clock. However, the more recent research of Torroni et al. (1993b), based upon the derivation of mtDNA haplotypes that are common to both Siberians and Amerindians, requires an initial migration of 17 000–34 000 years BP.

Virginia Morell, a science writer, described the 1990 Colorado conference in a Research News article in Science and concluded:

> '. . . there was indeed a pre-Clovis culture in the Americas and that
> the Greenberg hypothesis, which not long ago seemed to offer the
> hope of an overarching unification, has begun to show signs of age.'

As the conclusions of Greenberg and his colleagues slowly unravel, what is left in regard to the various theories concerning the peopling of the Americas? What conclusions can be reached as to the chronology and number of population movements into the New World?

Some pertinent evidence from linguistics, ecology, genetics and archeology has been uncovered since the initial formulation of Greenberg's three-migration theory. Rogers and his colleagues have combined ice-age geography together with linguistic data in order to assess the likelihood of human presence in the New World prior to the retreat of the Wisconsin ice mass. Rogers (1985) has argued in favor of an earlier settlement of the New World based upon the paucity of linguistic differentiation in the areas that were covered by the glaciers. His reasoning is based on the concept that geographical isolation and time are the essential elements in the differentiation of languages. The vast areas of North America that were covered by glacial ice exhibit fewer language groups than those regions that were ice-free. The areas not glaciated by 12 000 years BP contained 93% or 199 out of 213 Native American languages (see Fig. 9). This suggests that languages spread from North American areas free of ice into the areas deglaciated after 12 000 years BP. Meltzer (1989) has criticized this conclusion by pointing out that some of the languages in the previously glaciated areas became extinct and that is the primary reason for disparities in the number of languages present. Rogers also argues that Eskimo–Aleut and Na-Dene language phyla originated in Beringia, whereas the Amerinds originated south of the Wisconsin ice mass.

A later article by Rogers, Martin and Nicklas (1990) argues that the Na-Dene language group developed while isolated in a coastal ice-free refugium. This conclusion was based upon the evidence that maximum diversity within the Na-Dene family is found along the Pacific coast of Alaska. This area was a coastal ice-free refugium and gave rise to the Eyak, Tlingit and Athapaskan languages. This linguistic, geologic, and

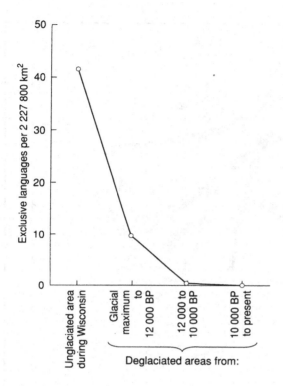

FIGURE 9 A plot of the percentage of exclusive Amerindian languages in unglaciated versus glaciated regions of North America (Rogers, 1985).

ecological evidence explains some of the observed linguistic variation and provides indirect support for a pre-12 000 year presence of Amerindians in the New World (see Fig. 10).

Additional evidence for the antiquity of human habitation in the New World comes from the Old Crow site, in northern Yukon territory, originally excavated by William Irving. In 1967 he uncovered two mammoth bones and a flesher that was dated to 27 000 years BP (Irving and Harrington, 1973). Recently, this flesher (often used as evidence in support of the early peopling of the Americas) was redated to about 1400 years by using the more advanced method of accelerator mass spectrometry of collagen. Yet this same technique yielded dates of 22 000–43 000 years BP for the human habitation site of Old Crow. These dates confirm the age and persistence of the Old Crow bone industry (Irving, 1987).

Additional genetic data have recently surfaced concerning the question of the number of population movements from Siberia into the New World (Bray, 1988).

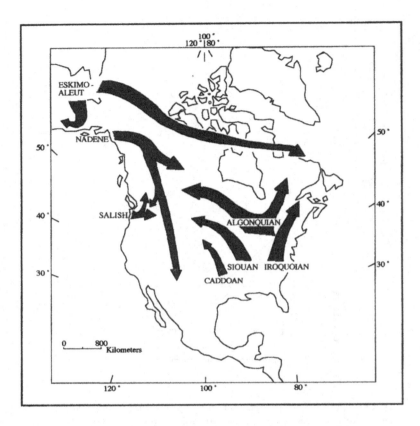

FIGURE 10 A diagrammatic representation of the origins of Na-Dene, Eskimo–Aleut, and other North American Indian languages (Rogers *et al.*, 1990).

The first data, the mitochondrial DNA work of Svante Pääbo, were mentioned earlier in this chapter. The existence of more than 30 major mtDNA lineages in the Amerindian groups tested to date has been utilized to argue either for more migrations or for larger migrating groups. Ward *et al.* (1991) after sequencing a 360-nucleotide segment of mtDNA among the Nuu-Chah- Nulth (Nootka) Indians, found a total of 28 molecular lineages within a single tribe. These data are consistent with my conclusion that the early Siberian hunters did not merely 'meander' into the New World in search of game but numerically and geographically expanded in a form of 'adaptive radiation' precipitated by the development of hunting technology and cultural methods of efficiently ameliorating the stress associated with severe cold. Given such a demic expansion into the Americas, a large number of 'migrants' would be expected.

The second bit of new genetic evidence comes from a publication on immunoglobulin variation in Siberian, Alaskan Eskimo and Amerindian populations. Schanfield *et al.* (1990) propose that the GM and KM markers support a four-migration pattern for the peopling of the New World. The earliest group, represented today by the South American Indians, had no GM*A T, only GM*A G and GM*X G. The second migrant group to enter was characterized by a high frequency of GM*A G and low frequencies of GM*X G and GM*A T. The Na-Dene speakers, who constituted the third migration, had a high frequency of GM*A G and moderate frequencies of GM*X G and GM*A T. The Aleut–Eskimo migrant gene pool probably contained GM*A G and GM*A T. This pattern of GM haplotypes of varying frequencies can be explained either by four distict migrations from Siberia or by fewer migrations but with the possible action of mutation, founder effect and/or selection. Given the paucity of evidence in support of natural selection operating at measurable levels on contemporary human populations, the more parsimonious explanation for the observed variation at the GM locus in Amerindian populations is one based upon separate migrations or numerical and geographic expansions of Siberian populations into the New World.

Recently, Szathmary (1993: 213) has criticized the use of immunoglobulin data on the basis that 'a dendrogram based on one locus produces only the history of that locus . . . and does not necessarily reflect the evolutionary history of populations.' She then goes on to construct a dendrogram using genetic distances based upon immunoglobulins. She notes that this dendrogram fails to show the three migrations into the New World purported by Williams *et al.* on the basis of GM haplotypes. The criticism involves the intermixing of Athapaskans and Eskimos with Siberians. Yet her solution to the problem (use of multiple loci) also resulted in dendrograms that distorted the ethnohistorically known relationships of the populations (Ferrell *et al.*, 1981). Granted, in a perfect evolutionary world, multiple loci should yield a more accurate representation of population affinities, but some loci are more informative than others. GMs have been extremely useful in studies of admixture because of the presence and absence of some haplotypes in parental populations. In addition, some of the information contained within a matrix is lost when it is used to construct evolutionary trees. Most topological methods provide much better graphic representations of the information contained in the genetic distance matrix.

1.8   SIBERIAN FOUNDERS

There has been dispute as to which Siberian groups have given rise to the New World populations. The Czech–American physical anthropologist Ales Hrdlička, on the basis of skull shape, linked the Neolithic specimens from eastern Siberia with some of the 'oblong-headed' tribes of the Americas. According to him, the southern Eneolithics

resemble the high-vaulted Algonkin type, the contemporary Chukchi relate to the
Bering Sea Eskimos and the round-headed Tungus were progenitors of the Aleuts
(Hrdlička, 1942). These typologically based conclusions ignore normal human variation,
as well as the concept of populations as the units of evolution, and have been disre-
garded by most scientists. Who were the carriers of the Dyuktai culture that most
likely expanded into the Americas 30 000 or more years ago? It is difficult to state with
great certainty because the present-day distribution of Siberian populations is vastly
different from the patterns encountered by early Russian explorers of the seventeenth
and eighteenth centuries. Ethnic groups such as the Yukaghirs were distributed widely
throughout northeastern Siberia and were one of the largest tribal groups of the seventeenth century. A combination of wars, severe epidemics of infectious disease and the
expansion of groups such as the Chukchi and Yakuts have brought this population to
the brink of extinction. Traces of Yukaghirs remain in the far northeastern section of
Chukotka, highly hybridized with the Chukchi. Crawford and Enciso (1982) were able
to detect the presence of Yukaghirs in a subdivision of a Chukchi population (see Fig.
11). Figure 11 provides a principal components genetic map based upon allelic frequencies in nine Reindeer Chukchi populations. Population 14, Rytkuchi, differs from the
other Chukchi groups because of admixture with the remnant Yukaghir elements. This
population also displays the closest affinity to the Samoyed populations. Did the
Yukaghirs give rise to the American Indians or did one of the ethnic groups from the
Kamchatka peninsula expand northward? The literature is replete with references to a
population from the Amur river region of Siberia that gave rise to Amerindians. This
place of origin was recently resurrected by Turner (1987), who bases his conclusions
upon a typological approach. Supposedly, the Mongoloids were being subjected to
severe cold and acquired their morphological traits while 'lounging' near the Amur
region, prior to their movement across Beringia and into the New World.

1.9   SETTLEMENT OF NORTH AMERICA

Most of the genetic, linguistic, and cultural data suggest an initial early radiation of
Asians into the Americas some 30 000–40 000 years ago. Minimally, three or four
migrations can explain the observed genetic variation when considered in combination
with the action of evolutionary processes and unique historical events. To date, much
has been written in the anthropological literature as to the 'exact' number of migrations
and the chronology of these movements. At present, it may not be possible to answer
these questions conclusively, particularly with the limited availability of the DNA data
representing the major Siberian indigenous groups. The analyses of mitochondrial DNA
from several Siberian groups by Torroni et al. (1993b) indicate that the Amerindians

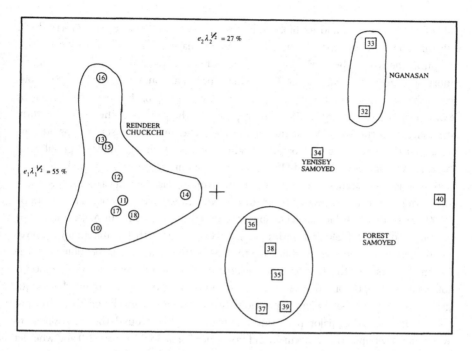

FIGURE 11 A principal-components analysis of allelic frequencies and a plot of the first versus second eigenvectors from Reindeer Chukchi, Samoyed, and Nganasan indigenous populations of Siberia. Note that the Reindeer Chukchi population no. 14 differs from the other subdivisions along the second eigenvector, showing some affinity toward the Samoyed cluster (Crawford and Enciso, 1982).

and Siberians share some of the same DNA markers and the distributions of the four mtDNA haplogroups are suggestive of four migrations or expansions (see chapter 4). However, Merriwether et al. (1995) believe that the mitochondrial DNA data are consistent with a single expansion into the Americas. There are major difficulties in attempting to reconstruct the genetics and the possible patterns of migration into the New World based upon contemporary populations, particularly in North America. Genetic admixture with European settlers, whalers in the Arctic regions, African slaves, and the possibly non-random decimation of the aboriginal populations by the introduction of Old World diseases may obscure the underlying genetic patterns.

The traditional interpretation of biological and archeological evidence suggests that the founders of Eskimo and Aleut populations constitute the last population movement into the New World and probably bore more extensive genetic affinities to Asian population than to the Amerindians. Hrdlička graphically characterized this late arrival of the Eskimo–Aleuts and its biological consequences in relation to the Amerindians. 'Picture the hand as representing all New World natives', he said, 'with their single origin in the wrist. The fingers represent the various kinds of Indians and the thumb, more separated, the Eskimo' (M.T. Newman, personal communication, 1968). However, this dogma about the separate Eskimo–Aleut grouping has been challenged by Szathmary and Ossenberg (1978) who propose that the peoples of the Bering Platform expanded into the New World during the Holocene and the coastal groups adapted to a maritime existence and became Eskimo–Aleuts, while the interior populations became Amerindians. Dumond (1987), after re-examining the archeological, linguistic and genetic data, argues that the remote ancestors of Na-Dene speakers moved from the Bering platform southward down to the Northwest coast approximately 10 000–8000 years BP. The Eskimo–Aleuts appeared about 4500 years BP in Alaska, according to Dumond. Most biological anthropologists would agree that a minimum of three population movements from Siberia into Alaska took place. It is possible that the Eskimo–Aleuts and the Na-Dene may represent a single expansion in the Holocene, followed by adaptation to diverse environments. However, based upon the unique genetic markers of the Eskimo populations that are absent from the northern Indians, that is unlikely. The major persisting argument revolves around the question as to when an early population movement did take place (circa 35 000 years BP) and whether all non-Na-Dene Amerindians constitute descendants from a single migration.

1.10   PEOPLING OF SOUTH AMERICA

Rothhammer and Silva (1989) have attempted to reconstruct the major migrational routes taken in the peopling of Andean South America. They utilized in their analysis archeological, craniometric, linguistic and genetic data. A total of 29 middle and late agricultural craniometric collections, 1239 skulls, were analyzed by principal component (PC) analysis. The first PC plot reveals a gradient of craniometric variation that extends from northwest to southeast South America. This first PC is probably due to a size factor. The gene-frequency-based genetic distances and, the craniometrically based isolines support Lathrop's (1970) claim (based on archeological reconstruction) that with the development of manioc cultivation the Proto-Arawakan and Tupian speakers (two of the linguistic groupings of Greenberg) expanded in separate directions. The Proto-Arawakans moved to the north and southwest, while the Proto-Tupi speakers expanded to the southeast. This population movement of approximately 5000 years BP was generated by population increases that followed the adoption of agriculture, and must have been preceded by the initial expansion of the hunters and gatherers who most probably came to South America more than 30 000 years BP. It has been suggested that the early Paleoindian colonists moved across the tropical forest along the major rivers and their tributaries. However, the recent excavations by Roosevelt *et al.* (1996) have revealed that the colonization of the Brazilian Amazon took place during the Pleistocene, from possibly 16 000 to 9000 years BP. These early foragers had a culture that was distinct from the Paleoindian big-game hunting tradition of North America.

There is evidence to suggest that once in the New World the Amerindians and Eskimo–Aleuts underwent genetic and morphological changes. Newman (1960a) presented data showing that some phenotypic changes took place over time. Stature increased in some Amerindians, and brachycephalization (the broadening of the cranium) and some clinal gradations in phenotypic characters could be observed in living Indians (see chapter 6 on the morphological variation). Similarly, the New World populations acquired unique genetic characteristics and frequencies most likely due to selection and/or unique historical events. For example, the South American Indians display a unique frequency distribution in the ABO blood group system with the O allele ocurring at 100% in many groups. There are numerous rare mutations that are found only in the New World, such as the group-specific component GC*ESK (found only in Eskimos) or Albumin Mexico variant, observed only in Central American groups (see chapter 4 on genetic variation). Thus, whatever the original biological ingredients, different environments in the Americas molded phenotypes differently (Newman, 1953). As Newman often told me, 'to a considerable extent American Indian morphology seems to be stamped "made in America".'

What follows in the next six chapters is a reconstruction of the New World population at European contact, the impact of this contact, and the demographic, genetic and morphologic characteristics of the contemporary populations. This characterization is interpreted on the bases of evolutionary theory and the often tragic unique historical events that have affected the New World peoples. The final chapter deals with the biological cost of surviving the conquest.

2 Population size and the effects of European contact

The Siberian populations that expanded into the New World some 30 000–40 000 years BP grew numerically, diffused geographically, and eventually reached Tierra del Fuego at the tip of South America. The total number of descendants of the original migrants is a function of the ecological factors in the diverse regions of the Americas and the extractive efficiency of the cultures of residents of these regions. For example, the population densities of hunting and gathering societies were much lower than those of the agriculturalists. As a result, the population densities must have differed in the various regions of the New World, with the highest densities occurring in the Highlands of Central America and Peru, where intensive agriculture could maintain populations of large size. In contrast,the Sonoran or Californian deserts could not have sustained populations of hunters and gatherers with densities greater than a few individuals per square kilometer.

How many Amerindians and Eskimos did inhabit the New World in 1492? Estimates of the size of the population have varied from Dobyns'(1966) 90–112.5 million to Kroeber's (1939) 8.4 million. Dobyns based his numerical estimates upon the reconstructions of population sizes by region when they were at their lowest and then corrected these figures by a 20 : 1 or 25 : 1 depopulation ratio. Kroeber, on the other hand, utilized a tribe-by-tribe 'dead reckoning' method (developed by Mooney in 1910) gradually building up totals for each geographical region.

2.2 METHODS OF ESTIMATION

Douglas Ubelaker (1988) has summarized the various approaches used to estimate the size of the aboriginal population of North America. These approaches can be grouped as follows: (1) ethnohistorical; (2) archeological; and (3) integrated, utilizing both kinds of data. The ethnohistorical approach is based on direct observations of explorers and settlers who made contact with the aboriginal groups. Thus, through successive descriptions of the total size of a tribe, the chronology, severity and mortality of an infectious epidemic can be ascertained. Similarly, early reports of the family size, political organization and social structure of native settlements have been used to reconstruct the demography of newly contacted Amerindian groups. However, these reports must be carefully scrutinized because they may include a number of biases (such as military commanders exaggerating body counts or priests providing inflated soul-conversion rates to their superiors), inadvertent miscounts (due to accidental omissions of settlements), or improper techniques of enumeration.

The archeological approaches reconstruct the settlement patterns, and the number and sizes of households, on the bases of excavated cultural artifact and skeletal remains. From these data it is possible to estimate the number of residents at a given time in a particular village, their form of subsistence and diet, and possibly their social organization. Osteological excavations provide some insights into the life expectancy, age–sex distribution, death rate and causes of death in specific sites (Ubelaker, 1974). However, extreme caution must be exercised in the interpretation of the archeologically based reconstructions because of possible sampling error. Burial practices, differential preservation of cultural and skeletal materials, and the accidental destruction of some sites may result in erroneous conclusions about demographic rates.

2.3   ESTIMATES OF REGIONAL POPULATION SIZE

North America

In 1910, James Mooney of the Bureau of American Ethnology, Smithsonian Institution, presented the first calculated estimates of the size of the aboriginal population of North America. After Mooney's death, John R. Swanton published Mooney's ethnohistorically based numerical estimates for the tribal areas north of Mexico and including Greenland. The overall estimate of a little more than one million pre-Contact Indians was made by a 'dead-reckoning' method of building up totals from tribe-by-tribe estimates. Table 1 provides a breakdown of the numbers of individuals living in geographically defined tribal areas.

Some of Mooney's estimates of tribal sizes were based upon exceptionally late reports (see Table 1) and may reflect populations already reduced in size by imported epidemic diseases and conflict due to the pressure of other tribes, themselves dislocated by White invaders. For example, Mooney used a 1780 total of 15 000 Blackfoot, and an 1845 figure of 4500 Western Shoshone. It would be impossible to be certain that neither estimate was affected by European contact. In the Northwest, Mooney estimated that the Makah numbered 2000 in 1780, but this was apparently based on counts provided by Lewis and Clark's expedition of 1805. It is also uncertain as to what proportion of Amerindians were simply left out of these estimates.

Table 1. *Mooney's population estimates for tribal areas north of Mexico*

Tribal area	Date	Estimate
North Atlantic States	1600	55 600
South Atlantic States	1600	52 200
Gulf States	1650	114 400
Central States	1650	75 300
North Plains	1780	100 800
South Plains	1690	41 000
Columbia Region	1780	88 800
California	1769	260 000
Central Mountain Region	1845	19 300
New Mexico and Arizona	1680	72 000
Greenland	1721	10 000
Eastern Canada	1600	54 200
Central Canada	1670	50 950
British Columbia	1780	85 800
Alaska	1740	72 600
Total	—	1 152 950

Source: Mooney (1928).

Ubelaker (1976) reviewed the original notes that formed the bases of Swanton's publication of Mooney's calculations and believes that Mooney was attempting to provide 'minimal' estimates for the specific dates. Thus, Ubelaker concludes that 'while the actual aboriginal number probably is no less than indicated by Mooney, it could be considerably higher' (Ubelaker, 1976: 287).

Mooney's North American population-size estimates heavily influenced those of later investigators. Kroeber (1939) utilized most of Mooney's reconstructions (with the exception of California, for which Kroeber substituted his own approximations) and believed that the total native American population fluctuated around 900 000 persons at contact. Dobyns and Swagerty (1966, 1983) painstakingly analyzed early documents concerning the Conquest and post-Conquest mortality rates among Indians and concluded that the earlier estimates were on the low side. He had seen the hundreds of thousands of acres of land in Peru that were cultivated in pre-Conquest time and abandoned, which suggested to him a much larger pre-Conquest population reduced because of higher post- conquest mortality rates. However, Dobyns re-evaluated the low estimates of Mooney and Kroeber; on the basis of a hemispheric depopulation ratio of 20 : 1 and the estimation of the populational nadir he came up with a hemispheric estimate of over 90 million. This is a population density of 2.1 people per square kilometer. Dobyns was not alone in these higher population estimates: Rivet estimated (1924) 40–45 million, Sapper (1924) 40–50 million, Spinden (1928) 50–75 million, and Borah (1962) 100 million.

Thornton and Marsh-Thornton (1981) noted that the pattern of the numerical decline of the natives of North America could be represented by linear regression of population size on time. A high correlation coefficient of −0.989 and the inspection of the plot indicate that this decline was in fact linear. They utilized Driver's (1968) estimate of the nadir of the Amerindians north of the Rio Grande by adding the tribal population sizes provided by Indian agents and published in the annual Reports of the Commissioners of Indian Affairs (RCIA). However, Thornton and Marsh-Thornton believed that Driver had overestimated the North American nadir by including freed-men and intermarried Whites living among the Amerindians. By subtracting the numbers of freedmen and Whites from the totals reported by the RCIA, Thornton and Marsh-Thornton (1981) provided an adjusted nadir of 228 000 for 1890. Extrapolating back, using a linear model, they concluded that the aboriginal population of North America consisted of 1 845 183 persons. Such an extrapolation is based upon the assumption that the decline could be characterized as a linear trend from the seventeenth to the nineteenth centuries. No evidence exists to support such an assumption; on the contrary, a curvilinear decline is indicated in most of the Central American reconstructions. Yet, despite this criticism, the estimate of 1.85 million is surprisingly close to Ubelaker's (1988) most recent figures based upon tribe-by-tribe totals.

Ubelaker (1988) updated the population size estimate for North America by compiling the tribal estimates from the specialists who wrote the 20-volume *Handbook of North American Indians* (Ubelaker, 1988) and other published data. He produced three estimates of population size, a minimum density (based on 7 persons km^{-2}), his own estimate (11 per km^{-2}) and a maximum density of 15 per km^{-2}. His total population size for aboriginal North America at initial European contact is almost 1.9 million with a possible range from 1.21 to 2.64 million (see Table 2). These appear to be reasonable sizes given the data available. However, it is possible that the 'dead-reckoning method' employed by Ubelaker may have missed some of the tribes that were inadequately described historically or became extinct owing to disease epidemics that preceded visits by Europeans. Ubelaker's estimates are considerably larger than either Mooney's or Kroeber's, but fall short of Dobyns' (1966) estimate of 9.8 million inhabitants in North America at Contact.

Table 2. *Ubelaker's (1988) estimate of North American population size subdivided by region at the time of European contact*

Area	No.	Estimate no. per km^2
Arctic	73 770	3
Subarctic	103 400	2
Northwest Coast	175 330	54
California	221 000	75
Southwest	454 200	28
Great Basin	37 500	4
Plateau	77 950	15
Plains	189 100	6
Northeast	357 700	19
Southeast	204 400	22
Total	1 894 350	11

Central America

Cook and Borah (1971; Borah and Cook, 1963) attempted to reconstruct the population size for Mexico by using an assortment of Spanish historical documents and reports. Most of the extensive native records on tribute have been lost. Two nearly complete codices contain summaries of the tribute levied by the Triple Alliance upon its subject provinces in the few years preceding the Conquest. The sources of the Spanish data that Cook and Borah utilized to estimate population size include records of tribute assessments and payment, census records, missionary reports, and military reports. They studied records of tribute assessed and paid by a 10% sample of 2000 towns of sixteenth-century Central Mexico (copied from the old *Matricula de Tributos* of the Audiencia de Mexico). There may have been some underenumeration because the earliest tribute count showed some gaps, resulting from exemption of certain classes and failure to cover large geographic regions. For example, the Tlaxcaltecan Indians, by virtue of their special alliance with the Spanish Crown, were exempt from tribute. Only in the middle of the sixteenth century did the Spanish adopt a more uniform system of tribute count and assessment with fewer exemptions and greater territorial coverage (Cook and Borah, 1971). Census records for 1547–1550 from 50% of the towns of Central Mexico were studied. These counts were embodied in the *Suma de Visitas*, a volume that contains reports of the inspectors who enumerated the inhabitants of each town. Missionary reports, when added to the information on

Table 3. *Population size estimates for Mex-ican and Caribbean Amerindians at Conquest*

Date	Population size (millions)
1518	25.2
1532	16.8
1548	6.3
1568	2.65
1585	1.9
1595	1.375
1605	1.075

Source: Cook and Borah (1971).

tribute, provided 90% coverage for Central Mexico. Military reports included infor-mation on the sizes of native armies met by the Spaniards in battle. These military reports undoubtedly exaggerated the numbers of enemies. For example, in 1519, on his march to the Aztec capital (Tenochtitlan), Cortez fought a battle with the Tlaxcalte-can army. He estimated its size to be 100 000 warriors. This figure contrasts greatly with the estimated population size of the entire territory of the Tlaxcaltecans, which contained approximately 300 000 persons at contact (Halberstein *et al.*, 1973). It would have been impossible for a population of 300 000 persons to field from its ranks an army of 100 000 adult males, when approximately one third of the population would have been adults and one half of the adults were women. Therefore, it is highly unlikely that the Tlaxcaltecans could have fielded an army of more than 50 000 able-bodied males. In Table 3 Cook and Borah (1971) summarize their estimates of Amerindian population sizes for Mexico and the Caribbean at Conquest.

Judging from these numerical reconstructions it appears that the Mexican and Caribbean population was reduced from 25 million to 1 million persons in less than one century. These survival rates (as posited by Dobyns) of 1 out of 20 or 25 for Mexico are not unreasonable. However, the survival rates varied in the different geographical regions of the New World and thus cannot be applied universally. These figures clash with the conservative estimates of Mooney and Kroeber, who disbelieved the so-called 'Black Legends', the reports largely by the Spanish clergy of a horrendous carnage wrought upon the Middle and South American Indians. Fray de las Casas, one of the top churchmen of New Spain (Mexico and Central America, minus Costa Rica and Panama), complained to the Spanish Crown about the total Indian death count of 24 million during the period between 1500 and 1540. For the area in which de las Casas estimated a total Indian mortality of 40 million by 1560, Kroeber (1939) reconstructed a pre-Conquest population of 3.3 million and Rosenblat (1945) 5.3 million persons, figures wildly inconsistent with the mortality levels reported by de las Casas.

Judging from the meticulous reconstructions by Cook and Borah (1971), the population size for Mexico and the Caribbean of 25 million at Contact may be close to the actual figure (see Table 3). A native population of this size could sustain a 20- to 25- fold diminution resulting from disease and war. Such a population size is almost consistent with Fray de las Casas' estimates of mortality, except for some exaggeration. To this figure, Cook and Borah estimate, through extrapolation, a high population size for Hispaniola of 7–8 million persons. They justify such a high estimate by pointing out the high population density at Contact due to favorable food resources (maize and cassava) available to the Indians of Hispaniola. Cassava gives a far greater yield per hectare than maize and has remarkable storage qualities. A more precise enumeration was not possible because disease epidemics devasted this population almost immediately after Contact.

Whitmore (1992) has attempted to add some rigor to the estimates of depopulation by computer-simulating the population decline in early colonial Mexico. He grouped the various estimates of depopulation into three clusters: 'mild,' 'moderate,' and 'severe.' The mild decline followed Cook and Simpson's (1948) estimate of 1.2 million persons with a proportional reduction of 25%. The moderate model contained a pre-Contact population of 1.5 million and a decline of 75%. The severe model assumed an estimate of 2.7 million persons at contact, followed by a proportionate decline of 94%. Whitmore was unable to eliminate any of the historical reconstructions, since all fit within the general envelope of possibility (± 25%). He tried to tighten his conclusions and reduce the 'possible' to the 'probable' by invoking historical documents and a 'feel' for the data. He concluded that the mild and severe models are less likely to have occurred and that the moderate group of curves better fit within the 25% confidence limits. Of all contributing factors, infectious diseases played the most significant role in the depopulation of Mexico. Using a 'scaling procedure,' Whitmore (1991) provides an estimate at Contact of 16 million residents for all of Mexico.

In a review of Whitmore's (1992) volume, Ramenofsky (1994: 383) concludes:

> 'If all estimates are possible and if a proportional decline of less than
> 90% is probable, then the acrimonious debates regarding Precontact
> population sizes and Postcontact declines are intellectual cul-de-sacs.
> We know that the collapse was catastrophic, and we know that infec-
> tious disease was the primary cause. It is time to turn our energy to
> other issues such as building theory of population collapse or tracking
> evolution that occurred following the collapse.'

South America

There is considerable disagreement as to the size of the South American aboriginal population at Contact. These estimates range from Kroeber's (1939) 4.5 million (Central America and the Antilles are included) to Dobyns' (1966) maximum projection of 48.75 million. The latter estimate is based on a survivorship rate of 1 out of 25 persons. Several other specialists have provided an assortment of projections and reconstructions of the population sizes: Rivet (1924) 25 million; Means (1931) 16–32 million; Rosenblat (1945) 4.75 million for the Andes and 2.035 million for the remainder of continental South America. Steward and Faron (1959) provided what they call 'approximations' of population densities that were in turn converted to population size per unit area. By summing the population sizes for each area, they calculated the approximate size of the indigenous South American population as 10.2 million in the year 1500 (see Table 4). The population densities that are utilized by Steward and Faron for South America are low when compared with those in other parts of the world. They state that much of the Central Andes region is exceptionally inhospitable because of the desert, tundra and the alpine zones of high mountains. These regions cannot sustain dense populations. Underlying these estimates are assumptions by Steward and Faron that the ecology acts in a rigid, deterministic way on population sizes and densities.

Some population estimates for the New World made by various demographers and anthropologists are presented as if they share an unrealistic level of numerical precision. The population totals, which range in the millions, sometimes include breakdowns to one or two individuals. Considering the imprecision of these estimates and the questionable assumptions made, we can only discuss these figures as rough approximations.

The total size of the population of the Western Hemisphere could not have been as low as Kroeber's 8.4 million. His low totals were in part a rejection of the 'black legends', and underestimates. Dobyns'(1966) estimates of from 90 million to 112 million are probably too large because he applied survivorship rates that may have been applicable to Mexico but not to other regions with lower mortality rates. He may also have overestimated the sizes of the populations at their nadirs. If some of the inflations and underestimates are adjusted for, a total hemispheric population of at least 44 million is derived, which may be close to the actual figure. I base this approximation upon the following regional estimates.

North America	2 million
Central America	25 million
Caribbean	7 million
South America	10 million
TOTAL	44 million

Table 4. *Population sizes and densities for South America near European Contact*

Area	Population size	Density (persons per square mile)
Inca Empire		
Central Andes	3 500 000	10.0
Chiefdoms		
Northern Andes	1 500 000	6.6
Central America	736 500	4.8
Northern Venezuela	144 000	1.1
Antilles	225 000	2.5
Southern Andes		
Atacama–Diaquita	81 000	0.38
Southern Chile: Araucanian	1 050 000	7.0
Tropical forests	2 188 970	0.6
Hunters and gatherers		
Chilean archipelago	9 000	0.2
Patagonia: guanaco hunters	101 675	0.12
Western Chaco	186 400	1.1
Eastern Chaco	80 250	0.5
Eastern Brazil	387 440	0.3
TOTAL	10 190 235	

Source: Stewart and Faron (1959).

The estimates that I used to create a population total are based upon several compilations from various researchers. The 2 million used for North America is derived from Ubelaker's (1988) estimate plus an addition of several hundred thousand in order to correct for any inadvertent omissions in the tribe-by-tribe reconstructions. The core of the figures utilized for Central America is Cook and Borah's (1971) estimate. However, they may have overestimated the population size for Mexico, so this figure has been scaled down before incorporation into the estimate for Central America. Steward and Faron's (1959) conservative estimates were utilized for South America. Cook and Borah have utilized extrapolation to estimate the population size of Hispaniola at AD 1500. Their numbers of 7–8 million for Hispaniola alone seems very high. Such a population size is more likely for all of the Caribbean islands combined.

Denevan (1992a,b) revised his original estimate of the size of the aboriginal American population, circa 1492. In 1976, he estimated the combined population of North, Central, Caribbean, and South America to have been 57 300 000. This original estimate was adjusted to almost 54 million in his 1992a,b publications. The major reductions in population sizes are for the Caribbean (2 850 000) and Central Mexico (4 461 000), which he believes were averages of a series of conflicting figures.

2.4   REGIONAL VARIATION IN DEPOPULATION

There has been considerable variation in the population response to European contact. Some societies became extinct, like the Chono of the Chilean archipelago (Steward and Faron, 1959) and many tribes of the eastern coast of the United States. Other groups suffered initial diminution, to a nadir, followed by recovery and eventual reattainment of their pre-Contact sizes. The magnitude of the survivorship varied by culture, the initial size of the population and various unique historical events. To flesh out the dynamics of depopulation, I provide some more detailed examples from my own fieldwork with three subdivided populations located in diverse environments and experiencing vastly different genetic and evolutionary repercussions from European contact. These three groups are the Tlaxcaltecans (from the Highlands of Central Mexico), the Eskimos of St. Lawrence Island (located in the Bering Straits), and the Caribs of St. Vincent Island of the Lesser Antilles.

Tlaxcala

The State of Tlaxcala is located in the Tlaxcalan–Pueblan Valley of the Central Mexican Highlands (see Fig. 12). The inhabitants of this valley played a pivotal role in the Conquest of Mexico. The Tlaxcaltecans, traditional enemies of the Aztecs, who lived in the adjoining valley, allied themselves with Hernando Cortez during his march towards Tenochtitlan in 1519. This military alliance, known as Segura de la Frontera, protected Tlaxcala from many of the ravages of early colonization and granted its residents a number of special privileges, including the right to bear arms, to travel on horseback and to be addressed by the honorific title of 'Don'. In return, the Tlaxcaltecans served the Spanish Crown in the capacity of loyal troops, helped pacify the marauding Chichimecs of northern Mexico, and accompanied the Spanish expeditionary forces to various regions of the empire such as Santa Fe, New Mexico.

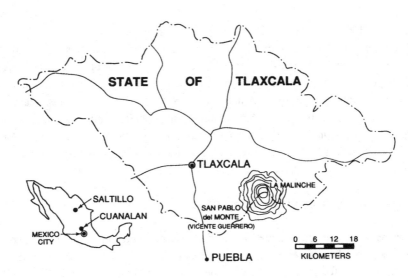

FIGURE 12 A map of the State of Tlaxcala, Central Mexico, locating the populations that were sampled during my field investigations in Mexico (1969–1975).

Initially, the Spanish presence in the Valley of Tlaxcala was limited primarily to the administrative and religious centers; thus, there was minimal contact with the conquistadores in the rural villages, particularly those on the slopes of the volcano LaMalinche. By 1580, the population of Tlaxcala was reorganized into a political entity with some degree of autonomy. Despite the fact that the Tlaxcaltecans managed to avoid many of the horrors of colonialization, the population in the valley was reduced significantly during early contact with the Spanish.

There is some disagreement between Halberstein *et al.* (1973) and Dumond (1976) over the extent of the Tlaxcaltecan population diminution during the sixteenth century. Dumond provides a maximum and a minimum population size at Contact for the Valley of Tlaxcala of 500 000 and 300 000 persons, respectively (Dumond, 1976). Halberstein *et al.* (1973) concur with Dumond but utilize the lower estimate for the size of the Tlaxcaltecan population at Contact. The disagreement concerns the population's nadir. Dumond projected a population reduction of Tlaxcala to less than 10% of its former size, whereas Halberstein and his colleagues (on the basis of historical records for San Bernadino Contla) contend that the population diminished only to 30% of its pre-Contact size. Figure 13 shows the reconstruction by Halberstein and his colleagues (1973) of the population decline in the Valley of Tlaxcala and the effects of various epidemics on this reduction. In addition to the effects of epidemic diseases, the population decreased because of the mortality from various military actions and

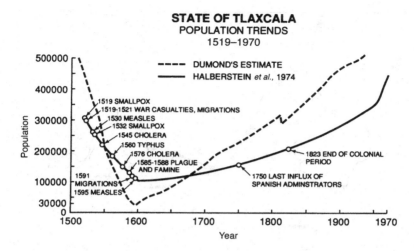

FIGURE 13 Reconstruction of the depopulation of the Tlaxcaltecans from the Valley of Tlaxcala. This plot compares the estimates of Dumond (1976) and Halberstein et al. (1973).

the transplantation of 400 families from the State of Tlaxcala to northern Mexico in 1591. Although a large number of Tlaxcaltecan Indians perished as a result of the epidemics introduced by the Spaniards, the survival ratio does not approach Dobyns' estimates of 1 out of 20–25 for Mexico. At the worst, 1 out of 3 (Halberstein *et al.*, 1973) to 1 in 10 (Dumond, 1976) survived in Tlaxcala. European Contact with the Tlaxcaltecans not only resulted in a genetic bottleneck but also caused the spread of the Tlaxcaltecan genes from Central Mexico to the northern regions of Saltillo, Parras and possibly Santa Fe. There is evidence to suggest that the Tlaxcaltecans also contributed genetically to the populations of Guatemala and the Valley of Mexico.

St. Lawrence Island

St. Lawrence Island is located in the Bering Sea approximately 40 miles from Siberia and 80 miles from Nome, Alaska (see Fig. 14). Although the existence of St. Lawrence Island was noted by a Russian explorer in 1658, European contact was limited until the development of the whaling industry, circa 1835 (Van Stone, 1958). This industry reached its peak between 1848 and 1885. During this interval the whalers made numerous stopovers on St. Lawrence Island and through intensive hunting destroyed most of the whales and walrus in the surrounding seas, the primary food resources of the islanders.

Before Contact, approximately 4000 people inhabited St. Lawrence Island. These

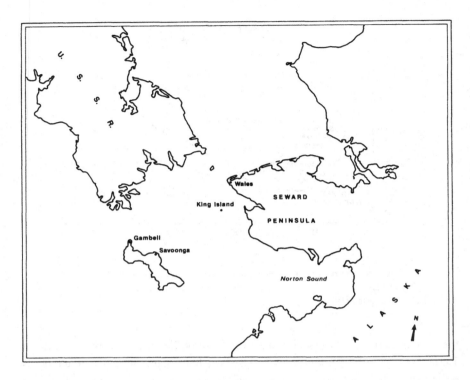

FIGURE 14 A map of the Seward Peninsula and the Bering Strait region of Alaska and northeast Siberia. The four villages included in this map were part of my research investigations of 1978.

inhabitants were distributed in small villages along the coast. In 1878, a combination of famine and epidemic struck the island with devastating effects. Approximately two thirds of the island's population perished, apparently from disease, starvation or both. Unusual weather conditions prevented the formation of solid pack ice near shore, thus limiting hunting. In addition there was a shortage of sea mammals because of the commercial whaling activities. These unique climatic and ecological conditions resulted in severe starvation. It is unclear which diseases may have been responsible for the epidemics observed on St. Lawrence Island during this period. However, dysentery, measles, 'black tongue' (interpreted as anemia), scarlet fever or vitaminosis (Milan, 1973) have all been blamed for the high mortality. Synergistic interaction between disease and famine may have produced the exceptionally high death rates. The population of St. Lawrence Island continued to decrease and in 1880 approximately 500 persons remained alive. By 1917, the population of St. Lawrence Island had reached its nadir: only 222 survivors were gathered at Gambell (see Fig. 15). The total popu-

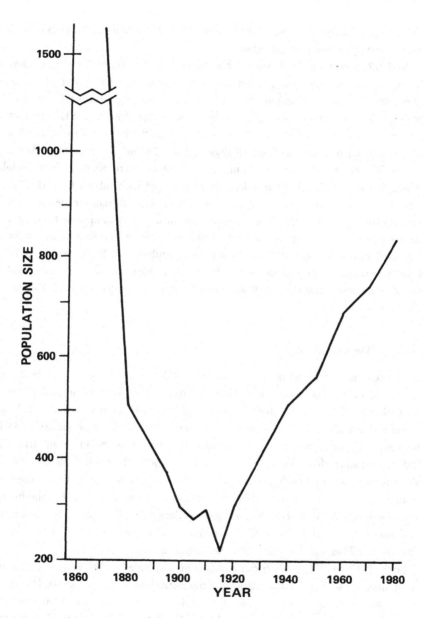

FIGURE 15 A reconstruction of the depopulation of St. Lawrence Island from 1860 to its nadir in 1920 (after Byard, 1981).

lation decreased from almost 4000 persons prior to 1835 to 222 survivors in 1917, a survivorship rate of 1 in 20 persons.

In 1917, a reindeer herd was established as a food resource for the St. Lawrence Island population. The best grazing land was located near the center of the island. Therefore, in order to maintain the herd, the surviving population at Gambell had to undergo fission and establish the village of Savoonga (Byard, 1981). Most of the younger men and their families relocated to be near the herd at Savoonga, while the older survivors remained in Gambell (Byard et al., 1984b). In order to further arrest the decline of the population, a number of orphans were adopted from mainland Alaska (Inupik speakers) and marriages were arranged with villages from the Siberian side. Thus, the genetic mixture of surviving adults with Siberian females and, eventually, the orphans from the Alaskan mainland, produced a unique melange of genes with frequencies that are intermediate between those of populations in Alaska and Siberia (Crawford et al., 1981b). As yet, the population of St. Lawrence Island has failed to reattain its pre-Contact size. This is due in part to insufficient time since the population's nadir and the fact that some of the St. Lawrence natives are emigrating to mainland Alaska.

St. Vincent Island

St. Vincent is a small island in the Lesser Antilles, located approximately 21 miles southwest of St. Lucia (see Fig. 16). Presently, it contains an admixed population consisting of Island Carib – Arawak Indian and West African components. St. Vincent Island was initially settled by Arawaks from South America in approximately AD 100. Between AD 1200 and European contact, another Amerindian group (the Carib Indians) expanded from Venezuela and intermixed with the original inhabitants of St. Vincent. Archeological evidence (ceramic motifs) has been interpreted by Rouse (1949) to indicate an invasion by the Caribs, extermination of the Arawak males, and hybridization with the Arawak females. However, Gullick (1979), on the basis of linguistic evidence, has argued in favor of a relatively small number of Island Caribs moving into the Caribbean and mating with the Arawaks.

From 1517 to 1646, an African component was added to St. Vincent's gene pool. A number of sources of this African component have been proposed: (1) runaway slaves from Barbados, an adjoining island and a center of the Caribbean slave trade; (2) raids on the European settlements by the Caribs, who then returned with African captives; (3) shipwreck of an eighteenth-century slave galleon destined for Barbados. This African component intermixed with the Amerindians to produce what has been called the Black Carib (or Garifuna) population, an exceptionally successful group that was deported by the British in 1797 to the Bay Islands. The Spanish navy further

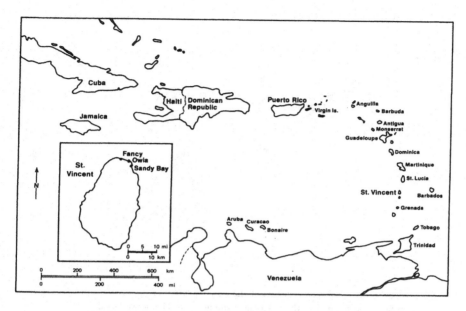

FIGURE 16 A map of St. Vincent Island and its location in the Lesser Antilles. The three villages on the north, windward side of St. Vincent were studied by my research team in 1979–1982.

transplanted the Garifuna to Honduras from whence they expanded and colonized most of the eastern coast of Central America (see Figs. 17 and 18).

There are no reliable estimates of the number of Amerindians living on St. Vincent Island prior to European Contact. Charles Gullick (1984) estimated the population of St. Vincent Island at approximately 5000 persons. This number includes both the Island Caribs and the Africans, even though the Amerindian component was put at a few hundred. After the second Carib war (1795–1805) approximately 2000 Black Caribs were deported to the Bay Islands. Gonzalez (1984) has argued that a total of 4200 Black Caribs (an admixed Afro-Amerindian population) surrendered to the British forces in July 1796, but only 2026 actually arrived in Roatan. She attributes this numerical discrepancy to the occurrence of a devastating epidemic (perhaps typhus) during the internment of the Garifuna captives on Baliseau Island.

Some of the Black Caribs who avoided British deportation in 1797 are ancestral to the estimated 1100–2000 Garifuna currently residing on St. Vincent Island. The contemporary population of Black Caribs of St. Vincent arose from a total of 45 individuals, some of whom were related (Gullick, 1984). Thus, this admixed population has experienced a considerable genetic bottleneck. Morphologically the Black Caribs appear to be from Africa. However, the frequencies of their gamma globulins (GMs)

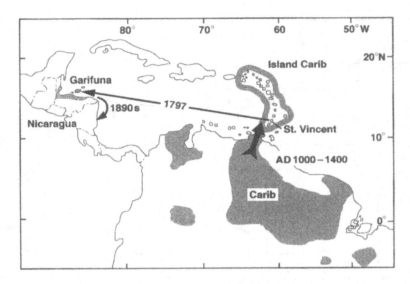

FIGURE 17 A visual reconstruction of the transplantation of the Black Caribs from St. Vincent Island to the Bay Islands and finally to Honduras (Davidson, 1984).

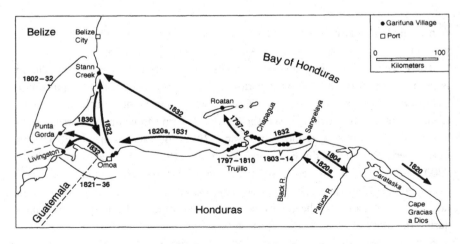

FIGURE 18 The pattern of the colonization of the coast of Central America followed by the Black Caribs (Garifuna) following their relocation in the Gulf of Trujillo in 1797 (Davidson, 1984).

reveal that from 32% to 42% of their genes have an Amerindian origin (Schanfield *et al.*, 1984). Thus, the only remaining evidence of the early presence of Caribs and Arawaks on St. Vincent Island is found among the genes of the highly admixed Black Caribs. The same is true for Dominica, where Afro-Amerindian populations exist to this very day.

There are a number of lessons to be learned from the three examples of New World depopulation provided in this chapter. The impact of imported diseases varied according to the form of subsistence of the affected population, the population density, the degree of geographic isolation, the nature of the conquest, and unique historical events that accompanied European contact. On the surface, the most severely affected population was on St. Vincent Island, where a combination of events including warfare, slavery, and epidemics almost annihilated the native population. However, some of the genes from the original Amerindian population were preserved in the Afro-Amerindian Black Carib population. In fact, St. Vincent's Amerindian component was transplanted to the coast of Central America, where it is currently found in almost 100 000 Garifuna distributed over a large geographic expanse. So, evolutionarily, St. Vincent's indigenous population was in fact successful, with the propagation and spread of its genes in new combinations through a colonizing group (see Figs. 17 and 18). Other island populations of the Caribbean were not as evolutionarily 'fortunate' and simply disappeared.

The Tlaxcaltecans were 'least affected' by their contacts with the Spanish, even though it is ludicrous to view the death of at least 200 000 Indians from a single valley as signaling the least affected! The mortality patterns of the Eskimos of St. Lawrence Island dramatically illustrate the results of synergistic effects of disease, starvation and cultural disruption all interacting and reducing the population from over 4 000 to 200 individuals in less than 50 years.

2.5   EVOLUTIONARY CONSEQUENCES OF POPULATION REDUCTION

The Conquest and its sequelae squeezed the entire Amerindian population through a genetic bottleneck. The reduction of Amerindian gene pools to from 1/3 to 1/25 of their previous sizes implies a considerable loss of genetic variabilty in New World populations. Who survived the epidemics? It is highly unlikely that survivorship was genetically random. Dubos (1965) reported that in the 1890s the annual death rate from tuberculosis was 9000 per 100 000 population. The Qu'Appelle Valley Indian reservation of Saskatchewan experienced a tuberculosis epidemic in which half the families were eliminated, with some 50% of all children dying. Those families who survived were likely to have been near the top of the normal range of variability in magnitude of protective response to the tuberculin bacillus. This epidemic is particularly perplexing, since recent molecular genetic evidence has demonstrated that precolumbian mummies from Peru contained in nodules of their lung tissue a DNA sequence that is found only in TB bacteria, *Mycobacterium tuberculosis*. Thus, tuberculosis was not a disease introduced into the New World but was present there at Contact (Salo *et al.*, 1994). In addition to the question of who survived the epidemic of Qu'Appelle, the most interesting question concerns what precipitated such an epidemic of this New World disease. Most likely the answer is contained in the living conditions experienced by the Indians on the Qu'Appelle reservation. In early colonial times, the Amerindians lived under crowded conditions, with their immune systems weakened or compromised by malnutrition and poor sanitation.

If Amerindians of today are different from their pre-Conquest ancestors with respect to many genetic systems, most likely those genetic traits that confered some selective advantage under the conditions of the Conquest are more numerous among contemporary Amerindians. Thus, the present gene-frequency distributions of Amerindian populations may be distorted by a combination of effects stemming from genetic bottlenecks and natural selection. In addition, the gene frequencies of the native populations were further modified by the massive gene flow or admixture with Europeans and Africans, thus possibly obscuring the pre-Conquest patterns. As a result, great care should be exercised in the interpretation of sophisticated multivariate analyses of gene-frequency distributions among New World populations based upon samples collected by various researchers utilizing a diversity of sampling techniques.

2.6   IMPACT OF OLD WORLD DISEASES

The Amerindian populations of the New World evolved for centuries in relative isolation from Old World populations, who numerically constituted the world's majority. The cultural and demographic encounter between the Old and New Worlds is one of the most dramatic events in the history of humanity (Ramenofsky, 1982, 1987). The population of the New World was reduced from over 44 million persons down to 2 or 3 million in fewer than 100 years and was eventually conquered by a small group of Europeans.

Given the immense size of the aboriginal population of the New World at Contact, how could comparatively small numbers of settlers or Spanish conquistadores subjugate it? The primary weapon was disease. This was one of the first uses of 'biological warfare' on a massive scale. As Stewart (1960) concluded, 'smallpox and not Spanish armor was the decisive factor in the fall of Mexico in 1520.' In many cases, the Old World diseases preceded organized European invasions by as much as a century. Cook

(1973) points out that the New England coast was known to explorers from 1497 on and that European fishermen were off the coast of Newfoundland and Nova Scotia throughout the sixteenth century. Indeed, as indicated by Cook, an epidemic in 1617, three years before the settlement at Plymouth, had 'softened up' the local Natives. Percy Ashburn (1947) has argued that English settlement may not have been possible had disease imports not paved the way. Without the effects of smallpox, Francisco Pizarro would probably not have succeeded in his conquest of the Inca Empire of Peru. The first smallpox epidemic started in Vera Cruz, Mexico, during Cortez' first contact in 1519. This disease spread into Guatemala, and then into what is now northern Peru in 1524–26. The Inca ruler and his entourage, including the only legitimate heir, all contracted smallpox and died. The result of their demise was the division of the Empire between rivals, thus lessening Inca resolve and facilitating the conquest of the Empire. Disease imports were thus the Europeans' best weapons against the indigenous populations of the New World and probably served as lethal 'advance men' time and time again in the Conquest of the Americas. Marshall Newman, in lectures at the University of Washington used to stress that 'the West was won in the United States by giving Indians blankets infected with smallpox.'

Were the Amerindians genetically more susceptible to Old World diseases? Crosby (1974) quotes a German missionary who stated in 1699 that the Amerindians die so easily that 'the bare look and smell of a Spaniard causes them to give up the ghost.' Centerwall (1968), during an expedition to the Amazon forest, observed a measles epidemic strike a 'virgin soil' Yanomama population of South American Indians. The researchers were able to monitor the effects of this epidemic and concluded that the death toll from measles was no different than what was observed in European populations that had not been repeatedly exposed to the same disease. Malaria and yellow fever were diseases that were equally fatal to both Europeans and Indians when these diseases were brought into the New World by African slaves. In Europe, epidemics caused by smallpox, yellow fever, and influenza were extremely severe with high mortality. The mortality was somewhat higher in the New World because the disease effects were further exacerbated by starvation, slavery and physical exhaustion. Thus, it has been argued that Amerindians did not have any special sensitivity or susceptibility to imported Old World diseases.

If there were no special susceptibility, why then did the Old World disease imports so ravage the indigenous populations at European contact? There have been several explanations offered for this phenomenon.

Cold screen

Stewart (1960) first explained the susceptibility of Amerindians in terms of his 'cold screen' theory. This theory suggests that the migration of Siberians through the Arctic into the New World prevented the flow of certain diseases. Newman (1976) agrees that there is validity to the cold screen theory, especially as applied to those diseases caused by pathogens with life cycles that require one or more developmental stages to take place outside the human body. He provides as an example hookworm, where neither the eggs nor the larvae can survive in soils with temperatures below 59 °F. Various vectors of pathogens, such as anophiline mosquitos, also cannot survive the cold of the Arctic. However, some diseases that are primarily vertically transmitted, such as T-cell leukemia caused by human T-lymphotropic virus type I (HTLV-1) have been carried across Beringia, despite the cold screen (Miura *et al.*, 1994).

Population base and disease maintenance

Crowd diseases, such as measles, smallpox, and rubella, often called density-dependent diseases, require large population bases for their permanent maintenance (Cockburn, 1963, 1967). Cockburn (1967) has argued that a population of about 1 million persons is required to maintain measles as an endemic infection. Populations of such size did not exist in most of the New World, with the possible exceptions of highland Mexico and Peru. However, St. Hoyme (1969) has argued that by the time Europeans arrived in the New World the conditions were ripe for the spread of epidemics. She points out that, with the advent of intensive agriculture, population centers and city states were coming into existence in North America, thus providing a reservoir for the maintenance of some of the crowd-type diseases. Most of these crowd diseases probably made their appearance in the Old World with the agricultural revolution of a little over 10 000 years ago. The Siberian immigrants into the New World were of insufficient numbers and density to maintain the crowd types of diseases on a year-around basis and these pathogens were 'lost' prior to crossing the Bering Strait.

Domestic animal reservoirs

Newman (1976) suggested that the relatively disease-free pre-Contact condition of New World populations was due to the paucity of domestic animal reservoirs for horizontal transfer of disease. He traced this idea to MacNeish's (1964) observation that the New World had many domesticated plants but few species of domesticated animal. The few domesticated animals of the Americas were the dog, the turkey and the South American llamas and related cameloids. Allison *et al.* (1982) described a number of pre-columbian dogs that were mummified in Chile. In contrast, the Old World had numerous species of domesticated animals but few species of plants. Thus, humans in the New World would have few zoonoses (infections transmitted to humans from animals directly or by ticks, mites, mosquitoes and other biting arthropods).

However, Cockburn (1967) pointed out that diseases such as botulism occurred in Alaskan Eskimos who consumed muktuk (skin and blubber of whales). Similarly, botulism can be acquired by eating utjak (seal flippers kept warm for several days until the skin falls off and the meat is ripe), which is often consumed by Labrador Eskimos. During fieldwork on St. Lawrence Island, I observed the continuation of the consumption of muktuk; however, great care is now practiced in its cold storage.

2.7   ABORIGINAL NEW WORLD DISEASES

Early travelers in the New World tended to romanticize the indigenous populations as being peaceful and disease-free. This romantic vision of the idyllic life of Amerindians prior to European contact persists even in scientific writings (see Denevan, 1992a). Dobyns (1993) state that the peoples of the Americas suffered none of the following diseases: measles, smallpox, chickenpox, influenza, typhus, typhoid, paratyphoid fever, diphtheria, cholera, bubonic plague, scarlet fever, whooping cough, and malaria. Therefore, if these populations were indeed so healthy, then conversely their life expectancy should have been higher. Yet numerous investigations of survivorship or life expectancy in pre-columbian Amerindian populations, based upon skeletal evidence, reveal a low life expectancy. For example, the average life expectancy at birth in the Pecos Pueblo in Southwestern United States was only 15–20 years (Ruff, 1981). Similar life expectancies have been reconstructed for the Larson Arikara site, with 13.2 years of life probable at birth, and the Indian Knoll site, with an 18.6-year life expectancy (Owsley and Bass, 1979). In addition, the skeletal evidence documents numerous instances of trauma and warfare, thus suggesting an existence far from the idyllic, peaceful one painted by the romantics (Larsen, 1994).

What diseases were present in the New World prior to European Contact? What evidence do we have for the reconstruction of pre-Columbian disease patterns?

The four primary sources of information on the disease state of pre-Contact Amerindians are: (1) observations and reports written by European settlers, priests and conquistadores; (2) native accounts; (3) paleopathological observations on mummified or osseous burials; and (4) molecular genetic evidence. The reliability of reports by Europeans is in question. Often, fevers and rashes were confused and some diseases went totally unrecognized. This is not surprising given the state of medicine in western Europe of the sixteenth century. A few native accounts, such as pre-Conquest Maya Indian life, contained in the *Book of Chilam Balam* (Roy, 1933), are almost idyllic descriptions reflecting nostalgia for the pre-Cortez period. According to this volume there were no sicknesses, no aching bones, no high fevers, no burning chests, no headaches. All of these problems were blamed on the foreigners and in many cases justly so! There is a large body of literature on paleopathological evidence for disease in pre-Columbian Americas. Some diseases, notably syphilis, tuberculosis, and arthritis, can leave identifiable markings on the bones of their victims and this evidence has been utilized to infer the presence of some of these diseases (Arriaza, 1993). Unfortunately, many diseases leave no imprint on bone. In particular, virally caused diseases either kill you rapidly or recovery follows without leaving any osteological marks (Ortner and Putschar, 1985). However, the study of mummified tissue permits the identification not only of microorganisms but also of the antibodies produced against various agents of disease. The recent development of techniques for the extraction and identification of DNA from contaminating pathogens permits more precise identification of some pre-columbian diseases of the New World. Recent molecular genetic investigations have revealed that tuberculosis is indeed a New World disease (Salo et al, 1994). In addition, the study of coprolites (desiccated feces) permits the recognition of intestinal parasites. Investigations have shown variation in parasites that is related to ecological and nutritional conditions (Reinhard, 1988).

The most likely aboriginal diseases of the New World (Merbs, 1992) (based in part upon Newman, 1976) include the following.

Infectious diseases

Treponematoses

These are caused by a group of spirochetes (genus *Treponema*) that cause infections and skeletal manifestations resulting from venereal syphilis, endemic and non-venereal syphilis, and yaws (Cockburn, 1961). There has been some controversy surrounding the origin of syphilis (*T. pallidum*) and other of these treponemal infections (Baker and Armelagos, 1988; Livingstone, 1991). Three hypotheses have been suggested as to the origin of syphilis in the New World: (1) syphilis was endemic to both the New and Old Worlds; (2) syphilis was brought to Europe, possibly by Columbus' crews; (3) syphilis was introduced into the New World from the Old World. Some authors have argued that, although there is some controversy about the origins of syphilis, treponemal infections in the form of Pinta (an infection of the skin caused by *T. carateum*, which enters the body through breaks in the skin) and possibly other infections existed in the Americas prior to Columbus' voyage (Hudson, 1965). These infections were non-venereal and caused mild infections of shorter duration. Hudson (1965) suggested that these various treponemal infections may represent adaptations of *T. pallidum* to differing environments. There is some skeletal and DNA evidence

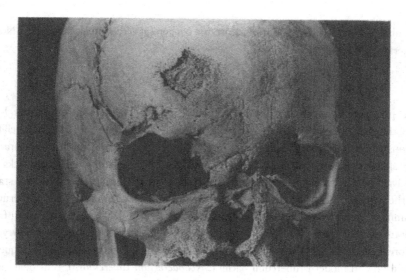

FIGURE 19 A cranium from an excavation in Norwich (England), bearing a typical treponemal lesion (kindly provided by Donald Ortner).

to confirm the pre-columbian presence of treponemal infections in the New World (Rogan and Lentz, 1994).

Reichs (1989) provides a description of a skeleton from the southeastern United States with both cranial and postcranial lesions that are strongly suggestive of treponematosis. This article, together with several others (Bullen, 1972; El-Najjar, 1979; Elting and Starna, 1984; Powell, 1986; Bogdan and Weaver, 1988), supports the presence of treponemes in prehistoric New World populations. The lesions derived from treponematosis are distinctive (proliferation and enlargement of bony tissue usually in the tibia diaphysis and calvarium (Ortner et al., 1992)). However, a recent excavation in Norwich, England, has revealed the presence of treponemal lesions on skeletons of individuals who died prior to Columbus' voyages of exploration. The photograph in Fig. 19 shows a lesion apparently due to a treponemal infection from the Old World, prior to contact. Perhaps treponemal infections existed concurrently on both sides of the Bering Strait.

Powell (1996) suggests that treponematosis was absent among North Amerindian populations prior to the appearance of horticulture. Evidence for treponematosis begins with high-density sedentary villages. The morbidity and mortality from this disease is much lighter than that observed with the Old World version of treponemal infections. It is likely that the form of treponematosis in the Americas was non-venereal and similar to modern yaws and endemic syphilis. Powell supports this conclusion with the observation that there is no evidence for congenital treponemal infection in North American Native groups.

Tuberculosis

The presence of tuberculosis in pre-Columbian New World populations has been hypothesized by several researchers (Allison et al., 1973). They observed the tell-tale lesions (involving loss of bone mass) in the lower thoracic and lumbar vertebrae of the spinal column in skeletal remains and the representation of 'hunchbacks' in Peruvian pottery, possibly depicting the destruction of the vertebrae in advanced stages of the disease. Allison et al. (1974b) found clinical signs of tuberculosis in Peruvian mummies. Recently, Salo et al. (1994) have sequenced the DNA fragments taken from lesions of lung tissue from the body of a female mummy who had died 500 years before Columbus set sail from Spain. These DNA fragments contained a sequence found only in modern tuberculin bacilli (Mycobacterium tuberculosis), thus definitively demonstrating the presence of tuberculosis in the Americas before European Contact.

Much like treponematosis, there is evidence to suggest that tuberculosis makes its appearance in North American Indians in agricultural communities (Powell, 1996). There is solid evidence for the presence of tuberculosis in the Lower Illinois Valley among Late Woodland settlements. This disease apparently requires high-density settlements for its transmission and maintenance.

Bacterial pathogens such as *Streptococcus* and *Staphylococcus*

Ortner (1972) has described osseous lesions that are suggestive of staphylococcal infection in pre-Columbian New World populations.

Diseases transmitted through an intermediate host

Typhus

There is evidence to suggest that typhus (a group of rickettsial diseases usually transmitted from infected rodents to humans by bites of lice, fleas, ticks or mites) was pre-columbian in origin. The Aztecs had a name for the disease (Matlazanahuatl) and they depicted Indians with nosebleeds and spots all over their bodies. Typhus is characterized by headaches, chills, fever and maculopaular rashes. Ashburn (1947) also noted the mild form of this disease in Peru, thus suggesting the build-up of immunity from long-term association with typhus. He states that typhus has been known in Peru since the sixteenth century and that it never caused epidemics as extensive as those in Mexico.

American leishmaniasis (protozoan)

Uta is an ulcerating disease caused by a protozoan (*Leishmania peruana*) transmitted by several species of sand fly. Uta occurs in the Andes of Peru and Argentina. The lesions of this cutaneous disease were graphically depicted in pre-Contact Mochica portrait pottery (100 BC – AD 700) from Peru's North Coast.

American trypanosomiasis (protozoan and insect-vectored)

Chagas' disease is, after malaria, the most prevalent vector-borne disease in South America (Rothhammer *et al.*, 1985). This disease is caused by the protozoan parasite *Trypanosoma cruzi*, which is commonly transmitted by triatomine insects (Chagas, 1946). Carpentero and Viana (1980) have suggested that these triatomine bugs adapted to life in human dwellings of South America circa AD 500 when guinea pigs were first raised for food. Before domestication, these small rodents probably served as hosts to Triatominae. Rothhammer and his colleagues describe the apparent presence of Chagas' disease in some of the 35 mummies from four sites with radiocarbon dates between 470 BC and AD 600. Some of the mummified bodies contained 'mega-syndromes,' an enlargement of hollow organs (such as colon, heart, and esophagus) caused by the destruction of their intramural nervous plexuses by the trypanosome (Koeberle, 1968).

Verruga or Carrion's disease

Carrion's disease (also known as bartonellosis) is an infection caused by the bacterium *Bartonella bacilliformis*, which is transmitted by the bites of sand flies, such as *Phlebotomus verrucarum*. This disease is characterized by fever, severe anemia, and bone pain. It is endemic in the valleys of the Andes in Colombia, Peru and Ecuador. Carrion's disease has two stages: (1) a severe, often fatal, fever and anemia (Oroya fever); (2) the verruga stage, in which the body is covered with wart-like nodules that vary in size from 2 to 50 mm. Evidence for this usually fatal disease was detected by Allison *et al.* (1974c) in a mummy from near Nazca, Peru. Ceramic vases dating back almost 2000 years depict the symptoms of Carrion's disease in South America.

Human fluke infection

A single fluke egg (possibly of the genus *Opisthorchis*, *Clonorchis* or *Heterophyes*) was contained in a coprolite from Glen Canyon, in southeastern Utah (Moore *et al.*, 1974). This infection usually requires two intermediate hosts, snails and fish. Infection results from eating the infected intermediary host. The flukes may invade the biliary and pancreatic ducts or the intestinal wall and cause diverse lesions and death. It is possible that this single egg was not the result of a human infection but rather was picked up by consumption of a bird or some other host that was infected.

Roundworms, especially ascarids and other endoparasites

Some of these parasites have been identified from pre-columbian mummies (Allison *et al.*, 1974b) and others have been studied in coprolites from various sites (Reinhard, 1988). Reinhard reports evidence for eight different species of helminth parasites, including nematodes, cestodes and acanthocephalans, in archeological sites on the Colorado plateau. According to Confalonieri *et al.* (1991), *Ascaris lumbricoides* is a North American Indian parasite that failed to reach South America.

Hookworm

Allison and his colleagues (1974c) have described the presence of *Ancylostoma duodenale* in a coastal Peruvian mummy dated to circa AD 900. However, there is some question concerning the origin of this parasite in the New World: the larval stage of this organism could not have survived the cold screen (Fonseca, 1969). In addition, Ferreira *et al.* (1980, 1983) uncovered hookworm eggs in human coprolites in an archeological site in Brazil dating to more than 7000 years BP. These findings are difficult to explain, since in order to be infectious, one part of the hookworm life cycle must be passed in soil. Araujo *et al.* (1988) have argued that this infection could not have crossed the Bering route and that its distribution can be explained only by trans-Pacific contact. An alternative explanation for the presence of hookworm is that it could have been a disease of animals prior to separation of land masses of New and Old Worlds.

The oldest documented human parastic infection in the New World involves a helminth, *Enterobius vermicularis*. Fry and Moore (1969) describe its presence in coprolites dating from 10 000 years ago. This parasite could have been brought across Beringia because its life cycle does not require development in the soil and it can be transmitted directly from host to host.

Food poisoning

Salmonella, botulism and parasitic agents acquired through the food chain have been reported in Eskimo and Amerindian populations. Most of these cases are transmitted through infected meat (Cockburn, 1971). Ferreira *et al.* (1984) found eggs of *Diphyllo-*

bothrium pacificum, a sea-lion parasite, in human coprolites from northern Chile. In all probability, this parasite was acquired by humans through the consumption of either sea lions or the fish that serve as intermediate hosts.

Viral diseases

Based upon minimal evidence, the possibility of influenza and other respiratory infections occurring in the New World, prior to European contact, has been suggested by Allison *et al.* (1974d) and Newman (1976). However, the best evidence for retroviral infection in the pre-Columbian New World comes from the recent DNA-sequence phylogenetic reconstructions of the genomes of human T-lymphotropic HTLV-I viruses, which cause adult T-cell leukemia/lymphoma and HTLV-associated myelopathy or tropical spastic paraparesis, HAM/TSP (Osame *et al.*, 1986). Three major types of HTLV-I virus are known: (1) the most divergent form, distributed in Melanesia, the Solomon Islands and Australian Aborigines; (2) the African type, which has been isolated from a person from Zaire; (3) a so-called 'cosmopolitan type', distributed over most to the world. Because the HTLV retrovirus is transmitted only by cell-associated infection and mainly by vertical infection from mother to child, its distribution reflects the phylogeny of human populations. The horizontal transmission of this virus is highly unlikely. Miura *et al.* (1994), using the neighbor-joining and unweighted-pair-group methods, constructed phylogenetic trees that reveal an African cluster with chimpanzee STLV-I, suggestive of interspecific transmission. The present distribution of the 'cosmopolitan type' reflects the movement of a single viral lineage from Asia into the New World, where it is found in Colombia and Chile.

In 1982 a related virus, HTLV-II, was recovered from a patient with hairy-cell leukemia (Kalyanaramen *et al.*, 1982). This virus can be transmitted by blood transfusions, but is most commonly passed on vertically from mother to infant. The HTVL-II virus has been detected in 11 of 38 North and South American tribes but is not present in any of the 10 eastern Siberian indigenous groups tested (Neel *et al.*, 1994). This virus has been reported in Mongolia, one of the sites where the mitochondrial B haplotype was also detected (Sambuughin *et al.*, 1991). These data prompted Neel *et al.* (1994) to hypothesize that the first migrants to the New World were derived from regions of Mongolia, Manchuria, and/or extreme southeastern Siberia.

Auto-immune diseases

Pre-columbian skeletal remains of Amerindians provide considerable evidence for the presence of arthritis in the New World prior to contact. For example, Ortner and

Utermohle (1981) reported polyarticular inflammatory arthritis in a 30–35-year-old female skeleton from Kodiak Island dating from before AD 1200. The skeletal lesions associated with the joints include porosity and destruction of articular surfaces. The largest New World prehistoric skeletal series was studied by Charles Snow (1948; Rothschild et al., 1988). He observed that approximately 60% of the 1234 skeletons from the Indian Knoll site (Kentucky) were afflicted with lumbar arthritis. Arthritis was therefore a common condition in some Amerindian groups throughout their history (Bourke, 1967).

Nutritional deficiency diseases

Iodine deficiency, leading to goiter, is among the most readily demonstrable of these diseases (Borhegi and Scrimshaw, 1957). Some of the disabilities and pathologies associated with nutritional deficiencies, such as goiter are graphically represented on the Mochica pottery of Peru. A study of a series of crania from both historic and prehistoric Anasazi Indians of the southwestern United States revealed the presence of porotic hyperostosis, a condition that apparently reflects increased bone-marrow activity. The research of El-Najjar et al. (1975) demonstrated that Amerindians who are highly dependent upon maize have a higher incidence of porotic hyperostosis than do groups who consume more iron-rich animal products. Heavy consumption of maize interferes with iron absorption. In addition, an amino-acid analysis of osseous material from an Indian child from the American Southwest with porotic hyperostosis revealed lower concentrations of amino acids containing hydroxyl groups and acidic side chains (Von Endt and Ortner, 1982). These data are suggestive of the absence of an enzyme cofactor, such as dietary ferrous ions, and provide a plausible explanation for the presence of a certain kind of osseous lesions that occur in Amerindian skeletal collections. The implication of this study is that porotic hyperostosis can result from iron-deficiency anemia, which might have been prevalent in pre-Contact populations.

Genetic (inherited) diseases

A number of inherited disorders leave some signs on skeletal remains. For example, Cybulski (1988) described brachydactyly (disproportionately short metacarpals or metatarsals) in a series of pre-columbian skeletons from Prince Rupert Harbour, Canada. Similarly, Danforth et al. (1994) described the presence of genetic syndromes in human remains from a small population (Pueblo III, circa AD 1100–1225) at Carter Ranch Pueblo, Arizona. They noted a number of dysplasias, including a hip syndrome similar to hip dysplasia in modern Navajo. There is evidence for the presence of Klippel–Feil syndrome (an inherited dysmorphology) in a young adult.

Most of the diseases listed by Newman are largely chronic (long-term and not highly contagious) illnesses that may have been endemic in small, scattered Indian villages. Apparently the Amerindians had little experience within their lifetimes of the swift ravages of the infectious epidemic diseases. They constituted 'virgin field populations' with regard to most of the Old World disease imports. Dobyns (1993) suggests the presence of some crowd-type diseases in the large populations of the Inca and Aztec Empires. He postulates that Uta and Carrion's disease may have reached epidemic proportions in the Inca Empire of Peru, while typhus may have ravished the Aztec Empire of the Highlands of Mexico.

2.8   OLD WORLD DISEASE IMPORTS

Ashburn (1947) listed three imported diseases 'in order of frightfulness: smallpox, typhus and measles.' He designated these diseases as the 'shock troops of the conquest.' Although the origins of typhus are controversial and may in fact be in the New World, this infectious trio caused havoc in the non-immune populations of the Americas. The severity of the epidemics associated with these diseases is described in the early parts of this chapter. In addition to these three deadly diseases, a large number of other contagions were brought from Europe and Africa into the New World. These include malaria, yellow fever, chickenpox, whooping cough, scarlet fever, diphtheria, plague, typhoid fever, poliomyelitis, cholera, and trachoma. Some of these diseases exacted higher mortality than did others. Among the deadliest of these diseases was falciparum malaria. Apparently this protozoal disease was brought into the New World by the West African slaves who were infected by *P. falciparum*, the deadliest of the malarial organisms. Other, milder forms of this disease can also be found in the New World: *P. malariae* causes quartan malaria; *P. ovale* causes mild tertian malaria; and *P. vivax* causes a common tertian malaria. This organism infected the most efficient New World anopheline vector, *Anopheles darlingi*, which in turn began infecting Amerindians and Europeans. The presence of a deadly malaria along the coast of Central America has extracted a phenomenal mortality from the Amerindians who have lived there historically. Unlike the Old World populations, such as the African populations or the Mediterranean peoples living in the malarial zones, the Amerindians possessed no genetic defense against malaria. They lacked any hemoglobin variants and any glucose 6-phosphate dehydrogenase deficiency (G-6-PD−), nor were any Duffy null alleles present at polymorphic levels among the Amerindians of the New World. Thus, the introduction of malaria devastated the Amerindian coastal populations and opened the door for the colonization of this region by Black Caribs and Creoles, whose gene pools contained alleles that offered a measure of protection against malaria and who also bore traces of the Amerindian culture, which permitted their efficient utilization of the region's natural resources.

2.9   CONCLUSION

In an earlier section of this chapter I discussed the evolutionary implications of the massive depopulation experienced by the peoples of the New World. Most of the depopulation was due to the high mortality associated with the various epidemics of infectious diseases. However, evolutionary effects of the epidemics can be ameliorated by replacement fertility. In a computer simulation of a Tlaxcaltecan population (Cuanalan) experiencing a series of epidemics, Kenneth Turner (1976) was able to demonstrate that in small villages where the epidemic is immediately followed by the replacement of those young who died, the genetic variances fluctuate less than in stable populations or in those without makeup fertility. Thus, the demographic patterns observed in Cuanalan, consisting of periodic mortality peaks due to epidemics followed a year or two later by fertility peaks, genetically minimize the effects of stochastic processes caused by the reduction of population through epidemics. Unfortunately, most of the aboriginal populations were initially faced by epidemics that were neither cyclic nor deadly to only the young and old. Diseases ravaged all age cohorts in many of the native societies because of the enslavement of that portion of the population that normally is most fit, the adults. Thus, the genetic effects of the epidemics were no doubt truly profound, and the contemporary Amerindian gene pool has been sculpted by the ravages of disease, the sword, and inhuman physical labors.

The introduction of the Christian religion to the native Americans further exacerbated the population decline. Larsen (1994) correctly points out that the concentration of Amerindians around missions promoted the spread of infectious disease owing to less sanitary conditions and higher population densities. Thus, the creation of an extensive network of Catholic missions in coastal Georgia and Florida increased the risk of transmission of crowd-dependent pathogens. Analysis of skeletal remains of mission Indians by stable-isotope ratios of carbon and nitrogen from bone collagen indicates a shift away from the consumption of marine foods to a greater reliance on maize (Larsen et al., 1992). Enamel defects such as hypoplasias in these Amerindians became more prominent during the missionary era apparently reflecting increases in infectious disease and starvation (Jones, 1978).

3 Demography of Amerindian populations

3.1 INTRODUCTION

The demography of human populations consists in part of their quantitative characteristics, including age/sex structure, size, fertility, mortality and migration patterns. The interrelations between fertility, mortality, emigration and immigration rates permit predictions of the numerical changes that a population may undergo during specific periods. For example, in reproductively closed populations such as species, the relative magnitudes of fertility versus mortality alone determine the size of the population. However, in most human populations it is the interaction of fertility and immigration versus mortality and emigration that determine the numerical trends in population size. This chapter considers the demographic structure of Amerindian populations both before European Contact and today. Given the broad scope of this volume, encompassing all of the indigenous peoples of the Americas and a heterogenous melange of populations of different sizes and subsistence patterns, it is not possible to reconstruct the demographic characteristics population by population. Instead, I shall endeavor to distill some of the common denominators of Amerindian populations. In addition, I shall present some examples that are particularly informative about the processes acting upon the demographic structures of these populations.

3.2 DEMOGRAPHY OF PRE-COLUMBIAN AMERINDIANS

The demographic characteristics of the populations of the Americas, prior to European contact, are often estimated from small amounts of information. There are no censuses or vital registers, both sources of data traditionally utilized by demographers. Instead, archeologists and biological anthropologists reconstruct the dynamics of prehistoric populations through the use of cultural artifacts, skeletal remains and the extrapolation of models derived from contemporary populations. Demographic data derived from archeological sites in prehistoric cemeteries often suffer from problems associated with sampling, which may be related to insufficient numbers, possible burial biases, and errors in the age and sex determination of skeletal materials. Often the data are based upon mixed groupings representing long or ill-defined periods and geographical regions of unknown size. There may be cultural constraints on who is buried in a specific cemetery. These problems have led demographers to doubt the validity of the estimation techniques commonly applied to prehistoric populations (Peterson, 1975). Also compounding the problems of prehistoric reconstruction is the reliance on small populations whose demographic parameters may fluctuate stochastically and which were altered by unique historical events. However, as the evidence grows, the techniques of aging and sexing are technically refined, and the distributions of vital statistics in small populations are better understood, the precision of the paleodemographic estimates improves.

Age and sex structure

Determinations of age and sex of individuals in skeletal collections are based primarily upon the morphology of the available bony structures. Recently, a number of DNA markers have been developed that can be used to determine the sex of osseous remains. Sex is usually assigned to a skeleton on the bases of the shape of the pelvis, particularly the breadth of the subpubic angle, the morphology of the pubic symphysis, the diameter of the acetabulum relative to the size of the innominate, and the lateral flare of the ischium. Cranial morphology also provides some indication of sex, which may be inferred from such traits as brow-ridge robusticity, frontal and parietal bossing, development of the mastoid processes, and the robusticity of the nuchal process. According to Buikstra (1976) the accuracy of age evaluation varies inversely with the maturity of the skeleton. Dental eruption patterns provide an accurate assessment of age up to approximately 12 years. Epiphyseal and basicranial suture closure permit the aging of adolescents and early adults. Similarly, the development of the third molar and the modifications of the pubic symphysis serve as age indicators for early adults. Standards of dental attrition and degree of endocranial suture closure provide for rough estimates of age of older adults. Age-related changes in osteon remodelling and bone mineralization can be assessed through photon–osteon analyses of cortical bone (Laughlin et al., 1979). Because of the absence of key bones from some of the excavated burials, differential training in the art of aging and sexing, and varying levels of expertise among archeologists and biological anthropologists, some error creeps into paleodemography.

The literature is replete with references to skeletal populations with unique age and sex distributions. For example, in the original description of the skeletal sample of 674 individuals from Pecos Pueblo, New Mexico, Hooton (1930) initially attributed a sex ratio of 150 for the skeletal remains from this site. He had utilized a visual

Table 5. *Age and sex distribution of the skeletal remains from the Larson Site*

Age interval (years)	Male	Female	Total
0–1	—	—	254
1–4	—	—	94
5–9	6	9	48
10–14	4	6	14
15–19	10	21	31
20–24	10	15	25
25–29	12	10	22
30–34	25	10	35
35–39	21	18	39
40–49	20	14	34
50–59	10	15	25
Sample total	118	118	621

Source: Owsley and Bass (1979).

inspection method for determining sex in this skeletal series. However, Ruff (1981), utilizing metric measurements, reassessed the sexes of the same sample and concluded that the sex ratio was 100. Hooton had, in fact, mis-sexed almost 25% of the total osseous sample. A large Arikara village and cemetery (circa 1750–1785) of the Northern Plains, called the Larson Site, contained a sample of 621 individuals. The sex ratio for adults buried at Larson does not differ significantly from 100 (Owsley and Bass, 1979). Unequal sex ratios were observed for the subadults, suggesting different sex-specific mortality rates during childhood.

The age and sex distribution of the Arikara skeletal remains from the Larson Site suggests a population with high neonatal and childhood mortality and a preponderance of female deaths in the 15- to 25-year-old cohort. Overall, the observed age and sex distribution is consistent with the patterns predicted for small Amerindian populations.

Table 5 summarizes the age and sex distribution of the skeletons excavated from the Larson Site.

According to ethnographic reports, Eskimo sex ratios should show a preponderance of males in the various age categories. This sex ratio reflected systematic female infanticide, practiced until recently in the Arctic. Eskimos preferred male children to females, presumably because the males are providers of food whereas the females might constitute an economic burden. Early ethnographic reports on the Netsilik

Eskimos of the Canadian Arctic suggested female infanticide rates as high as 66% (Rasmussen, 1931). However, skeptics of these high mortality rates argue that Eskimo populations would have become extinct if such high rates persisted over an extended period of time. Using computer simulation, Schrire and Steiger (1974) modeled the growth of a hypothetical Eskimo population utilizing mortality and fertility data derived mostly from field observations. They concluded that even for rates of female infanticide as low as 8%, the population tended toward extinction in several hundred years. Chapman (1980) criticized this model for ignoring an important effect of infanticide. Schrire and Steiger had assumed in their computer simulation that all Eskimo women were subject to identical fertility rates regardless of infanticide. However, cessation of lactation following infanticide should decrease the minimum interval before the next pregnancy. Thus women whose previous infants had been killed could be more fertile than women whose children had lived. When this factor was included in Chapman's computer simulation, it was observed that an Eskimo population could sustain a rate of 30% female infanticide and still survive. Thus, some pre-Contact Eskimo populations contained a preponderance of males as a consequence of female infanticide. Some such rate of female infanticide in Eskimo populations would have both maintained these groups in numerical equilibrium and preserved a favorable proportion of providers for the community.

Population density

Buikstra (1976), using the numbers of mounds per unit area and the average numbers of burials per mound, attempted to calculate the population density in the Middle Woodland region of Illinois. She summarized the surveys of the area and found on average 12 mounds per mile of bluff. Over 140 miles (225 km) of bluffline she estimated a total number of Middle Woodland mounds to be 840. With no fewer than 30 individuals buried per mound, a median age of 28, and a 550-year occupation in a region of 2800 square miles (7250 km²), Buikstra computed a population density of 0.46 individuals per square mile (0.18 km⁻²). She based these calculations on the formula

$$P = \frac{SNA}{m^2 T},$$ (1)

where the population density (P) is based upon the product of S (minimum number of skeletons per totally excavated mortuary unit), N (estimated number of mounds in the region), and A (average age at death), divided by the product of m^2 (number of square miles inhabited) and T (length of time). These sorts of estimates are in all probability highly inaccurate, but are one of the few means available for estimating the population densities of pre-Columbian populations of Amerindians.

Mortality

There is considerable variation in the mortality patterns of pre-columbian Amerindian populations reconstructed by using skeletal and archeological evidence. This variation should not be surprising given the tremendous diversity of the subsistence patterns and living conditions of New World groups. However, some of the diversity in mortality reported in the literature may be due to less than rigorous applications of demographic methods to small populations. For example, Nancy Howell (1976) raised doubts about the conclusions of W.S. Laughlin (1975) with regard to the apparent lower mortality of the Aleuts when compared with Eskimos as being an indication of ecological adaptation. She criticized Laughlin et al. for failing to provide information about the size of the population from which these deaths were drawn. Howell correctly chastized these researchers for ignoring the likely under-representation of infant deaths and for not providing adequate information on the age assessment of the skeletal materials. Laughlin et al. (1979), in a later publication, acknowledged the severe underestimate of mortality among infants and defined age 10 as the earliest that could be used for comparisons of mortality. They continued to champion the greater life expectancy of the Aleuts over the Eskimos (by their calculations a 15-year-old Aleut will on average outlive his Eskimo counterpart by 11–12 years) and explained this differential mortality in terms of the accessibility of reefs, which permit elderly Aleuts to provide for their own dietary needs. Laughlin et al. (1979), in their comparison of Eskimos and Aleuts, concluded that the greatest life expectancy was found in pre-contact Paleo-Aleut males who at age 10 appear to have lived an average of almost 26 years. Based upon historical documents, a decline in life expectancy was observed in historic Aleut populations.

Storey (1986) constructed life tables for a pre-Columbian urban center, Teotihuacan. This preindustrial city (circa 150 BC – AD 750) covered 8 square miles (20.7 km²) and had a population of between 125 000 and 200 000 residents (Millon, 1973). Storey analyzed the burials from an apartment compound (named Tlajinga 33) located on the southern edge of the city and apparently occupied by full-time craftsmen. Out of a total of 166 skeletons unearthed at Tlajinga 33, 33.1% were perinatal and 35.5% of the population died at less than one year of age (Storey, 1986). According to Storey 'the most dangerous point in the lifespan, according to the life table, was around birth.' She notes that the Tlajinga infants were shorter at birth than modern and Arikara infants (Northern Plains of United States) and interprets these data to signify low birthweight due to maternal nutritional deprivation. The result of this apparent

Table 6. *Life expectancy at birth in diverse Amerindian skeletal populations*

Populations	Reference	Dates	Life expectancy (in years)
Lawson	Owsley and Bass (1979)	AD 1750–1785	13.2
Tlajinga 33	Storey (1986)	150 BC–AD 750	17.1
Indian Knoll	Ubelaker (1974)	3000 BC	18.6
Ossuary I	Ubelaker (1974)	AD 1500–1600	20.9
Ossuary II	Ubelaker (1974)	AD 1500–1600	22.9
Hopewell	Buikstra (1976)	50 BC–AD 400	29.5
Texas Indians	Ubelaker (1974)	AD 850–1700	30.5
Pecos Pueblo	Hooton (1930)	AD 800–1700	42.9[a]

[a]Ruff (1981) has recalculated the life expectancy for this skeletal population. After correcting for aging errors the apparent life expectancy is 25 years.
Modified from Owsley and Bass (1979).

nutritional stress is a population with high perinatal, infant and juvenile mortality and short life span. The population from Tlajinga 33 at birth appears to have had a mean life expectancy of about 17 years.

Owsley and Bass (1979) summarized the life expectancies at birth for an assortment of skeletal Amerindian populations (see Table 6). A comparison of the Tlajinga 33 urban population with groups that were rural indicates a lower life expectancy in urban populations. This interpopulational comparison of life expectancies at birth indicates that the people at Larson Site had the lowest, followed by the urban population from Central Mexico. The Lawson residents were Arikara Indians who relied on a mixed subsistence of hunting and horticulture. This post-Contact community was undoubtedly exposed to several infectious diseases such as smallpox, measles, chickenpox, whooping cough and cholera. Thus, the Lawson Site provides insight into populations undergoing epidemics, but is not representative of a pre-Contact Amerindian community with regard to life expectancy. The Tlajinga 33 urban community of craftsmen probably approximates the lowest life expectancy observed in the New World, owing to the crowded living conditions, chronic undernutrition and poor sanitation. The survivorship patterns observed in Tlajinga 33 are similar to those seen in pre-industrial urban populations of Europe and suggests that 'crowd' diseases existed in Tlajinga as well.

There is no reason to suspect that the aboriginal populations of the Americas differed demographically from other groups at similar levels of technological develop-

Table 7. *Life expectancy estimates for extant hunting and gathering populations and pre-Columbian Natives of North America*

	Life expectancy at age		
	Birth	15	50
North American Natives	23	20	5
Average hunter-gatherers	22	22	13

Source: Weiss (1973).

ment in other parts of the world (Howell, 1976). Estimates of life expectancy of pre-Columbian Amerindian populations were compared by Weiss (1973) with those of average hunters and gatherers elsewhere and found to be similar (see Table 7).

Various measures of mortality have been estimated for a number of different skeletal populations of North America. Owsley and Bass (1979) computed crude death rates (m) from North American skeletal collections and found a range of 44–76 deaths per 1000 people per year. Again, the Larson and Leavenworth Sites of South Dakota exhibited the highest crude death rates and suggest populations in numerical decline. Buikstra (1976) computed death rates of 2.5–5.0% per 10 years for the Illinois Hopewell population.

Arriaza *et al.* (1988) examined the nature and frequency of childbirth-related deaths in a sample of female mummies from the pre-columbian population of Arica, Chile. They found that 14% (18/128) of those had died between the ages of 12 and 45 years from childbirth-complicated death (CCD). Although the sample of mummies is large, the site covers a span of 2700 years. Therefore, generalizations as to the exact rates and causes of the birth complications cannot be made. However, only three mothers died without completing delivery. Arriaza and colleagues suggest that septic conditions, acute diseases such as pneumonia, circumstances of delivery and methods of delivery were likely causes of death of pre-columbian mothers in labor.

Migration patterns

Lane and Sublett (1972) proposed that past human residential patterns could be inferred from the analyses of male and female morphological variability. They suggested that decreased between-group variation and increased within-group variation for one of the sexes would be expected if that sex had a higher mating migration rate as observed in patrilocal or matrilocal systems. These decreases in between-group variation can be assessed by calculating distance measures between groups (Lane and Sublett,

1972). Konigsberg (1988) extended this model theoretically from one based upon Wright's island model of gene flow (Wright, 1951) and used a migration matrix method. He demonstrated that the usual measures of standardized genetic variance can be decomposed into female/male and male/female components, which in turn can be used to assess the effects of differential residence on the within-group genetic diversity of the two sexes. Konigsberg applied some of these techniques to the analysis of non-metric cranial traits from pre-Columbian archeological sites of Illinois. The females exhibited higher morphological variability than males for most of the archeological sites, suggesting a patrilocal residence pattern. These results parallel the findings by O'Rourke and Crawford (1976) in contemporary Tlaxcaltecan Indian populations of Central Mexico. Odontometric variation was always higher in females than in the males; this phenomenon was explained on the bases of patrilocality and the recruitment and movement of females to their husbands' villages of residence.

3.3   DEMOGRAPHY OF THE LIVING

After initial contact with Europeans and Africans, the Amerindian populations underwent a cataclysmic reduction in size (Johansson, 1982). According to the census, this reduction continued in the United States until approximately 1900, when the population reached its nadir of 237 196 persons. From 1900 to the present, the Amerindian population of the United States has rebounded to well over one million persons (IHS, 1989).

Census sizes

North America

For the 1990 fiscal year, the Indian Health Service (IHS) provided a total count of 1.105 million native Americans residing in the United States. This figure is an underestimate because those Amerindians no longer living on reservations are not enumerated. From these estimates, it appears that the present population is increasing at about 2.7% per year with 32% of the population being younger than 15 years of age. Given these figures, IHS has projected, on the basis of fertility and mortality rates, a population size of 1.450 million by the year 2000 and almost 2 million by 2010. The average birth rate of American Indians and Alaskan Natives was 28.0 per 1000, 79% greater than the birth rate for the total United States population in 1985. The Amerindian maternal mortality rate has dropped from 82.6 per 100 000 live births in 1957–59 to 8.2 in 1984–86 (a decrease of 90%), contributing to growth of the population of Amerindians. Thus, the contemporary Amerindian population of the

United States can be characterized as a young group that is growing rapidly (IHS, 1989). For example, in the Norton Sound Region of Alaska, the Eskimo population has been rapidly increasing numerically (Norton Sound Health Corporation, unpublished report, Nome, 1978). Between 1950 and 1970 the overall native population grew by 44.5% as a consequence of a high birth rate (due to a high fertility rate of 126 births per 1000 women aged 15–44 years of age) coupled with a decreasing death rate. The birth rate in 1970 was 23.3 per 1000 population, compared with a rate of 24.9 for all Alaska and 18.0 for the whole United States. In 1950 the infant mortality rate for Alaskan natives was approximately 4% (Alaska Bureau of Vital Statistics). The overall age composition at Norton Sound shows a young population with 48% under 18 years of age and 52% under 20 years of age.

Price (1988) summarized the sizes of Canadian Indian populations and the proportions living on and off the reserves. On the basis of the 1976 enumeration, he estimated that a total of 288 938 Indians reside on reservations. A total of 570 bands (a band is an administratively defined group of Indians who have the right to specific lands) are distributed on 2301 reserves and 66 settlements (Price, 1988). About 28% of Canadian Indians live off the reserve lands. When these non-status Indians (i.e. those without rights to specific lands) are included, there are approximately 600 000 people with a Native heritage living in Canada (Price, 1988). Out of the total of Native people living in Canada, approximately 20 000 are Inuit (Eskimo) who reside in the Arctic and Subarctic regions. Like the Amerindians of the United States, the Natives of Canada are the fastest-growing ethnic minority of their country.

Salzano (1968a) attempted to enumerate the sizes of the unacculturated Indian populations of Central and South America. He estimated the total Indian population of Central America to be approximately 8 million persons, of whom fewer than 1 million were unacculturated. This figure of 1 million unacculturated Indians, living primarily in Guatemala and Mexico, was probably a gross overestimate and was challenged by the moderator of the Pan-American health conference at which Salzano provided this estimate. For South America, Salzano estimated 10 million to be Amerindian, of whom he judged 980 000 were 'unacculturated.' One must keep in mind that these calculations were based upon data collected 22–25 years ago and that much has happened in the Amazon basin and Yucatan to the so-called unacculturated populations. I suspect that at present there are few remaining South Amerindian groups that have been left alone by the expansion of Mestizos and Europeans into the rain forest.

Of the roughly 20 million Natives living in the Americas, over 1.6 million reside in the United States and Canada. Most of those in the USA and Canada live on reservations, which themselves produce significant effects on population structure. Reservations bring together different tribal units and, evolutionarily, result in the elimination of some interpopulational genetic heterogeneity. Some reservations are listed as multitribal; in Oregon, for example, the Warm Springs Reservation is inhabited by Paiute, Tenino, and Wasco Indians, and the Umatilla Reservation includes Umatilla, Walla Walla, Cayuse and Paiute Indians.

Hulse (1960) explored the genetics of the Hupa Reservation in Northwest California. The BIA Guide Map listed a total of 952 Hupa residing in an area of 200 square miles (518 km²); in addition, 365 Yurok lived on a 7000 acre (2833 ha) extension of the reservation. Hulse collected blood specimens from 282 Hupa, 27 Yurok, 19 Karok and 28 Indians of other tribal affiliations including the Sioux, Choctaw, Wintun, Shoshone, Cherokee, Chippewa and Oklahoma. This sample revealed the heterogeneous nature of the Reservation's population, created as an amalgam of various tribes. Of the entire group, only 31 individuals showed no evidence of admixture with Europeans. Many of the younger Indians of both sexes had frequently left the Reservation for work and returned with spouses and children of uncertain parentage. These waves of genes from the outside created the 'ripples on a gene pool' that Hulse referred to in the title of his paper.

Not all reservations have experienced considerable gene flow; some have been highly endogamous. For example, at Acoma, near the Arizona – New Mexico border, no Europeans were allowed to stay overnight on the mesa housing the pueblo. Until the construction of roads into the Papago reservation, 99.5% of the children born prior to 1900 were fullbloods (Smith, 1980). Thus, some of the reservations were reproductively isolated and were made up of a single tribe or nation, whereas other reservations pushed together an assortment of tribes who did not share recent historical roots.

Latin American

Marvin Harris (1964) divided Latin America into three subdivisions.

(1)   Highlands, which stretch from Mexico south through Central America along the Andes to northern Chile. The populations inhabiting these regions are made up of Indians, Europeans and Mestizos.

(2)   Tropical and Semi-Tropical Coastal Lowlands, including the Caribbean Islands. The ethnic makeup of these populations includes Africans, Europeans and Creole hybrid groups.

(3)   Temperate South. This region is inhabited primarily by Europeans. Most of the indigenous populations became extinct centuries ago.

During the 1960s, within the highlands of Latin America, three countries remained predominantly Amerindian. They were Guatemala, Bolivia and Peru. Four countries, namely Mexico, Ecuador, Paraguay and Chile, were essentially Mestizo but with medium-sized (10–28%) Indian populations. Honduras, Costa Rica, Nicaragua, Panama, Colombia and Venezuela showed even less Indian representation, although enclaves of Amerindians could be found. By the 1980s, Mestizos had largely replaced the Indian communities of Latin America.

Fertility

The fertility rates of populations are of great importance evolutionarily: the interaction of fertility and mortality are determining factors of the size and genetic composition of subsequent generations (Stinson, 1982). Achieved reproduction or fertility is a result of both biological and cultural factors operating on societies to produce the next generation. Among the most significant biological factors influencing fertility are: disease, incidence of multiple births, sterility, age at menarche and menopause (which define the length of a woman's reproductive career), sexual desire (libido) and the effect of breast-feeding on ovulation. Some of these apparently biological factors, such as nutrition (which may affect the age at menarche and susceptibility to disease) and the duration of breast-feeding, may be influenced by cultural or environmental factors. Cultural factors that influence fertility include culturally constituted or legally defined age of marriage, coital rates, use of contraception, sexual taboos, education, and even occupation. All of these factors have been associated with differential fertility in some human cultures. These factors have influenced and continue to influence the fertility patterns of humans and particularly of Amerindian populations.

Given the complexity of the interactions between cultural and biological factors, it is not surprising that there is extreme variation in achieved reproduction in the human populations widely distributed from Tierra del Fuego to Barrow, Alaska. The highest recorded fertility in New World populations was reported by Brennan (1983) for the oldest generation of Black Caribs of Sambo Creek, Honduras. She noted that the mean number of liveborn children in women 45 years of age or older was 10.9. However, later cohorts of Black Carib women showed a considerable decrease in fertility. This mean number of liveborn children in the Black Caribs is the highest reported for any human population, and surpassing even the fertility rates computed for the Hutterites (Eaton and Mayer, 1953) and Irish Tinkers (Crawford and Gmelch, 1975). Given fertility rates of this magnitude, it is not surprising that the Black Caribs constitute one of the most successful cases of biocultural adaptation (Crawford, 1983). Other researchers have failed to confirm such high fertility rates in this Afro-Indian population (Firschein, 1984; Custodio and Huntsman, 1984). Firschein reports the mean number of liveborn children for Black Carib mothers who had completed their reproductive careers (45 years of age or older) as 5.40. He observed a significant generational trend in women 65 years of age or older, who produced an average of 6.04 children. Custodio and Huntsman confirmed Firschein's findings in the Black Carib populations of Honduras. They observed that females (40 years of age or over) with normal hemoglobins produced a mean of 6.14 children. This discrepancy in the fertility rates of Black Carib populations may relate to other differences between the more rural, village environment of Sambo Creek and the towns of Stann Creek and Punta Gorda studied by Firschein.

McAlpine and Simpson (1976) report an exceptionally large average number of pregnancies for postreproductive Inuit women of Igloolik and Hall Beach. The 11.1 ± 0.9 average includes both livebirths and stillbirths. The birth rate in Igloolik was 51.4 per 1000, compared with 36.3 per 1000 in Hall Beach. Given the small sizes of these populations, 564 in Igloolik and 248 in Hall Beach (i.e. 89 and 43 mothers, respectively), the unusually high reproductive performance reported for these Eskimo groups may be a chance event.

Yanomama

Among the least acculturated populations of South America to be studied demographically were the Mucajai Yanomama. This small group lives along the Mucajai River in northern Brazil and subsists through a combination of horticulture (plantain or manioc), hunting and foraging. They were first contacted in 1958 by missionaries who established a permanent mission there. The missionaries registered births and deaths, collected periodic censuses and constituted genealogies (Early and Peters, 1990). Yanomama females claim that pre- and post-Contact fertilities were much the same, with an average of 7.9 live births per woman who has completed her reproductive career. According to Early and Peters, the population has grown from 121 in 1958 to 319 in 1987, an increase of 198 people, or an annual growth rate of 3.5%. This high figure conflicts with the more modest estimates (0.5–1% annually) of population growth rates for the Yanomama made by Neel et al. (1977b). These Yanomama are highly fertile, with a crude birth rate of 50 per 1 000 population. This high fertility results from an early age of menarche (12.4 ± 0.18) and short intervals between pregnancies. Early and Peters report that females become active sexually (as evidenced by cohabitation) at 13 or 14 years of age and that on average the first child is born when the female is 16.8 years old. Some forms of population control are practiced among the Yanomama such as induced abortion and prolonged breast-feeding, in some cases for 2.25 years after birth of a child. The effective live birth rate for an adult Yanomama woman has been estimated by Neel and Weiss (1975) to be 0.23, or one birth every four years. The mean number of children produced by those women who reach the age of menopause is 8.2 births (Neel et al., 1977b). However, Chagnon

(1968) indicated that the fertility levels for the Yanomama were relatively low, with an average of 3.8 births per woman. Early and Peters believe that Chagnon's low estimates indicate deficient data. Salzano et al. (1967a) provide the following numbers (means ± 1 SD) of live births to women who have completed, or nearly completed, their reproductive careers (over 39 years of age).

Xavante	5.7 ± 2.4
Caingang	6.6 ± 3.8
Upper Xingu	5.3

Similarly, Salzano and Callegari-Jacques (1979) summarized completed mean family sizes for a number of Brazilian populations.

Cashinawa	6.9 ± 1.6
Katukina	10.0 ± 1.5
Kanamari	6.4 ± 0.8
All villages	7.4 ± 0.7

These data suggest that the Indian populations of the Amazon forest are highly fertile and rarely, if ever, employ contraceptive techniques.

The crude birth rate of Yanomama females has been estimated as 0.059; the male crude birth rate is slightly lower at 0.056. The crude death rates of the Yanomama are 0.050 and 0.048, respectively; thus it is clear that the Yanomama rates are high when compared with national standards (Neel et al., 1977b).

Demographic studies of small populations such as the Yanamama are subject to two general types of error, namely methodological error (ascertainment bias and misreporting) and statistical error, arising from sampling variance (Leslie and Gage, 1989). In the Yanomama, stochastic fluctuations in rates can occur even with complete and accurate enumeration. Thus, some of the parameters should be estimated by indirect approaches, thereby discounting the stochastic fluctuations (Leslie and Gage, 1989).

Tlaxcaltecans

The Tlaxcaltecans of Central Mexico offer an opportunity to observe populations in demographic transition. Halberstein and Crawford (1975) compared the demographic structure of a Tlaxcaltecan enclave (transplanted from the Valley of Tlaxcala to the Valley of Mexico-Cuanalan) with that of two communities from Tlaxcala (San Pablo del Monte, primarily an Indian village, and the City of Tlaxcala, a Mestizo community). The populations of Cuanalan and San Pablo del Monte were more fertile than that of

Table 8. *A comparison of the achieved reproduction rates for Tlaxcaltecan populations of Central Mexico*

	Populations			
Parameters	Cuanalan	San Pablo	Tlaxcala City	Mexico City
Total number of prolific women	415	205	129	1622
Mean number of liveborn children	5.51	4.71	4.22	4.07
Variance in number of offspring	10.54	12.35	10.56	—
Average number of reproductive years	25.63	26.60	25.20	—
Mean number of livebirths per prolific woman (40 years of age +)	6.50	6.10	5.36	5.11
Percentage of married childless women over 40	7.60	4.35	4.10	—

Source: after Halberstein and Crawford (1975).

the town of Tlaxcala (see Table 8). Reproductive wastage, including sterility, occurs more frequently in Cuanalan than in San Pablo or the City of Tlaxcala. Despite this greater reproductive wastage, the average Cuanalan woman who has children produces more children than do her counterparts in Tlaxcala and San Pablo. The City of Tlaxcala exhibits the characteristic reductions in fertility and mortality that are diagnostic of populations undergoing a demographic transition.

Cultural practices may have reduced the fertility of some Amerindian populations. For example, the Havasupai Indians of Arizona inhabit a well-watered side canyon of the Grand Canyon. Despite agricultural abundance, the Havasupai population has been numerically stable throughout historic times, although there have been some minor fluctuations due to disease epidemics. Married couples produce an average of 4.4 liveborn children but lose 1.1 to childhood mortality (Alvarado, 1970). According to Alvarado, a number of cultural preferences apparently limit the fertility of the Havasupai: (1) age at marriage (mean age of first marriage is 24.6 for males and 18.9 for females); (2) age differential between spouses; (3) high percentage of unmarried adults; (4) spermatocidal effect of prolonged use of the steam lodge. The Havasupai men customarily utilize the sweat lodges four times each afternoon for periods of approximately 10 minutes each. Each session in the sweat lodge is followed by a swim in the icy stream. Temperatures of the sweat lodges range from 118 to 157 °F; the heat, followed by the cold of the stream, may serve as spermatocidal agents. (5) Despite recent denials, infanticide, particularly in the case of female twins, may have been practiced historically by the Havasupai. Infanticide has been used to explain the high sex ratio (with a range of 101 to 151) observed among the Havasupai. (6) Induced abortions have been linked to the rare occurrence of illegitimate pregnancies in this population.

Birth seasonality

Seasonality of births was reported by Malina and Hines (1977a) for a rural Zapotec-speaking municipio in Oaxaca, Mexico. They noted that the monthly distributions of births deviated from random expectation in that the highest frequencies of births occurred during the rainy season. Seasonal distributions of births are probably influenced by cultural patterns associated with the annual agricultural cycle. In contrast to this finding, Cowgill (1966) has reported that the highest number of births occur during the dry season in the Southern Pacific regions of Mexico. In her data, the month of December shows a major birth peak and is followed in frequency of births by February and January. She suggests that deliberate efforts are made to avoid birth during the rainy season because of disease and known high infant mortality.

Rural versus urban fertility

A survey in 1972 of two different urban and two rural Amerindian populations revealed apparently contradictory results. A comparison of urban and rural Seminole Indians of Florida suggested that the rural reservation Seminoles had higher achieved reproduction than those in more urban enclave (Liberty et al., 1976b). The reservation population exhibited not only slightly higher fertility (3.0 living children per woman 20–40 years of age, as opposed to 2.2 children per urban counterpart), but showed higher reproductive wastage. The authors explained the differential fertility on the basis of the greater use of birth control by the urban women (74% used birth control in the urban enclave, compared with 32.5% on the reservation). A similar comparison (Liberty et al., 1976a) was made between two other groups of Indians, this time from Nebraska. Comparison of the fertility of urban and rural Omaha Indians showed the relationship between these variables reversed. Larger numbers of children were produced by urban women than by their reservation counterparts: 3.89 versus 3.24. Unfortunately, the authors failed to statistically test the significance of their findings. Given the small sample sizes and the fact that the authors interviewed women in the 20–40 age category, the differences in the two sets of results may reflect different age distributions within the samples. The question of whether the fertility of urban Amerindian women is reduced after they move from the reservations is yet to be answered.

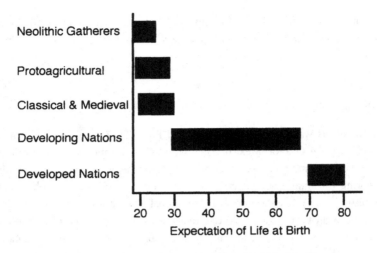

FIGURE 20 The range of variation in the expectation of life at birth in societies of varying socioeconomic levels and cultural complexities (Gage *et al.*, 1989).

Mortality

There is considerable variation in the observed patterns of mortality in contemporary New World populations. Both the level of mortality (as measured by death rates and life expectancies) and the shape of the age-specific mortality curves differ greatly among Amerindian populations. These mortality characteristics are thought to be statistically independent of each other (Gage *et al.*, 1989). Life expectancy is usually measured as the average number of years remaining to an individual of a particular age in a given population. Figure 20 illustrates the range of variation in the expectation of life at birth in societies at various socioeconomic levels and of different cultural complexities. This figure suggests that in pre-industrial societies the expectation of life at birth was between 20 and 30 years of age (Gage *et al.*, 1989). Contemporary Western societies that are economically developed exhibit life expectancies in the range of 70–80 years of age. Among contemporary Amerindians, the Yanomama have life expectancies at birth of 21.5 years for males and 19.8 years for females (Neel and Weiss, 1975). The Trio Indians of Surinam exhibit higher life expectancies at birth of 42 years of age for males and 59 for females, a considerable sex-related difference (Gage *et al.*, 1984). The age pattern of mortality among the Yanomama suggests high childhood mortality and an intermediate rate of mortality in adulthood (see Fig. 21). Neel and Weiss (1975) estimated infant and childhood mortality, including infanticide, to be about 50% for the Yanomama. Survivorship to age 15 is about 50%, and a little higher in males than in females. This slight difference in survivorship between the sexes was

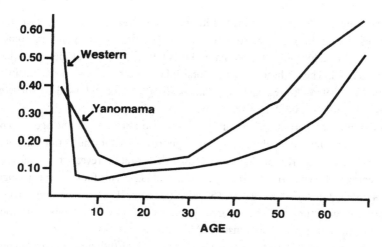

FIGURE 21 Comparison of a Western life expectancy curve with that of the Yanomama (Gage et al., 1989).

due to female infanticide (Neel *et al.*, 1977b). Neel and his colleagues also note a high level of mortality throughout the reproductive period. Figure 21 compares a western life table with that of the Yanomama (Gage, 1988). In contrast to the exceptionally low life expectancy at birth of the Yanomama, the United States' Indians and Alaskan Natives exhibit life expectancies at birth of 67.1 years of age for males and 75.1 for females. In 1980, the average life expectancy of Amerindians was more than 3 years shorter than that of the US Whites (IHS, 1989). The Indian life expectancy has increased dramatically since the 1940 census, which indicated an expectancy of 51.3 and 51.9 years of age for male and female American Natives, respectively. Kenen and Hammerslough (1987) compared mortality rates, subdivided by age and sex, of American Indians living in reservation and non-reservation counties for 1970 and 1978. They detected an improvement in mortality for the decade that includes the age group 0–4 years. In 1970, non-reservation death rates were not different from those on reservations. By the end of the decade, non-reservation death rates diverged downward from the reservation rates.

Infant mortality is much higher in South American Indian populations than it is among US Indians on Reservations. The total Peruvian infant mortality rate for 1960 was 103.4 per 1000 births, over four times the US figure of 25.3 (all races). For US Indians on Reservations, the infant mortality rate at about that time, 1954 and 1962, was 65 and 41.8 per 1000 births, respectively. There are complex reasons for the greater infant death rates in Peruvian populations. Mazess and Mathisen (1982) and Haas (1980) demonstrated that neonatal death rates in Andean regions of Peru and

Bolivia were considerably higher at high altitudes. The Indian babies on US reservations receive much better health care from the Indian Health Service with, according to the 1964 Health Education and Welfare (HEW) Report, 97% of deliveries being physician-attended. The highest causes of total infant death for American Indian and Alaskan Natives, 1984–86, was sudden infant death syndrome, followed by congenital anomalies and respiratory distress syndrome.

Evolutionarily, the percentage of children born who fail to reach the age of reproduction, 15 years, may be indicative of the opportunities for the action of natural selection. Salzano's research on Caingang Indian demography reveals the failure of a high percentage of newborns to reach 15 years of age. For four of the Caingang villages, 41% died before the age of 15 years (Salzano, 1961). Neel and Chagnon (1968) provide an estimate of the mortality among Xavante and Yanomama Indian children and adolescents. Approximately 33% of Xavante and 16% of Yanomama newborns fail to reach 15 years of age. In contrast, 17% of the Ramah Navajo females born from 1890 to 1909 failed to reach the age of reproduction and 13% of those born from 1945 through 1954 died before attaining their fifteenth birthday (Morgan, 1973). However, in the Ramah Navajo a high proportion of females (28%) who survived to age 15 failed to produce any children and thus did not contribute genetically to the next generation. Among the Canadian Eskimos, approximately 25% of all liveborn children died prior to 15 years of age (McAlpine and Simpson, 1976). Thus, the Inuit hunters are intermediate beween the Amazonian groups and the reservation Navajo in their survivorship to the age of reproduction.

Opportunity for selection

In 1958, James Crow introduced an index of total selection, later called the index of opportunity for selection. This index is defined as

$$I = I_m + (I_f /p_s),\qquad (2)$$

where $I_f = V_f/ X^2$. X is the mean number of offspring produced by women 45 years of age, V_f is the variance in the number of births among these women, and p_s is the percentage of women surviving to the age of reproduction. The mortality component $I_m = p_d /p_s$, and $p_s = 1 - p_d$ (p_d is the proportion of women who fail to reach the age of reproduction). Crow (1989) recently refined his original derivations. He subdivided the index into components acting at different stages of the life cycle.

Although there is some dispute as to what this index actually measures, there have been numerous estimates of the index of opportunity for selection in Amerindian populations of both hemispheres. In North American Indian populations, the index

Table 9. *The index of opportunity of natural selection in New World populations*

Population	Reference	I_m	I_f	I
Navajo 1844–1924	Morgan (1973)	0.27	0.50	0.91
Cuanalan	Halberstein and Crawford (1975)	1.47	0.28	2.16
San Pablo	Halberstein and Crawford (1973)	1.63	0.31	2.46
City of Tlaxcala	Halberstein and Crawford (1973)	0.59	0.35	1.14
Mexico	Spuhler (1962, 1963)	0.49	0.61	1.41
Cashinahua	Johnston and Kensinger (1971)	0.79	0.11	0.98
Maca (Paraguay)	Salzano et al. (1970)	0.56	0.21	0.88
Yanomama	Neel and Chagnon (1968)	0.22	0.66	0.88
Xavante	Salzano et al. (1970)	0.49	0.41	0.90
Chile: towns	Crow (1966)	0.15	0.45	0.67
Chile: villages	Crow (1966)	0.33	0.22	0.62
Chile: nomads	Crow (1966)	1.38	0.17	1.78
Terena	Salzano and Olivera (1970)	0.27	0.28	0.63
Cayapo	Salzano (1971)	0.34	0.38	0.71
Caingang tribes ($n = 6$)	Salzano (1961, 1964 1965)	0.49–0.78	0.26–0.81	0.90–2.20

falls in an intermediate range, I ranging from 0.4 to 1.0. This range applies to Navajo age cohorts with births between 1844 and 1924 (Morgan, 1973). Neel and Chagnon (1968) provided measures of the index of potential selection among the Yanomama and Xavante Indians of South America, for whom I is 0.90 and 0.88, respectively. However, these measures of the opportunity for selection were based upon the extremely low estimate of achieved reproduction among Yanomama females of 3.8 liveborn children. Furthermore, they judged the indices for populations of 'early man' as being of intermediate values. The highest opportunities of selection occur in agricultural populations, with I values as high as 3.16 among the Banyoro Bantu of Uganda (Neel and Chagnon, 1968). The Tlaxcaltecan communities of Cuanalan and San Pablo del Monte exhibit the characteristically high total index for the opportunity of selection observed in agricultural populations (see Table 9).

Longevity

The literature is replete with references to hidden valleys where populations with extremely high life expectancies not only exist but flourish (Leaf, 1973). Among these supposedly longevous populations, a New World community, Vilcabamba, Ecuador, has been championed (along with Abkhasia, Georgia, and Hunzas, Pakistan) by physician Alexander Leaf. Soviet researchers reluctantly drained the fountain of youth in Abkhasia when they demonstrated that two thirds of the 'dolgozhitili' (literally translated the term means long-lived) had in fact inflated their ages significantly. Further studies indicated that when corrections are made for these age exaggerations, life expectancy in Georgia is no different from that in Miami, Florida.

Mazess and Forman (1979) examined the anecdotal claims of extreme longevity in Vilcabamba and found, as in Abkhasia, age exaggeration among the supposed centenarians. There was a traditional pattern of exaggerating ages from 60 or 70 years on (Mazess and Forman, 1979). In a later study, Mazess and Mathisen (1982) developed lifetables for Vilcabamba, based upon a synthetic cohort method that solely utilized mortality data. They also computed life expectancies using data corrected and uncorrected for age exaggeration. The analysis revealed that rural Ecuadorean populations do not show unusual longevity and that life expectancy at birth in Vilcabamba is similar to that in the USA about 60 years ago.

Causes of death

The list of the leading causes of death for US Native Americans for all ages is headed by the diseases of the heart, accidents, malignant neoplasms, cerebrovascular disease, chronic liver disease and cirrhosis, diabetes mellitus, pneumonia and influenza, homicide and suicide. The high incidence of liver disease and diabetes are characteristic of Amerindian populations but not of US Whites. Subdividing Native mortality by sex reveals a higher level of death due to accidents (112.8 per 100 000) in males than in females (41.6 per 100 000). Accidents are the second highest cause of death of Amerindian males.

It is difficult to ascertain the leading causes of death among the Indians of the South American rain forest because of the paucity of local laboratory facilities and of personnel trained in medical diagnosis. Thus, most information about causes of death comes either from missionary stations or from visiting anthropologists or physicians who may stumble upon an epidemic of infectious disease in progress. Some evidence of disease exposure and probable cause of death has been unveiled in newly contacted Indian tribes of the Xingu National Park in Brazil through their reactions to tuberculin tests (Nutels, 1968). The effects of the introduction of measles into newly contacted populations have also been described (Centerwall, 1968). In the Mucajai Yanomama, Early and Peters (1990) described the age, sex and possible causes of 129 deaths (assigned by missionaries) that have occurred since Contact. Infectious diseases were the principal causes of death (approximately 50%), followed by infanticide (14%), homicide, and accidents; for 28% of all deaths no specific cause was given. The frequencies and the exact causes of death in these newly contacted populations should be viewed only as the roughest approximations since the missionaries lacked proper training to diagnose and even if they had this training there were no facilities for technologically based diagnoses.

In Tlaxcaltecan populations of Mexico, the leading causes of death differ significantly from those observed in United States natives. For example, the two leading classes of killer in the Indian communities are respiratory infections (pneumonia, pneumonitis, and bronchitis) and gastrointestinal disorders (gastroenteritis, gastritis, dysentery, enterocolitis, diarrhea, parasitosis and related diseases). However, secondary respiratory infections often accompany major systemic infections (Halberstein and Crawford, 1973). The major factor contributing to the high rate of gastrointestinal disorders in San Pablo and Cuanalan is the consumption of contaminated water. In the Mestizo town of Tlaxcala, cardiovascular dysfunction replaced the gastrointestinal disorders as the second leading cause of death (see Table 10). Accidents and violence combined are the fourth or fifth leading cause of death. The causes of mortality in the three Tlaxcaltecan populations reflect the socioeconomic changes taking place in Mexican societies as the 'Indios' and their children assume the Mestizo lifestyle. These changes result in the lowering of mortality (due to better health care, vaccinations and sanitation) and a reduction in fertility through the use of contraceptives.

The leading causes of death among the inhabitants of Mexico in 1960 (Gabaldon, 1965) were gastrointestinal disorders (a death rate of 170.5 per 100 000), pneumonia (141.4 per 100 000), and diseases of early infancy, including prematurity (134.00 per 100 000). The two leading causes of death are due to infectious agents; diseases of the cardiovascular system are the fifth most common source of mortality. In this regard, both San Pablo and Cuanalan fit the Third World pattern, whereas the City of Tlaxcala begins to show the higher incidence of degenerative diseases characteristic of more developed nations.

Migration

Migration patterns are influenced by a number of social, demographic and ecological factors. These factors include the age and sex structures of populations, social organization, geographic distribution of aggregates, and mating patterns. For example, in traditional matrilineal societies such as the Navajo, a preference for matrilocality

Table 10. *Leading causes of death (percentages) in three Tlaxcaltecan communities of Central Mexico*

San Pablo	Cuanalan	City of Tlaxcala
1. Respiratory infections 32.65	1. Respiratory infections 21.05	1. Respiratory infections 25.89
2. Gastrointestinal disorders 17.55	2. Gastrointestinal disorders 19.85	2. Cardiovascular dysfunction 17.98
3. Childhood infections 11.42	3. Childhood infections 12.67	3. Childhood infections 12.22
4. Accident or violence 10.61	4. Cardiovascular dysfunction 8.37	4. Accident or violence 8.63
5. Cardiovascular dysfunction 3.67	5. Accident or violence 7.17	5. Gastrointestinal disorders 5.75
6. Tuberculosis 1.63	6. Cancer 7.17	6. Cancer 5.03
7. All others 22.47	7. Cirrhosis of liver 4.54	7. Tuberculosis 4.31
	8. All others 15.54	8. All others 20.19

Source: after Halberstein and Crawford (1975).

usually exists. As a result, higher rates of in-migration and out-migration are expected, particularly for the males. The data on Navajo migration presented by Morgan (1973) are consistent with these expectations.

The primary reasons for in-migration among the Mucajai Yanomama Indians are marriage, captives seized in raids and incorporated into the society, and flight from the village of birth in order to escape revenge (Early and Peters, 1990). Almost 50% of the immigration is for purposes of marriage. Marriage with outsiders is often a necessity for Mucajai men because of the absence of females within the group who have the prerequisite kinship status. In these cases, social organization and kinship are the primary determinants of the observed migratory patterns.

Salzano *et al.* (1980b) described the demographic structure of a group of Brazilian Indians (the Ticuna) who were first contacted in the seventeenth century. The Ticuna (numbering approximately 11 000) constitute one of the largest groups of descendants from a single Brazilian tribe. Approximately 8% of the Ticuna gene pool is mixed; less than 1% of this admixture is with individuals of non-Indian ancestry. There is considerable exogamy among Ticuna settlements and the distance between the birthplace and place of residence varies from 0 to 608 kilometers. Most movements take place within a radius of 150 km; yet in one community, 58% of the adults were born more than 200 km away. The distances traversed by these acculturated New World

Table 11. *Migration matrix based upon birthplaces of the parents versus the birthplaces of the children in six Maya Indian villages near Lake Atitlan, Guatemala*

	Birthplace of parents					
	SA	CDO	SAP	SCP	SCL	JAI
Birthplace of children						
SA	298	0	0	0	0	0
CDO	6	134	0	0	0	0
SAP	0	0	543	2	0	0
SCP	0	1	7	298	0	0
SCL	0	0	0	0	176	3
JAI	0	0	0	0	2	26
Population size	3000	300	548	309	182	32
Outside migration	0.051	0.014	0.005	0.010	0.003	0.125

Source: after Cavalli-Sforza and Bodmer (1971).

natives are surprisingly long. Undoubtedly, the effects of rubber exploitation have contributed much to the mobility of these Amazonian groups.

In many agricultural communities of indigenous peoples of the New World there is great sedentism and villages are often separated from each other by large expanses of unoccupied land. This relative geographic isolation encourages local endogamy and few individuals or family members tend to migrate. Cavalli-Sforza and Bodmer (1971), in their classic volume, reprinted some previously unpublished data of their Stanford colleagues on the migration patterns of agricultural Maya Indians residing near Lake Atitlan, Guatemala. These data were assembled in the form of a migration matrix consisting of the birthplaces of the parents in the columns and the birthplaces of the children in the rows (see Table 11). The villages were highly endogamous, with 93–100% concordance between the birthplaces of the parents and children. Outside migration (systematic pressure) varied, with 0.003–0.051 of the marriages coming from outside the region, indicating that the villages were relatively closed, small populations whose gene pools were subjected to stochastic processes. The high degree of isolation between the Guatemalan villages results in attainment of predicted genetic equilibria in about 200 generations or 4000–5000 years (Cavalli-Sforza and Bodmer, 1971). This contrasts markedly with other less endogamous populations that require few generations to achieve equilibrium. Using this matrix, the predicted kinship ϕ and F_{st} values were computed for these populations.

The predicted F_{st} value (a rough measure of genetic differentiation of population subdivisions, defined more completely in chapter 5) of 0.08 suggested that the structure of this subdivided Mayan population is conducive to the genetic microdifferentiation of the isolated villages. Stochastic processes would tend to prevail in these small, reproductively isolated villages that experience low levels of systematic pressure.

The study of Tlaxcaltecan communities permits the examination of the effects of acculturation and urbanization on population movement. Both Cuanalan and San Pablo del Monte are rural, agricultural communities in the Mexican altiplano. The City of Tlaxcala is a small town of approximately 15 000 residents. The migration patterns of these three communities indicates that the City of Tlaxcala has the highest rate of movement both inward and outward. In contrast, San Pablo exhibits the greatest endogamy, with only 2% movement outside the municipio. Over the past four generations, the in-migration rate for San Pablo was 4.1%, compared with 18.8% for City of Tlaxcala. Mate selection is endogamous not only at the level of the municipio but also at that of the barrio or neighborhood. Almost 75% of the marriages in San Pablo are contracted between partners who are born in the same barrio. In the City of Tlaxcala the same pattern of barrio endogamy exists, but with a higher proportion of mates (almost 20%) coming from outside the municipio.

In addition to the extreme endogamy of most native agricult- ural populations, one off-reservation community in Arizona has been described as a 'genetic cauldron.' This predominantly Papago Indian community, Ajo, is a copper-mining town with Anglo, Indian and Mexican residents. The community is located ten miles west of the largest Papago reservation and has been characterized by Lamb (1975) as young and highly unstable. Approximately 40% of the population originated outside of Ajo; 70% of those over 20 years of age are immigrants. Outmigration is also high, with more than 30% of the 20–44 age group leaving the community. Lamb analyzed the mating patterns for Ajo (based on a total of 161 matings) and found that the most frequent mating type was between full Papagos (53.4%). Almost 37% of the matings were heterogeneous with respect to Indian ancestry. The availability of work in the mines has attracted temporary residents and fostered admixture of Papago Indians with other ethnic groups.

3.4   CONCLUSION

Salzano and Callegari-Jacques (1988) summarized the demographic and population structures of Amerindian populations, by designating each population one of three types, each corresponding to an evolutionary stage. These are: (1) hunters and gatherers with incipient agriculture; (2) advanced agriculturalists and fishermen; (3) pastoralists and populations living in densely inhabited areas and industrialized centers. They consider the Yanomama, Trio, Cayapo, Xavante and Warao representative of the hunters and gatherers with incipient agriculture. According to Salzano and Callegari-Jacques, these groups follow the fusion–fission model of population structure, with moderate mortality and fertility; migration is along kinship lines, resulting in a non-random partitioning of gene pools. The more advanced agriculturalists and fishermen are sedentary and live in groups separated from each other by large tracts of unculti-vated land. Thus, migration involves individuals or small families and there is consider-able endogamy. These agricultural groups tend to experience high fertility and high mortality. The migration matrix for six Mayan villages (discussed earlier in this chapter) illustrates a relatively low migration rate and high endogamy. Similarly, the Indian Tlaxcalatecan villages, such as San Pablo del Monte, exhibit high mortality and high fertility in contrast to the more 'Mestizoized' communities such as those of the City of Tlaxcala or Saltillo. Finally, the population size increases dramatically, reproductive isolation decreases and there is a tendency towards a uniform density over a large area. Fertility, mortality and morbidity rates all decrease with the implementation of public-health practices, the availability of Western medicine and the use of contracep-tion. The population structure resembles the isolation-by-distance model of Malecot (1959). The Inca, Aztec and Maya empires best approximate this stage of evolution.

Consideration of New World populations from the perspective of demographic structure provides a context for fitting the enormous amount of observed variation into a single model. However, there are numerous exceptions to this construct: examples are high fertility in the Yanomama, instead of the predicted moderate levels, and failure of Arctic hunting groups to follow the fission–fusion model. The use of a unifying model, as suggested by Salzano and Callegari-Jacques (1988), is useful as long as the so-called stages are not viewed solely from a cultural evolutionary perspective. Each stage does not necessarily have to evolve into the next stage but rather is the result of ecological constraints acting on unique historical events and through the action of evolution. For example, the Mennonite Anabaptist populations practice intensive agriculture but also fulfill the prerequisites of the fission–fusion model. In other words, populations with structures paralleling this model appear historically and evolutionarily at various levels of technology.

4 Genetic variation in contemporary populations of the Americas

Before we delve into the genetic variation observed in the contemporary native populations of the Americas, some causes of this phenomenon must be explored. It is highly unlikely that the patterns of variation observed in today's Amerindian populations resemble those that existed in pre-Columbian times. The present patterns of observed genetic variation are a result of a melange of factors. These are:

(1)   The numbers and sizes of migrations (expansions) across Beringia. As discussed in chapter 1, there is considerable controversy as to the number of migrations or demic expansions, from Siberia to Alaska, which added genes to the populations of the Americas. What proportion of the genetic variation is due to the heterogeneity of the founding populations and what proportion is made in America? If the ancestral groups were small in size, then stochastic processes would have contributed much to the observed variation, especially if these groups were large in number.

(2)   The continuation of gene flow into the Americas after the last glaciation. My fieldwork with the Eskimos of St. Lawrence Island and Wales, Alaska, revealed that despite the severest impediments to gene flow, such as political borders and hostile troop concentrations, contact between the Old and New Worlds continued. Up to World War II, Alaskan Eskimos crossed the winter ice pack into Siberia to obtain wives. It is my contention that social contacts persisted in the Norton Sound region between the Eskimo groups of both sides of the Bering Strait and that complete reproductive isolation between the Old and New Worlds is a myth. Similarly, there is evidence based upon parasitology (see chapter 2) that the presence of some helminths in South America can be explained by means of trans-Pacific contacts. One species of helminth could not have accompanied the early migrants across the Bering Strait because part of its life cycle requires the parasite's survival in the soil. This parasite cannot be directly transmitted from host to host but could have arrived in the New World with shipwrecked fishermen from Asia.

(3)   Population reduction due to epidemics, warfare and slavery. At the time of

European contact, the total population of the New World has been estimated to be from as low as 8.4 million persons (Kroeber, 1939) to as high as 90 to 112 million (Dobyns, 1966). The truth probably lies somewhere in between; I have suggested 44 million as a reasonable approximation (see chapter 2). This population was reduced to but a few million survivors at its numerical nadir in the two continents. The depopulation varied from region to region with the total extinction of some populations and individual rates of survivorship of up to one in three in areas like Tlaxcala, which managed to avoid some of the ravages of colonization. Fitch and Neel (1969: 395) warn that

> ' . . . a number of tribes . . . have undergone marked non-random post
> Columbian decimations as a result of disease, warfare, or persecution;
> as a consequence, indeterminate distortions may have been introduced
> into the pre-Columbian picture.'

Genetically, the important question is 'who survived?' Are the survivors a random sample of the pre-Contact populations, or are they the result of intensive selection and unique historical events? Most likely, the frequencies of the many gene products, be they variants of blood groups, enzymes or other proteins, have experienced selection and are not representative of the frequencies in native populations at contact. Similarly, conclusions as to the exact number of mtDNA lineages and their bearing on the peopling of the New World may prove to be elusive because of the probable extinction of numerous maternal lines during depopulation.

Although the frequencies of the alleles at various genetic loci in Amerindian populations may not be representative of pre-Contact values, it is essential to our understanding of Amerindian evolution that the contemporary variation be assessed. Measurements of gene frequencies of hematological markers constitute only a rough sampling of a population's gene pool. However, this genetic characterization, repeated over generations, often permits quantification of evolutionary changes that occur in a population.

4.2   ACCURACY AND RELIABILITY OF SAMPLES

Given possible laboratory errors and the use of different sampling techniques in characterizing gene pools, how reliable or accurate are evolutionary studies of human populations? Most of the research is conducted as a single sampling of the population and blood specimens are usually analyzed in a single laboratory. The scientific community accepts these results unless there is lack of verification by subsequent studies. There have been a few cases where vastly different results were obtained when two or more sets of results from the same population were compared. Usually, in follow-up studies the samples are collected from different individuals and/or the original research was conducted years or decades earlier, leaving questions as to whether the observed differences were due to discrepancies in sampling, evolutionary changes or laboratory or methodological error.

During almost a quarter century of field research, I have encountered two cases where the initial genetic characterizations failed to be verified. One is the unusual series of gene frequencies reported by Yuri Rychkov and his associates in indigenous Siberian populations. Rychkov and Sheremetyeva (1977) asserted that the average frequency of FY*A in 38 Siberian populations was 0.2319. Because of the high frequency of the FY*A allele throughout Siberia, Ferrell et al. (1981) raised a 'red warning flag' concerning the low frequencies reported by Rychkov and Sheremetyeva. The probability of these low frequencies being due to sampling error is less than 10^{-52} (Ferrell et al., 1981). I had suggested that the most probable reason for these low frequencies is faulty antisera (Crawford et al., 1981b). Studies of the same populations by Sukernik and his colleagues failed to verify other observations by Rychkov, particularly the frequencies of RH haplotypes in Chukotka populations. Another perplexing series of results reported by Rychkov et al. (1969) concern the unique gene frequencies in the populations of the Altai. This genetic distinctiveness is in part due to the high incidence of the NS chromosomal segment, with a frequency reaching 53% in one of the Tophalar villages. When compared with the incidence of NS in other Siberian populations the high frequency in the Altai appears to be anomalous. Similarly, two Touvan communities display an unusual incidence of CDE when compared with surrounding groups. Measures of genetic microdifferentiation between the Siberian groups, based upon Wright's F_{st}, reveal that if the Touvinian and Tophalar groups are included in the computation, there is virtually as much genetic differentiation in Siberian indigenous groups as among all the races of the world. The F_{st} value for the circumpolar groups (including the Altai) is 12.2%; the differentiation of world populations, measured by F_{st} values, is approximately 15%. However, with the exclusion of the Altai populations, the F_{st} values for the circumpolar groups appear to be more similar to values reported for populations distributed in comparable geographical regions, such as the Amazonian populations.

The population of St. Lawrence Island was genetically sampled twice within a two-year period. The first was in 1977 to investigate possible effects of reproductive barriers such as languages on the genetic divergence of Eskimo populations on the Seward Peninsula. Two Yupik villages from St. Lawrence Island (Savoonga and Gambell) and two Inupik communities (Wales and King Island) were selected. In addition to linguistic differences, there were a number of historical factors that made the genetics of St. Lawrence Island populations of great interest. The following year, Laughlin and his team initiated another project on St. Lawrence Island, which included blood genetic characterization of the populations. This created a rare opportunity to study sampling methods and the accuracy of laboratory techniques on samples collected from the same villages one year apart. Here was a chance to determine how good a job scientists do when genetically characterizing human gene pools. The blood specimens were analyzed by two first-rate research laboratories, the War Memorial Blood Bank of Minneapolis and the Center for Demographic and Population Genetics, University of Texas Health Sciences Center, Houston. The immunoglobulins were typed by two other laboratories, one at the American Red Cross, Washington, D.C., and the other at the Department of Human Genetics at the University of Michigan Medical School. In the case of the immunoglobulins, Moses Schanfield from our group was trained by Henry Gershowitz from the University of Michigan. Thus, the same techniques were used to type immunoglobulins.

The two investigations of St. Lawrence Island Eskimos focused upon slightly different questions and these affected the choice of genetic markers employed. Our group, from the University of Kansas, was interested in the genetic structure of Eskimo populations and how unique historical events and cultural factors such as social organization and language influence allelic distributions. The University of Texas group appeared to be more preoccupied with populational affinities and with screening for rare variants (Ferrell et al., 1981). As a result, they examined more proteins and blood-group systems. There were slight differences in sample sizes, with 172 vs. 222 sampled in Savoonga and 81 vs. 73 in Gambell. There were also some technical differences in the storage and preparation of the specimens for shipment, with the Kansas group separating the erythrocytes from the serum and packing the blood components separately in wet ice. The Texas group transported their specimens in liquid nitrogen.

Although most of the results by the two groups were similar, there were a number of significant differences. Ferrell et al. (1981) reported that the frequency of ABO*A1 in Gambell was 0.183 whereas our group calculated a frequency of 0.334. The apparent difference appears to be due to non-random sampling, with twice as many individuals (possibly related) with A1 participating in our project. The results for Savoonga are close, with the two studies showing insignificant differences: A1, 0.268 vs. 0.251; B, 0.094 vs. 0.103; O, 0.638 vs. 0.646. I assume that in Table 1 of Ferrell et al. (1981) the second of the B-gene frequencies is actually the O blood group frequency. There are at least eight typographical errors or miscalculations contained in their Table 1 that require interpretation, including an NS frequency of 0.619 instead of 0.062 and ESD*1 frequency of 0.068 instead of 0.932. Another major difference between the results of the two analyses is at the Diego locus. We failed to detect any DI*A reactions for the two communities whereas Ferrell and his colleagues found 14 individuals who were DiA+. This difference may have stemmed from the way we shipped blood specimens, which may have affected the Diego reaction. Although both groups utilized the same five antisera for the RH system, Ferrell reported that RH*RO appeared at a frequency of 12.3% in Savoonga; our group failed to detect any RO phenotypes.

The results of the immunoglobulin testing were highly similar, with frequencies often varying only in the second decimal place. There is one important difference: our group described the presence in a Savoonga family of GM*A,Z- (GM*A-), which is reported by Ferrell et al. (1981) as the haplotype GM*A,Z B. The availability of solid pedigree data allowed us to identify the unequal crossover (Schanfield et al., 1990).

This comparison of two independent research groups' genetic characterizations of the same population one year apart reveals that even under optimal conditions some differences in gene frequencies and detection of a few specific alleles will be seen. Unlike some studies in cultural anthropology, where the results of different investigators working in the same communities sometimes bear little resemblance (for example the Mead vs. Freeman controversy surrounding growing up in Samoa: Freeman, 1983), the differences in gene frequencies are relatively small and have little effect on the measures of population affinities. Both of the formal analyses, one based upon dendrograms and the other on an R-matrix, reveal that St. Lawrence Island populations cluster closely together and show proximal affinities to other Siberian Eskimo groups. This should not be construed as an attempt to sweep the differences away. On the contrary, I believe that greater rigor is required in the sampling methods employed and in the replicability of laboratory tests. The best solution for improving the accuracy of experimental work in the laboratory is (funds permitting) to send blood duplicates to two laboratories and to compare results. Any tests showing differences in characterizing the phenotypes should be repeated. Although one can attempt to institute random sampling methods, it is difficult to impose such schemes, because human research subjects have the right to refuse to participate in projects; elegant sampling designs often deteriorate to 'get-what-you can' samples!

The remainder of this chapter summarizes the observed genetic variation in New World populations as measured by markers found in blood. Within the constraints of this volume, gene frequency distributions of the New World cannot be summarized population by population, locus by locus. As a result, I shall endeavor to highlight the patterns that are of greatest interest evolutionarily and/or those that pertain to the peopling of the New World.

Table 12. *Blood-group and protein-marker frequencies in New World populations*

The ISGN (1987) notation system is used for describing alleles

High-incidence markers	Low-incidence or absent markers
ABO•O	ABO•A2
MN•M	ABO•B
RH•R1	RH•RO
FY•A	LU•A
DI•A	K•K
ABH•SE	LEA+
	Abnormal hemoglobin

Source: After Layrisse (1968).

4.3   AMERINDIAN AND SIBERIAN MARKERS

William Boyd (1952) believed that Amerindians as a whole were distinct from other major continental populations in their blood- group frequencies. He proposed a single American Indian serological race, one of seven such major groupings. In the volume *Biomedical Challenges Presented by the American Indian* (1968) Miguel Layrisse summarized the distinctive patterns of frequencies in Amerindian populations (see Table 12).

During the past two decades, innovations in biochemical genetics and serology have produced a plethora of genetic markers that can be utilized to evaluate populational affinities and movements. Since Layrisse's compilation of 1968, a number of new genetic markers have been identified through electrophoresis, isoelectric focusing and immunologic techniques. Two of the most informative sets of genetic markers encode the immunoglobulins (GMs and KMs) and the human leukocyte antigens (HLA system). The newer markers that distinguish Amerindian populations are listed in Table 13.

According to Lampl and Blumberg (1979), the HLA•A2 allele has the highest incidence among Amerindians. In addition, HLA•A9, HLA•W28 and HLA•W31 are common alleles. Bodmer and Bodmer (1973) note the absence of HLA•A1, HLA•A3, HLA•A10, HLA•A11, HLA•W29 from Amerindian populations. North and South American Indian populations can be distinguished on the bases of HLA•AW31 and HLA•W15, which occur at high frequencies among South American Indian groups, whereas HLA•W28, HLA•A9 and HLA•W5 are North and Central American markers.

Table 13. *Newer genetic alleles which, by virtue of their frequencies, distinguish New World populations*

High-incidence	Low incidence or absences
HLA•A2, HLA•A9, HLA•W28	HLA•A1, HLA•A3, HLA•A11
HLA•BW15, HLA•BW16, HLA•BW40	HLA•B29, HLA•B18
GM•A G, GM•A T	GM•F B, GM•A,F B
GC•1S, GC•IGLOOLIK	BF•F
GC•CHIP	
ALB•MEX, ALB•NASK	
TF•DCHI, TF•BO-1	
CHE1•S, CHE•2+	

Asian populations are characterized by high frequencies of HLA•A2 and HLA•A9, HLA•BW15, HLA•BW22, and HLA•BW40. Other genes, such as HLA•AW31 and HLA•AW32 and HLA•B14 and HLA•BW21, are absent from Asian populations. Although Amerindians share a number of HLA haplotypes with Asian groups, thus further documenting their Asian ancestry, some unique haplotypes exist in both groups. To date, there is a paucity of comparative HLA haplotype information available from Siberian indigenous populations. The available data on the HLA frequencies of the New World are summarized later in this chapter.

In contrast to the HLA system, a greater corpus of information is available on the world distribution of the immunoglobin haplotypes. The New World populations share with Asia the highest incidence of GM•A G in the world. This haplotype varies in frequency between 0.35 and 0.86 in Siberia but has a higher incidence among Amerindians and Eskimos. Although GM•A T is the second most frequent haplotype among the Chukchi, Athapascan- speaking Indians, and Eskimos, it is of negligible incidence in or absent from Mesoamerican and South American Indian groups. In addition, GM•A T is maintained at moderate to high frequencies in all Siberian and North American Arctic populations tested to date (Sukernik and Osipova, 1982).

Sukernik and Osipova (1982) argued that the immunoglobulin haplotype distributions of Siberia and the New World support Szathmary's conclusions that Eskimos, Reindeer Chukchi, and North Athapaskan groups cluster together, as indicated by similar allelic frequencies. They suggested that the Reindeer Chukchi and the Athapaskan Indians formed a continuum along the land bridge, but were separated from other groups that expanded into the New World during the Wisconsin glaciation. This conclusion is further supported by the observed gradual increase in the KM*1 allele from Siberia, across Western Alaska to inland North America.

Other genetic systems that can be used to establish population affinities between Siberian and New World groups include: group-specific component (GC), serum pseudocholinesterase (CHE1 and CHE2), properdin factor B (BF), transferrin (TF), and albumin (ALB). The group-specific component locus provides a number of marker alleles, such as GC*IGL among Eskimo populations and GC*CHIP in Amerindians. New World populations can be distinguished from all others by the high incidence of the GC*1S allele, identified by isoelectric focusing, and the low frequency of GC*2. The properdin factor locus exhibits a low frequency of the BF*F allele and a high incidence of BF*S in Amerindian and Asian groups. In addition to these gene products, DNA RFLPs and sequences of mitochondrial and nuclear DNA are providing finer discrimination among populations and individuals.

4.4   TRADITIONAL MARKERS OF THE BLOOD

Blood group systems

ABO

In the year 1900 Landsteiner discovered the ABO blood group system, the first of its kind identified in the human population. He described three of the most common blood types, A, B, and O; while the fourth phenotype, AB, was identified by Decastello and Sturli (1902). Bernstein (1924) showed that the four most common phenotypes of the ABO system were the result of three alleles, A, B, and O. Biochemical studies later indicated that the blood-group antigen is a secondary gene product made up of complex mucopolysaccharides. According to Roychoudhury and Nei (1988), the genetic locus controlling the ABO system is located on the long arm of the ninth chromosome (9q34).

An assumption made by early physical anthropologists and geneticists was that the A and B alleles were introduced into the Americas by gene flow from Europeans (Coca and Deibert, 1923). One suggestion was that the Amerindians separated from all other human populations prior to the evolution of the A and B antigens. This assumption was based in part upon the absence of both ABO*A and ABO*B alleles from Amerindians dwelling in the South American tropical forests. This assumption led some researchers to compute European admixture in Mesoamerican Indian populations based upon the incidence of the ABO*A and B alleles (see, for example, Matson, 1970). However, Matson and Schrader (1933) found in the northern Plains that the Blackfeet and Blood Indians exhibited blood type A at a high frequency. This evidence argued against the introduction of the A allele into North America by Europeans.

Based upon genetics, morphology and geographic location, there is a consensus among scientists that Siberian hunters and gatherers peopled the Americas. Obviously, these migrants brought their genes with them. There are two major sources of information that bear upon the hypothesis of the presence or absence of the A and B antigens in New World populations prior to European contact. The first is the contemporary distribution of the ABO blood types. The geographic distribution of these antigens is one indication of their presence in the founding populations. All of the Siberian populations are highly polymorphic at the ABO locus, with the average frequency of the ABO*A1 allele (type A$_1$ is a subtype of A) being almost 20% among the reindeer Chukchi of Chukotka Peninsula and 14% among the Siberian Eskimos (Sukernik et al., 1981). Both the ABO*A and ABO*B alleles are relatively common among Samoyed groups, with frequencies for the ABO*A1 allele of 12% among the Forest Nentsi and more than 20% among the Nganasan of the Tamyr Peninsula (Karaphet et al., 1981; Sukernik et al., 1981). Similarly, the ABO*B allele occurs at a frequency of 23% among the Nentsi and 18% in Siberian Eskimo populations. In a collaborative research program (US–Canada–Russia) focused upon the Tungusic-speaking Evenki, it was found that the combined frequencies of ABO*A and ABO*B total 27% (Crawford et al., 1992, 1994). Assuming that the gene frequencies observed in contemporary Siberian populations approximate those of the founding groups that peopled the Americas, at least one in five founders would have carried an ABO*A1 allele and approximately one in five an ABO*B allele. Unless the effective sizes of the populations that crossed the Bering Strait were exceedingly small, considerable genetic drift would have been necessary to 'lose' these two alleles in all populations by the time the Amerindians reached South America. If the populations were larger, then selection would have to be be invoked to explain the loss of the genes.

The second source of information is the blood grouping of mummified tissues from South America. Llop and Rothhammer (1988) have tested 54 mummies from northern Chile for the presence or absence of A and B antigens and demonstrated the presence of the ABO*A gene in pre-Columbian populations. The typing of skeletal or mummified materials for blood group antigens has been in some dispute because Thieme and Otten (1957) demonstrated that the bacterium Clostridium tertium (located in the soil) can decompose the A antigen and produce false negatives. Llop and Rothhammer did find clostridial spores in the eight mummies that lacked ABO antigenic determinants. Other researchers have found evidence for the presence of both the ABO*A and ABO*B alleles in prehistoric South American populations (Allison et al., 1978; Boyd and Boyd, 1937).

Geographic distribution: The ABO•A and the ABO•B alleles are distributed throughout indigenous Siberian groups and in Eskimo populations on both sides of the Bering Strait. Among Eskimo populations, the ABO•B allele is concentrated mainly in Alaska and in Angmagssalik in eastern Greenland, with frequencies reaching 18% in Point Barrow and in some Siberian groups. However, most of the North American Indian tribes exhibit only the ABO•A and ABO•O alleles, with an occasional low frequency of ABO•B, which is usually attributed to admixture with Europeans and Africans (Mourant *et al.*, 1976a). South of the Mexican border, both the ABO•A and ABO•B alleles are rare and the presence of these genes has been explained by European contact. Some of the Amazonian tribes lack either antigen; the ABO•O allele is fixed in their gene pools. However, Andean populations (with no known history of European admixture) possess both ABO•A and ABO•B genes. Judging from the recent evidence from pre-Columbian mummified tissue, it is likely that both the A and the B antigens were brought into the New World across the Bering Strait and were lost in some populations through the founder effect, genetic drift, and possibly selection. In Central America, both ABO•A and ABO•B alleles are present in most of the populations but at relatively low frequencies. In the 12 Indian tribes of Mexico sampled for the blood-group antigens, Cordova *et al.* (1967) observed variation in frequencies of both the ABO•A and the ABO•B alleles ranging from 0 to more than 5%. Unfortunately,the samples were small, with the Chinantecs represented by only 21 individuals. The Tarahumara Indians of northern Mexico lack the ABO•B allele (Rodriguez *et al.*, 1963). Matson (1970) compiled the blood-group gene frequencies for a large number of Mesoamerican populations. He also computed levels of European admixture for these groups, using Bernstein's formula and the combined frequencies of A and B alleles as indicators of outside admixture. Matson (1970: 109) states that 'a high degree of **racial purity** is suggested by these high values' (emphasis added). This comment suggests a typological genetic approach to the concept of race.

In those Eskimo and Amerindian populations exhibiting the ABO•A allele, subtyping reveals the predominance of the ABO•A1 type. Newman (1960b) argued that the ABO•A1 allele is aboriginal in the New World because of its high frequency in northwestern North America, especially among the Blackfoot and Blood Indians. The ABO•A1 allele occurs at a frequency of about 0.45 in the Blackfoot and Blood and clines radiate out sharply in all directions. In a ground-breaking compendium Cavalli-Sforza, Menozzi and Piazza (1994) graphically documented the primary focus of the A1 allele in north-central North America (see Fig. 23). Brues (1963) argued that most Indian populations could have had ABO•A, and lost it through drift as a consequence of its low frequency in the founding populations. Computer simulation confirmed this likelihood, particularly for the small tropical forest Indian populations of eastern South America. Gene loss is a more parsimonious explanation than trying

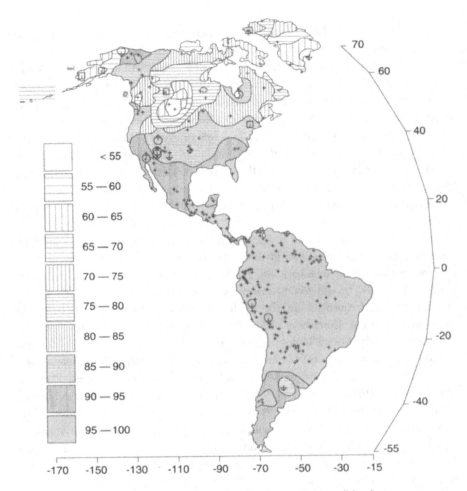

FIGURE 22 A gene map of the geographic distribution of the O allele of
the ABO blood-group system in the indigenous populations of the Americas
(Cavalli-Sforza *et al.*, 1994).

to explain the observed variation entirely through migration. There is evidence for the
fixation of ABO*O and other alleles in the Xavante, Yanomama and other small
groups. The presence of ABO*A2 among Amerindians and Siberians is often attributed
to admixture with Africans and Europeans.

Rhesus (RH)

The Rhesus blood group system was first detected by Landsteiner and Weiner (1940)
when they injected Rhesus monkey cells into guinea pigs and produced antibodies that

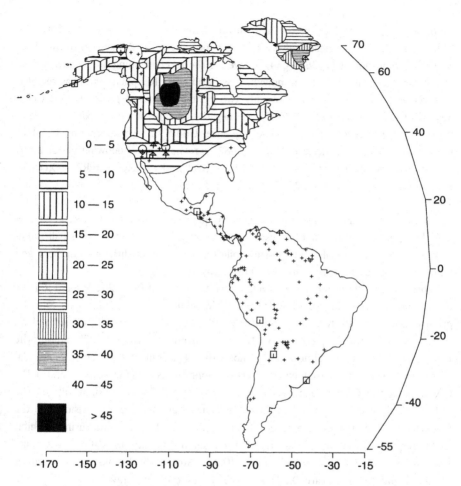

| 0 — 5 |
| 5 — 10 |
| 10 — 15 |
| 15 — 20 |
| 20 — 25 |
| 25 — 30 |
| 30 — 35 |
| 35 — 40 |
| 40 — 45 |
| > 45 |

FIGURE 23 A gene map of the geographic distribution of the A1 allele of the ABO blood-group system in the indigenous populations of the Americas (Cavalli-Sforza *et al.*, 1994).

would agglutinate some human red cells. There has been considerable controversy over the number of loci involved in the regulation of gene action for the Rhesus system. Initially, Weiner hypothesized that the various RH phenotypes resulted from a single locus with multiple alleles, such as r, R1, R2, RO, at a single locus. However, after examining the transmission of RH phenotypes, Fisher *et al.* (1944) suggested that the most parsimonious explanation of the observed variation required three closely linked loci, C, D, and E, each with two common alleles. The segregation of a -D-chromosomal deletion in a Tlaxcaltecan family was interpreted as being supportive of the three-locus hypothesis (Turner *et al.*, 1975). In addition, Turner *et al.* concluded that the most probable sequence of loci on the first chromosome was not CDE (as predicted by Fisher) but ECD. This research also verified the location of the RH system to the short arm of chromosome 1 (1p36.2–p34). However, the recent cloning and sequencing of cDNAs and genomic DNA from individuals with different RH phenotypes has revealed that the RH locus is composed of only two related structural genes, D and CcEe, instead of three (Mouro et al., 1993). RH-negative phenotypes are apparently homozygous for the deletion of the D gene sequence. Under the An International System for Human Gene Nomenclature (ISGN, 1987) agreement, the following designations are utilized for the RH system: r (cde) = R, R1(CDe) = R1, R2(cDE) = R2, RO(cDe) = RO, r′ (Cde) = R′, r″(cdE) = R″, RZ(CDE) = RZ, and r^y (CdE) = RY (Shows et al., 1987).

There is considerable variation in allelic frequencies at the Rhesus loci in Siberia and the Americas. Some of this observed variation is a reflection of the small populations studied, the small sample sizes collected, the use of differing numbers and kinds of antisera, and admixture with Europeans and Africans.

Siberia can be characterized as having a high incidence of RH*R2 (cDE) and RH*R1 (CDe) and a low incidence of RH*R (cde). In addition, some indigenous Siberian groups have a relatively high frequency of RH*RO (cDe), normally considered an African marker. The frequency of RH*R2 varies from an average of 22% among the Chukchi subdivisions to almost 60% among the Nganasan of the Tamyr Peninsula. This haplotype occurs at frequencies of 0.46 among the Nentsi (Samoyeds), and 0.21–0.26 among the Chelkanians, reaching a low of 0.097 in one Eskimo village. The RH*R2 haplotype is observed at high frequencies among the subdivisions of the Chukchi of Chukotka, with a range of 0.66–0.72. The Evenki from Surinda exhibit allelic frequencies that are characteristic of indigenous Siberian populations, with high frequencies of RH*R1 (0.60) and RH*R2 (0.37). Among the Evenki, the frequencies of RH*R and RH*R1W are 0.025 and 0.004, respectively (Crawford et al., 1994). The Chelkanians and the Kumandinians show a narrow range in frequency (RH*R1, of 0.50–0.55). The haplotype RH*RO appears to be common in southern Siberia, with frequencies ranging from 11% among the Yakuts to 66% among the Tophalars (Rychkov et al., 1969). However, no other study verifies the high incidence of RH*RO reported in the Altai groups.

The New World Eskimos exhibit a high frequency of RH*R2 (as do a number of indigenous Siberian groups), surpassing 50% in some Eskimos of Alaska and Canada. The two Eskimo villages on St. Lawrence Island have RH*R2 frequencies of 0.45 and 0.37 (Crawford et al., 1981b). The RH*RO haplotype was either absent from or at a low frequency in Eskimo groups prior to admixture with Europeans.

Amerindian populations of North, Middle and South America have Rhesus-system haplotypes similar to those observed in Siberian and Eskimo groups. The RH*R1 and

RH•R2 gene complexes are common in native populations, frequencies of the former often exceeding 50%. For example, among the Cherokee, Creek and Choctaw of Oklahoma, the frequency of RH•R1 ranges from 42 to 57%, while that of RH•R2 ranges from 25% to 38% (Kasprisin et al., 1987). The haplotype RH•RZ is usually found at lower frequencies, approximately 1–5%, in the various tribes of the Americas. Judging from their low frequencies and patchy distributions, it is likely that RH•RO and RH•R were introduced into the New World by Europeans and Africans (Mourant et al., 1976a). Throughout the Americas, there is considerable variation in the frequencies of the common RH haplotypes, particularly in small South American isolates and population subdivisions. For example, Neel et al. (1977a) show that, among 11 Wapishana tribal villages, the frequency of the RH•RZ haplotype ranges from 0 to 17%, and that of RH•R2 varies from 12 to 46%. The frequency of RH•RZ varies in Mexican Amerindian populations from 0.009 in the Cora to 0.136 in the Mixe Indians (Cordova et al., 1967). In sum, the Rhesus gene markers observed in the Americas are, in general, similiar in frequency to those found in Siberia. The one ambiguity concerns the RH•RO haplotype. Its presence the New World has always been attributed to African admixture (Mourant et al., 1976a), yet it occurs in Siberia at a relatively high frequency and in the absence of African gene flow. If this haplotype is native to Siberia and was brought across the Bering Strait, then measures of African admixture based upon RH•RO are overestimates.

RH•R (cde) has its highest frequency in Europe (up to 40% among the Basque) and this marker has been used as an indicator of European gene flow into Amerindian populations. Cavalli-Sforza et al. (1994) plotted the distribution of RH•cde (RH•R) in the Americas and found its highest concentration amongst Amerindian populations located on the northeastern coast, with gradients emanating out (see Fig. 24). These populations had the longest history of contact with European settlers and are undoubtedly the most admixed of the native peoples of North America.

MNSs

The MN system was described in 1927 by Landsteiner and Levine (1927a,b, 1930). They immunized rabbits with different human erythrocytes in order to identify antibodies that reacted to antigens carried by people with identical ABO phenotypes. Walsh and Montgomery (1947) found an antigen that was eventually called S. Sanger and Race (1951) demonstrated that the S allele belongs to an MNSs system and that the M and S loci are closely linked. The MN blood groups serologically define glycophorin A; S, s define glycophorin B. In ISGN, 1987 (Shows et al., 1987) the system is referred to as MNS with two loci, MNSM and MNSS, and M = 1, N = 2 and MN = 1, 2. Under this system S = 3 and s = 4. However, this nomenclature is confusing to

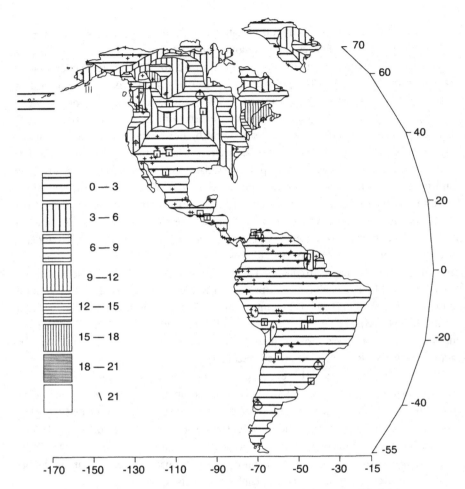

FIGURE 24 A gene map of the geographic distribution of the RH•R allele (cde) in the indigenous populations of the Americas (Cavalli-Sforza *et al.*, 1994).

those of us who were brought up on the MNS nomenclature. A concession to ISGN nomenclature is used in this chapter, namely the separation of the loci by a space, and the use of M, N, S, LS = s (Shows *et al.*, 1987). Thus, to avoid confusion I have utilized a system resembling the old nomenclature, i.e. MNS•M LS (Ms), MNS•M S (MS), MNS•N LS (Ns), and MNS•N S (NS). These loci are located on the long arm of the chromosome 4 (4q28–q31).

The MNS system and the geographical distributions and frequencies of its constituent alleles are often presented in the literature locus by locus. That is, frequencies of the M and N alleles are considered separately from those of the S and s alleles. The highest frequency of the M allele occurs in Amerindian and Eskimo populations. The

Sarcee of Canada and the Naskapi of eastern United States have frequencies of M in excess of 90% (Mourant *et al.*, 1976a). Samoyed-speaking Siberian populations have the lowest frequencies of M in Siberia: the Nganasan, 0.22–0.34; the reindeer Chukchi, 0.42; and the Nentsi (0.50). The Chelkanstii and the Kumandintsii (from the Altai) have the highest frequencies of M (0.72–0.8). The Siberian Eskimos of Siryeniki (0.57) and Chaplino (0.59) have intermediate frequencies. The Evenki of Surinda have an intermediate frequency of 0.62, which can be apportioned among MS (0.077) and Ms (0.542) haplotypes. Because of linkage between the MN and S loci, N must be viewed in conjunction with the S allele, i.e. the frequency of NS is 0.006 and that of Ns 0.375 among the Evenki (Crawford *et al.*, 1994). There is little comparative information in indigenous Siberian populations on the distribution of the MS chromosomal segments. In some Siberian populations (Nentsi, Chukchi, Nganasan and Eskimos) the s allele is more often linked to the M than is the S allele. Siberian populations also resemble Amerindians in a higher incidence of MNS*N LS than of MNS*N S. A comparison of the Siberian (Yupik-speaking) Eskimos with the Alaskan Inupik-speaking Eskimos reveals a higher incidence of the M allele in the Inupik group. For example, the populations of Wales and King Island have frequencies of M of 0.86 and 0.73 versus 0.58 and 0.46 for those of Savoonga and Gambell, respectively. There is a division between the high frequencies of M alleles from Wales, Alaska, to Greenland, and the lower frequencies of Siberian Yupik populations. The high frequency of MNS*M LS in New World Eskimos is in contrast to the low frequency of MNS*N LS. The MNS*N LS haplotype is the least common of the MNS haplotypes in North American Indians, with frequencies ranging from 0 to 17%. The highest incidence of MNS*N LS occurs among the Ojibwa of northern Ontario (17%) and the Blackfoot of Alberta (16%).

The M allele is common in North, Central and South America, with frequencies reaching 90% in regions of Ecuador and Peru. However, there is considerable variation in frequencies of M, N, S, and s throughout the New World, with few clines or clearcut structures apparent. According to Kasprisin *et al.* (1987), Amerindians of Oklahoma, such as the Cherokee, Creek and Choctaw, exhibit high frequencies of M (71% to 81%) and S is found most commonly with the M allele. In Mexican Amerindian populations, the M allele varies in frequency from 68 to 80%. Little information is available on the MNSs haplotypes in Mexico, but the frequency of MNS*M S is 37.6%, of MNS*M LS 40.6%, of MNS*N S 7.1% and of MNS*N LS 14.7% in San Pablo del Monte in the State of Tlaxcala (Crawford *et al.*, 1974).

Diego

The Diego system and the anti-DIA antibody were identified in 1955 by Layrisse *et al.* in a family with hemolytic anemia. They found that the DIA antigen was infrequent in families of European origin but was common in South American Indians. Further studies revealed that the DIA antigen was present in Asians and other Amerindian populations (Gershowitz, 1959). Thompson *et al.* (1967) identified a second antibody, anti-B, and demonstrated the presence of the alleles DI*A and DI*B.

The Diego DI*A allele has been used as a marker gene for Asian and Amerindian populations. In Siberia, this allele occurs at low frequency among the Nganasan (Karaphet *et al.*, 1981). Reindeer Chukchi and Siberian Eskimos have the DI*A alleles at frequencies of 2–4%. The Eskimos of mainland North America appear to lack this Diego allele, while Ferrell *et al.* (1981) have detected it at low frequency on St. Lawrence Island. According to Mourant *et al.* (1976a), the Eskimos differ from Siberians and Amerindians in their lack of the DI*A allele.

The DI*A gene is present in most Amerindian populations but at frequencies lower than those observed in eastern Asia. The frequencies of this allele is between 2 and 6% in North America, with slight regional variation. For example, the tribes in Oklahoma, Cherokee, Creek and Choctaw, exhibit the national range of variation. Mexican Indian populations vary in frequency of the DI*A allele from 1% among the Tarahumaras (Rodriguez *et al.*, 1963) to a high of 22.7% among the Huichol (Cordova *et al.*, 1967). I suspect that the unusually high frequency of the DI*A allele among the Huichol is an artifact of sampling, since only 72 individuals were used to represent the entire population. South American tribes, such as the Wapishana and Macushi, have relatively high frequencies of the DI*A allele, from 8 to 29%, in their populations subdivided by villages.

Duffy

The Duffy blood group system was first reported by Cutbush *et al.* (1950) when they discovered an antibody in the serum of a man (named Duffy) who had received multiple transfusions in the course of treatment for hemophilia. This antigen was given the designation of FY*A. Another allele, FY*B, was detected by Ikin *et al.* (1951). Shortly thereafter, Sanger *et al.* (1955) showed that the majority of Africans did not react with either anti-FY A or anti-FY B and the allele they carried was eventually termed FY (now FY*3). Roychoudhury and Nei (1988) locate the Duffy system on the long arm of chromosome 1 in the region q12–q21.

Geographically, the FY*A allele occurs in European populations at frequencies ranging from 40 to 50%. Higher proportions are generally observed among Asian groups, but the highest frequencies are found among Melanesians and Australian Aborigines (Mourant *et al.*, 1976a). With few exceptions, Asian, Siberian and Amerindian populations have higher frequencies of FY*A than of FY*B alleles. The average FY*A frequencies of 0.232 for 38 Siberian populations were reported by Rychkov and

Sheremetyeva (1977). However, their results were not verified by other researchers even when the same populations were sampled. Ferrell et al. (1981) report that the probability of obtaining the results published by Rychkov and Sheremetyeva is less than 10^{-52}. The most likely explanation for the anomalous findings is that the Duffy antisera were faulty (Crawford et al., 1981b). In Siberia, the FY*A allele varies in frequency from 0.95 or 0.96 in the Reindeer Chukchi and Siberian Eskimos to about 0.60 among the Chelkanians and Kumandinians (Sukernik et al., 1978). The Evenki population of Surinda exhibits an FY*A frequency of almost 83% (Crawford et al., 1994). The FY*A frequency varies in Alaskan Eskimo populations from 85 to 100% (Crawford et al., 1981b).

In North American Indian populations the FY*A allele is found in the 50–80% range in populations represented by large sample sizes. Oklahoma tribes have a range of FY*A frequency of 57–62% (Kasprisin et al., 1987). Tribes in the Southwestern United States, such as the Maricopa, Pima, Papago, and Zuni, exhibit a relatively narrow range of variation in the frequency of FY*A, from 0.71 to 0.82 (Workman et al., 1974). South Amerindians show considerable variation, with some studies reporting frequencies of FY*A lower than 50% (Post et al., 1968). However, frequencies as high as 0.916 and 0.919 have been described for the Trio and Wajana of Surinam (Geerdink et al., 1974).

Kidd (JK)

The Kidd blood group system, discovered by Allen et al. (1951), is located on the short arm of the second chromosome [2p(I)]. An antibody was found in the serum of Mrs Kidd, whose newborn suffered from hemolytic anemia (Plaut et al., 1953). The frequencies of the JK*A allele vary among populations from 0.135 among the Guaymi of Costa Rica (Barrantes et al., 1982) to 0.946 among the Khoi of South Africa (Jenkins, 1972). The JK*A allele occurs at highest frequencies in African groups (an average of 75%), lower in European groups (an average of approximately 50%) and lowest in Asian populations (an average of 30%).

The JK system is highly variable among the New World populations, with frequencies of JK*A reaching as high as 0.796 among the Eskimos of Savoonga, St. Lawrence Island (Crawford et al., 1981b) and as low as 0.135 among the Guaymi. The Cavalli-Sforza et al. (1994) plot of the JK*A alleles in the Americas fails to show the focus of high JK*A alleles among the St. Lawrence Island populations (see Fig. 25). There appears to be no clear-cut pattern in the geographical distribution of the JK*A allele in the native American populations, with a few unique clusters of high or low frequencies surrounded by populations with frequencies of JK*S of 40–50%. Mourant et al. (1976a) have argued that the Kidd (JK*A) gene frequency varies widely but averages 45–50%

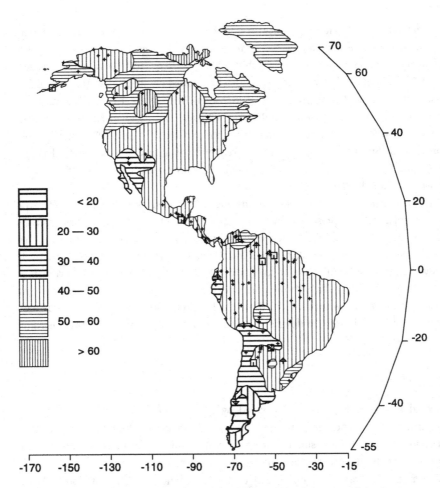

FIGURE 25 A gene map of the geographic distribution of the JK∗A allele of the Kidd blood-group system in the indigenous populations of the Americas (Cavalli-Sforza et al., 1994).

in both Amerindians and Eskimos. However, in Mexico the JK∗A frequency falls outside Mourant's expected average, dropping to 0.36–0.37 in the State of Tlaxcala (Crawford et al., 1974). Similarly, this allele is found at relatively low frequencies among the Papago (0.360) and Pima (0.359) Indians (Niswander et al., 1970).

The only gene-frequency data published on the Kidd blood-group locus in Siberian indigenous populations suggest that considerable variation exists. Although the average frequencies for two tribes (Reindeer Chukchi and Forest Nentsi) are similar (JK∗A frequencies of 0.54 and 0.41) there is considerable intratribal variation. In particular, the JK*A frequencies in Forest Nentsi range from 26 to 56% while those in the Chukchi vary from 47 to 69%. These are small subdivisions of the populations and are more sensitive to stochastic fluctuations. The Nganasan of Tamyr exhibit the highest frequency (83%) of the JK*A allele. However, the sample size is small, which may account for the uniquely high frequency measured among the Ngansan as opposed to the other Samoyed-speaking groups (Sukernik et al., 1978).

P blood group

The P blood-group system was discovered by Landsteiner and Levine in 1927. Sera from rabbits immunized with human cells agglutinated red cells from some people, even when the ABO and MN types were kept constant. These antibodies were intially termed P+ and P− and the phenotypes were eventually called P_1 and P_2. Landsteiner and Levine later demonstrated that the P_1 antigen was inherited as a dominant allele and that this system was polymorphic (Landsteiner and Levine, 1930). This P locus appears to be carried on chromosome 6 (Roychoudhury and Nei, 1988).

The P locus appears to be polymorphic in Siberian populations, with frequencies of the P*P1 allele varying between 20% (Eskimos of New Chaplino) and 42% (Chukchi of Chukotka). This difference of 22% between Siberian Eskimos and Chukchi in the frequency of the P*P1 allele make this locus useful in distinguishing these indigenous groups of Chukotka. The Eskimos on the Alaskan side of the Bering Strait exhibit relatively low frequencies of P*P1, with the Inupik Eskimos having the lowest frequencies: 9% in Wainwright (Corcoran et al., 1959), 8% in Wales, 17% in King Island (Crawford et al., 1981b) and 19.7% among the Thule of Greenland (Gurtler, 1971). The Yupik Eskimos exhibit a slightly higher frequency (21%) of P*P1 on St. Lawrence Island. Amerindian populations exhibit higher frequencies of P*P1, ranging from 30 to 40%. There are some exceptions: namely the Ojibwa Indians have a P*P1 frequency of 76%, while the Dogrib in the North West Territories exhibit a frequency of 18% (Szathmary et al., 1975; Szathmary, 1983). The reasons for such disparities in frequencies should be investigated further.

Other blood groups

There is a large number of other blood-group systems and rare familial antigens that occur in human populations. These systems include: Colton, Dombrock, Gerbich, Henshaw, Kell, Hunter, Penny, Lewis, Lutheran, Miltenberger, Scianna, Stoltzfus, Sutter, Vel and XG. Miltenberger and Henshaw have been shown to be part of the MNS system, representing different groups of rare substitutions in glycophorin A or B or possibly a recombination between the two. Kell and Sutter are substitutions within the KEL system. Some of these blood-group systems, such as Hunter and Henshaw, are found at high frequencies within the Black population. Other alleles, such as Penny, are rare, often restricted to specific families, or are low-level polymorphisms in Whites. Statistically, some of these systems are less informative because of the low incidence of one of the alleles.

I will briefly mention a few systems that are polymorphic, but have not been studied extensively in either the New World or Siberian populations. Although the Kell (KEL) blood group system has been studied widely, it appears not to be particularly informative for New World populations. The KEL*LK allele occurs in a narrow range of frequencies (94–100%) in the Americas. The majority of Amerindian and Eskimo groups have no KEL*K alleles (Roychoudhury and Nei, 1988). The KEL*K allele is absent from the gene pools of the Papago, Maricopa, Tlaxcala, Yaqui, Tarahumara, Chontal,Totonac, Zapotec, Mixtec, and Nahua Indians, to name a few South West United States and Mexican populations. The Lewis blood group system (LE), although highly polymorphic, has been studied in few Amerindian and Siberian groups. It appears to be a promising marker for South American Indian groups, among which the LE*LE allele varies in frequency from 0.52 to 0.20. The Sutter blood group system (JS*A) was discovered by Giblett (1958) and is an excellent marker in populations of African origin. However, this system seems to be of limited usefulness in Amerindian populations, where the JS*A allele has been described only among the Aymara (at a frequency of 0.011) by Ferrell et al. (1978). The only sex-linked blood type, XG*A, occurs at high frequency among the few Amerindian groups tested to date. Dewey and Mann (1967) found the XG*A allele at 91% frequency in the Zuni Indians, 85% in the Cheyenne, and 77% in the Navajo. A lower frequency of 56% was described by Lucciola et al. (1974) in the Cree of Manitoba.

4.5   ELECTROPHORETIC MARKERS

Proteins of the blood provide a sampling of the genetic material maintained in the gene pool of a population. These proteins are primary gene products and reflect some of the variation found in the DNA. Rapid screening techniques became available in the 1950s, first with the development of filter-paper electrophoresis, which allowed the separation of proteins on the basis of molecular charge. This was followed by the development of starch gel electrophoresis, which added another component, i.e. molecular size, to the separation of proteins. Other techniques, such as isoelectric focusing (IEF), in which proteins with different isoelectric points are separated, constitute a further refinement and reveal the existence of greater and greater variation in the human genome.

Serum proteins

Albumin (ALB)

Albumin is a serum protein whose locus resides on the long arm of chromosome 4, in the q11–q13 position. Its physiological functions include pH buffering and the maintenance of the viscosity of the plasma. Scheurlen (1955) reported the presence of a second, slow-moving albumin band in a diabetic patient. This research was followed by Wieme's (1959, 1962) demonstration that bisalbuminemia was inherited as a codominant allele. Melartin and Blumberg (1966a,b) detected a relatively common, electrophoretically fast albumin variant among the Naskapi and Montagnais Amerindians, and a slow-moving albumin variant was described in Mexican Indians (Melartin and Blumberg, 1966a,b). Albumin has been particularly useful for the investigation of New World populations because it provides markers for Mesoamerican and North American Indian populations. Few albumin variants with widespread geographical distribution have been detected in the New World (Arends and Gallengo, 1972). However, familial mutations were characterized among the Makiritare (Arends et al., 1970), Wapishana (Neel et al., 1977a) and Yanomama (Weitkamp et al., 1972).

  Albumin Mexico (ALB*MEX) is an electrophoretic component found in serum samples from populations descended from the Aztecs of Mexico and Piman groups of SW United States (Johnston et al., 1969). It is absent from the South and most of the North American populations. The Pima, Papago, Cocopah and Maricopa all exhibit ALB*MEX at frequencies ranging from 1 to 5% (Workman et al., 1974). The presence of ALB*MEX has been reported in two Highland Guatemalan samples (Johnston et al., 1973). This albumin variant has proved to be useful in my study of Black Carib geographic expansion and admixture in Central America, because it is found in mainland Garifuna populations but is absent from the founding communities of St. Vincent Island. The presence of ALB*MEX in the Black Caribs of Belize and Guatemala appears to be a result of gene flow from Central American Indian groups into the Black Carib gene pool. ALB*MEX could not have been introduced by the Carib or Arawak Indians, since this allele is a Central American marker and is not present among indigenous South American groups. Thus, the presence of this variant allowed me to propose a resolution of the ethnohistorical controversy about the degree of reproductive isolation of Black Caribs of Central America, and ultimately led to the measure of the rate of gene flow from surrounding highland Indian populations (Crawford et al., 1984).

  The second common albumin variant, distributed among Eskimos, Athapaskan, and Algonkian-speaking groups, is termed Naskapi (Szathmary et al., 1974; Szathmary, 1983; Crawford et al., 1981a). It occurs at its highest frequencies among the southern Athapaskan groups such as the Navajo (among whom the frequency of AL*NA is

0.032) and Apache (0.015). This allele among Eskimos may have been acquired through gene flow. The Naskapi variant is found at a frequency of 0.098 among the Slaves and Beaver Athapaskans and constitutes 2–3% of albumin alleles among the Algonkian Crees. The frequencies of the Naskapi and Mexico alleles are summarized for North and Central America by Schell and Blumberg (1977).

Haptoglobins (HPA and HPB)

Polonovski and Jayle (1938, 1940) discovered that a serum protein, alpha-2-glycoprotein or haptoglobin, could bind to hemoglobin. Further research revealed that haptoglobin is a group of closely related molecules (Jayle and Moretti, 1962). Two loci are involved in the synthesis of haptoglobin, one for alpha chains and the other for beta chains (Bias and Migeon, 1967). The alpha locus is found on the long arm of the 16th chromosome (16q22.1). Javid (1967) described a genetic variant of the beta polypeptide chain of haptoglobin and suggested calling this chain the binding peptide. Smithies (1955), using starch-gel electrophoresis, described three haptoglobin phenotypes: Hp 1–1, 2–2 and 2–1. Smithies and Walker (1956) demonstrated that these three phenotypes were determined by two autosomal alleles, Hp^1 and Hp^2. These common alleles are distributed in varying proportions in world populations.

The indigenous populations of the New World show considerable variation at the haptoglobin (HPA) locus. The HPA*1 allele ranges in frequency from 0.26 among the Ojibwa Indians of Canada (Szathmary et al., 1974) to 0.83 among the Yanomama of Venezuela (Weitkamp et al., 1972). Salzano et al. (1986) demonstrated considerable variation at this locus within the various villages of the Icana River Indians (the HPA*1 frequency was 0.27–0.52). The Eskimos of Alaska and Siberia have the lowest frequencies of HPA*1, similar to those observed in Asian groups. Siberian populations vary in the frequency of HPA*1 from 14% among the Koryaks to 52% among the Yakut (Spitsyn, 1985). However, because many of the Siberian sample sizes are relatively small, caution should be exercised in interpreting these results.

Simoes et al. (1989) have recently applied isoelectric focusing (IEF) techniques to the haptoglobins in blood samples from two Brazilian tribes. The IEF method separated both HPA*1 and 2 into slow and fast components. They observed significant intratribal variation of alleles HPA*1S and HPA*2FS among the Macushi but not among the Icana River Indians. The application of IEF techniques to the haptoglobins of Amerindian and Siberian populations should allow finer discrimination of affinities among populations, as well as of their origins.

Transferrins (TF)

Transferrins (TF) are a beta-globulin fraction of the serum and transport iron from the liver to the receptors or cells of the bone marrow and tissue storage compartment (Giblett, 1969). Apparently, the liver is the primary site of synthesis, which begins at an early stage of embryogenesis (Scheidegger et al., 1956). Polymorphism at this genetic locus was first described by Smithies (1957) using starch-gel electrophoresis. The transferrin locus is believed to be located on the long arm of chromosome 3 (3q21), closely linked to the ceruloplasmin (CP) and pseudocholinesterase loci.

The most common transferrin phenotype contains a single, darkly staining (with amido black) electrophoretic band of medium mobility. This phenotype, produced by homozygosity for the TF*C allele, occurs in all human populations at high frequency, ranging from 0.85 to 1.00. Variants in the New World include TF*B, TF*B0-1, TF*D and TF*DCHI. The TF*D and TF*B variants occur among populations of Mexico. The TF*B allele is shared by the Lacandon Maya (where its frequency is 0.19), the Mestizos of Tlaxcala (0.23), the Chinantecos (0.28), the Cora (0.05) and the Pima of the southwestern United States (0.32). The Nganasan of Siberia (Sukernik et al., 1978) and the Blackfoot Indians of Montana (Rokala et al., 1977) apparently share the same HP*B0-1 mutation at polymorphic levels (its frequencies are 0.48 and 0.017, respectively). These data do not imply common origin, but most likely the chance occurrence of the same mutation acted upon by either the founder effect or selection. In addition, the TF*DCHI variant occurs in several North American tribes (Montgnais, Naskapi, Cree and Ojibwa), in Costa Rica (Guaymi) and the Jandu Cachoeira, Asurini and Ticuna of South America.

Recently, transferrins have been subjected to isoelectric focusing and this technique has shown variation in what was originally judged to be a single allele, TF*C. This method utilizes the presence of pH gradients in the gel media to separate proteins with slightly different isoelectric points. One of the subtypes, TF*C4, appears to be particularly informative in the separation of Amerindian populations from either European or African contributions. Whereas the TF*C4 allele is absent from both the U.S. European and Afro-Americans, it occurs at a frequency of 8–18% in North American Indian tribes such as the Apache, Blackfeet, Navajo, Pima and Walapi (Dykes and Polesky, 1984).

In indigenous Siberian populations, both TF*B0-1 and TF*DCHI have been observed. The most complete summary of transferrin variation in indigenous Siberian populations is provided by Spitsyn (1985). He indicated that TF*DCHI varies in frequency from 0 to 2.3% in Siberia. Alexseyeva et al. (1978) described the presence of the TF*DCHI allele (at low frequencies of 0.004–0.02) in Aleuts of Bering Island, the Chukchi of Uelen and the Eveni of the Bastrinskii region. In his vast surveys of

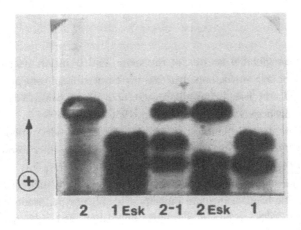

FIGURE 26 A photograph of the group-specific component (GC) Eskimo variant and its electrophoretic mobility compared with common phenotypes.

native Siberian populations, Spitsyn (1972) has documented the presence of this variant among Eskimos, Khanti, Nentsi and Buryati. Considering the lack or rarity of the TFD*CHI in populations other than Asian or Amerindian ones, this marker provides further evidence for the Siberian origins of New World peoples. Unfortunately, no information on IEF-subtyped transferrins is available for indigenous Siberian populations.

Group-specific component (GC)

Group-specific component, also known as the vitamin D-binding protein (DBP), is located on the long arm of chromosome 4 (4q12). It was first found to be polymorphic by the immunoelectrophoretic methods of Hirschfeld (1959). Familial studies revealed that the GC patterns are determined by a pair of codominant alleles called GC*1 and GC*2 (Hirschfeld et al., 1960). Population studies of the geographical distribution of the GC alleles revealed that, with few exceptions, the GC*1 allele is more common than the GC*2. The GC*1 allele varies in frequency from 0.37 to 0.94 among the populations of the Americas. However, the GC*1 frequency for the Parakana of Amazonia appears to be anomalous in the region (Black et al., 1980b). Aside from a few rare variants in specific populations, such as GC*IGL (initially termed Gc Eskimo) in Eskimos (see Fig. 26) and GC*CHIP among the Ojibwa Indians (Szathmary et al., 1974), there is no clear-cut pattern for Amerindians. Similarly, the few electrophoretic surveys of Siberian native populations fail to reveal north–south clines for the GC system. They do demonstrate that GC*1 is more common than GC*2 and that GC*1 frequency varies between 0.53 and 0.81 (Spitsyn, 1985). We found GC*1 among the Evenki populations of central Siberia at still higher frequencies of 0.884 and 0.967 in Surinda and Poligus, respectively (Crawford et al., 1994). In his volume, Spitsyn indicated the existence of a gradient for the GC*1 allele radiating from central Siberia both east and west. Mourant et al. (1976b) have demonstrated a north/south geographic gradient and an association between GC alleles and the intensity of sunlight.

The application of isoelectric focusing (IEF) to the GC proteins has yielded a bounty of genetic variation that is useful in establishing affinities between human populations. To date, IEF has revealed the existence of three common alleles (GC*1F, GC*1S and GC*2) plus almost 90 rare variants. Dykes et al. (1983) screened a total of 11 682 samples representing 20 different populations from the Americas and demonstrated that the Amerindians tend to cluster because of their high frequency of the GC*1S allele and their low incidence of GC*2 (see Fig. 27). Figure 27 contains a bivariate plot of the frequencies of GC*1S and GC*2 from an assortment of continental populations. The Eskimo groups tend to cluster with Asian populations. The only Siberian sample, from Touva (Altai), is intermediate between the Asian and hybrid African/Amerindian groups (see Fig. 27). Constans et al. (1988) attempted to reconstruct the peopling of the New World using the distribution of rare GC variants. They argued that since the GC*1A2 and GC*1A4 mutations have been detected only among Eskimos from North America then the ancestors of these people must have entered the New World with the last migration, after the arrival of the ancestors of the Amerindians. Obviously, the late arrival of the Eskimos into the New World should not surprise; however, the presence of the two mutations could have originated de novo in Alaska. Constans et al. (1985) also suggest that Amerindians can be clustered into three groups: (1) Bolivian Indians; (2) Gorotire, Kraho and Caingang from Brazil and Lumbees from the United States; (3) Pima, Macushi and Ixils from North and South America. Constans and his colleagues suggest on the basis of GC*2 allelic frequencies (0.10–0.20 versus 0.30–0.40) that the Americas were peopled by two distinct groups. These conclusions should be criticized because: only one locus was utilized, and the presence of a north–south gradient in the frequencies of GC alleles might be expected because of the GC proteins' association with vitamin D and, hence, sunlight. However, Corvello et al. (1989) argue that the proposed sunshine hypothesis may not be applicable to South America because of the potential involvement of founder effect and/or genetic drift.

Properdin factor B (BF)

Variation in a glycine-rich beta-glycoprotein in the serum, also known as protein properdin factor B (BF), was first demonstrated by Alper et al. (1976). Linkage of BF

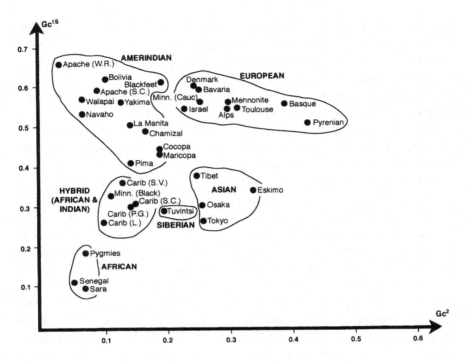

FIGURE 27 A bivariate plot of the frequencies of GC•1S and GC•2 alleles from an assortment of human populations. It is interesting to note that these two alleles are sufficiently informative to separate major human geographic groups.

to glyoxalase (GLO), phosphoglucomutase (PGM3) and the histocompatibility (HLA) complex led to its localization on the short arm of chromosome 6 (6p21.3).

To date, few studies of the BF locus have been performed on New World populations. The most informative investigation, by Dykes *et al.* (1981), revealed an exceptionally high frequency of BF*S in Amerindian and Eskimo groups and a low frequency, even absence, of BF*F. This association between BF*S and BF*F is shown in Fig. 28, which indicates that little variation exists between New World tribes. In a study of the Mapuche of Argentina, Haas *et al.* (1985) show similar BF patterns for a South American tribe with a high frequency of BF*S (0.91) and a low frequency of BF*F (0.077). In contrast to the Amerindians of the southwestern United States, the Mapuche have a polymorphic level of variant BF*S1. The Navajo share with the Columbian Indians of South America a frequency of BF*S0.7 in excess of 1% (Bernal *et al.*, 1985). The presence of this allele may be due to admixture with either Europeans or Africans, who possess BF*S0.7 at frequencies as high as 8%.

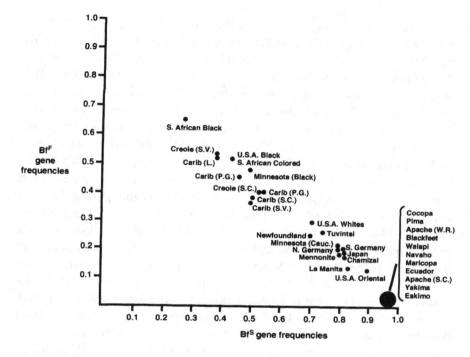

FIGURE 28 A bivariate plot of properdin factor BF alleles BF*F versus BF*S. The Amerindian populations have a limited distribution of these alleles compared with other populations of the world.

The only Siberian populations investigated for BF variation are the Tuvintsi (Touvinians) from the Altai and the Evenki from the central plateau. The Touvinians constitute an amalgam of European, Mongolian and possibly Chinese admixture (Schanfield *et al.*, 1980). As a result of this hybrid origin, the Tuvintsi, on the basis of their BF frequencies, cluster with European instead of with Asian or Amerindian populations. Although the Evenki are not included in Fig. 28, the frequencies of 0.920 for BF*S and 0.080 for BF*F would place them in close proximity to the Amerindians.

Cholinesterase 1 and 2 (CHE1 and CHE2)

Human blood contains two kinds of cholinesterase, acetylcholinesterase in red cells and pseudocholinesterase in the serum. The serum cholinesterase can be differentiated from the red cell enzyme by its ability to split benzoylcholine but not acetyl-beta-methylcholine (Giblett, 1969). The function of serum cholinesterase is unknown.

Two loci, CHE1 and CHE2, control the inheritance of serum cholinesterase; CHE1 appears to be located on the long arm of chromosome 3 (3q25.2). The CHE1 locus has four known alleles: the normal (CHE1*U), atypical (CHE1*A), fluoride inhibition (CHE1*F), and silent (CHE1*S). The second locus, previously called C5, has alleles that generate two electrophoretically discernible phenotypes, C5+ and C5–. These phenotypes are defined by the presence or absence of an 'extra' electrophoretic band. Harris et al. (1962) reported four characteristic bands of electrophoretic activity in most British subjects; in about 5% of the blood samples they detected a slower-moving band, which they called C5. The silent and atypical cholinesterase alleles were of medical importance because of the role of this enzyme in the hydrolysis of succinylcholine, a muscle relaxant that used to be administered in conjunction with anesthesia. Those individuals homozygous for the silent allele may experience prolonged apnea, a consequence of muscle paralysis, and are unable to breathe without mechanical help. Individuals who are homozygous for the silent allele lack the pseudocholinesterase enzyme, yet they lead long and healthy lives as long as they avoid contact with succinylcholine. The presence of the C5 band increases the mean cholinesterase activity by about 30% (Harris et al., 1963). In addition to the usual allele (CHE1*U), the dibucaine inhibition-resistant or atypical allele (CHE1*A) occurs at polymorphic levels in some human populations. The CHE1*A frequency ranges from 0% in Eskimos (Motulsky and Morrow, 1968) to 3% in some European groups (Altland et al., 1969). The CHE1*S allele is relatively rare in most human populations, with an average frequency of 0.0032, and approximately 1 in 100 000 people are homozygous for it.

Overfield (1975) has described a polymorphic level of the CHE1*S allele in Eskimo populations of southwestern Alaska. She found a range of frequencies from 8–22% in eight villages, with an average frequency of 12%. Most of the other studies of Eskimo populations in Greenland, Alaska and Canada fail to report the presence of the silent allele. Several screenings of Middle and South American Indian populations (Gutsche et al., 1967) (Nahua, Atacameno and Makiritare) for cholinesterase variation have failed to detect the silent allele but have observed the atypical allele at frequencies from 1 to 3% (Lisker et al., 1967; Goedde et al., 1984; Arends et al., 1970). Salzano et al. (1991), in their genetic analyses of four Amazonian populations of Brazil, tested for CHE1 alleles and described no variation at that locus. However, they did note that Amazonian Indians in general exhibit the usual allele at a frequency of 99.8%. There is no further discussion of the variants that constitute the remaining 0.2% of the total. Black et al. (1988) found no variants at the CHE1 locus in three Brazilian tribes.

More data from Siberia and the New World are available on the genetic variation at the second cholinesterase locus (CHE2). This locus appears to be polymorphic in a number of indigenous populations. Spitsyn (1985) summarized the literature on CHE2 variation in Siberia. The frequency of the C5+ allele ranges from 0 among the Koryaks

(Alexseyeva *et al.*, 1978) to 0.133 among the Mongols (Spitsyn, 1985). Spitsyn (1985) located the highest frequency of C5+ on the border of Mongolia. In her review, Szathmary (1981) chose not to include some of the populations screened by Spitsyn that were of small size. (For example, Spitsyn (1985) reported C5 alleles at frequencies of 11% and 12% among the Yakuts and Chukchi.) Scott (1973) and McAlpine *et al.* (1974) report frequencies of C5+ alleles among the Eskimos of Alaska and Canada at 3% and 7%, respectively. The C5+ allele appears at a frequency of 7% among the Cree Indians of Canada (Simpson, 1968). Three studies of South American Indians have detected the C5+ allele at polymorphic levels. Of seven Amazonian tribes studied, four exhibited the extra electrophoretic band at frequencies from 2 to 14% (Arends *et al.*, 1970; Black *et al.*, 1988; Salzano *et al.*, 1991).

Pseudocholinesterase is a fascinating enzyme with, apparently, no indispensible function, yet it is distributed polymorphically throughout the New World. Individuals who lack this enzyme appear to be unaffected unless they are injected with the muscle relaxant succinylcholine. From my dissertation research, it became clear that this enzyme was found not only in humans but also in pongids, and at even higher levels (Crawford, 1967). What is the evolutionary history of this genetic system of apparently no known physiological function?

Red-cell proteins

Phosphoglucomutase (PGM)

Phosphoglucomutase (PGM) is an erythrocytic enzyme that catalyzes the transfer of phosphate from the first to the sixth position of glucose (Giblett, 1969). Spencer *et al.* (1964) demonstrated the existence of a polymorphic forms of PGM. Subsequently, the enzyme was shown to be under the control of three different loci (Hopkinson and Harris, 1965; Harris *et al.*, 1968). These three loci, PGM1, PGM2, and PGM3, are located on three different chromosomes, 1p22.1, 4p14–q12, and 6q12, respectively (Roychoudhury and Nei, 1988). All three of these loci have at least one common allele plus a number of rare ones, most of which are limited to specific populations or regions.

There is a great deal of genetic variation in frequency of the PGM*1 allele among indigenous Siberian and American populations, ranging from 0.59 among the Mapuche of Argentina to 0.98 among the Parakana of Brazil. Spitsyn (1985), in his summary of the frequency of the PGM*1 allele in Siberia, provided a range from 100% among the Evens to 54% among the Nentsi. However, the significance of this range is suspect because of the small sample sizes. When Spitsyn's frequencies for the Evens are compared to those frequencies measured by Posukh *et al.* (1990) using larger samples, it is clear that the frequencies of the PGM*1 allele vary between 83% and 92% among the Evens. The Eskimos and Chukchi of Chukotka have frequencies of PGM1*1 of 0.92 and 0.94, which are higher than those in most of the New World, while the Samoyedic populations, such as the Nganasan and the Nentsi, have lower frequencies, 0.655 and 0.833 (Sukernik et al., 1978, 1980, 1986b). The Chelkenians and the Kumandinians exhibit intermediate frequencies of the PGM1*1 allele (0.70–0.80). The average of PGM1*1 allelic frequencies in Siberian Eskimos is 0.846. However, this average figure has been lowered by the frequency observed in Naykan, which is 0.689. If this small settlement is excluded from computation, then the average frequency for Siberian Eskimos of Chukotka (0.90) is closer to that of PGM1*1 for the Alaskan Yupik speakers. The Savoonga and Gambell populations of St. Lawrence Island display PGM1*1 frequencies of 0.90 and 0.93 (Crawford et al., 1981b). This allele is less frequent among the Inupik Eskimos of Igloolik and Barrow.

The application of IEF techniques to products of the PGM locus has revealed a wealth of genetic variation that standard electrophoresis failed to disclose. Bark et al. (1976) demonstrated that the PGM alleles 1 and 2 could be subtyped by IEF into four common types: PGM1*1+, PGM1*1−, PGM1*2+, PGM1*2−. For purposes of standardization, the numerical designation has recently been replaced with an alphanumeric one, and the alleles are now referred to as PGM1*1A, PGM1*1B, PGM1*2A, and PGM1*2B. Dykes et al. (1983) applied the IEF techniques to 11 North and Central American populations. A consistent pattern was noted in the frequencies of PGM1*1A and PGM1*1B genes. The Amerindian populations from Central America, Saltillo, Chamizal, and LaMinita and the Pima Indians, cluster together, while the Apache, Eskimos and Asians form a separate cluster (see Fig. 29). The most dramatic interpopulational differences were based upon the relative frequencies of the PGM1*2A and PGM1*2B, which produced a clustering of New World Amerindians and Eskimos as a result of their high frequencies of the PGM1*2B allele (Dykes et al., 1983).

The two Evenki populations, Surinda and Poligus, exhibit frequencies of PGM1*1A of 0.767 and 0.62, respectively. The second most common allele in these populations is PGM1*2A, which we found at frequencies of 0.19 and 0.27, respectively (Crawford et al., 1994). These frequencies place the Evenki within the Amerindian cluster (Dykes et al., 1983).

Uridine monophosphate kinase (UMPK)

The enzyme uridine monophosphate kinase (UMPK) is involved in the phosphorylation of uridine monophosphate (UMP) to uridine diphosphate (UDP) in the synthesis of pyrimidine nucleoside triphosphate, which is involved in RNA and DNA production

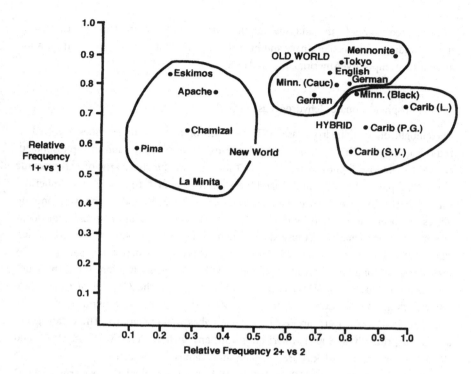

FIGURE 29 This is a plot of phosphoglucomutase (PGM) isoelectric focusing subtypes 1+ vs. 1− against 2+ vs. 2− (Dykes *et al.*, 1983).

(Giblett *et al.*, 1974). This enzyme is located on the short arm of chromosome 1 (1p32). Giblett *et al.* (1974) developed an electrophoretic and staining procedure for the detection of UMP kinase variation. They also explored populational differences in the frequencies of the three common UMPK alleles and found that the UMPK*3 allele occurs at a frequency of over 10% among the Cree Indians, whereas it is absent from Asians, and Africans, and rare in Europeans. Scott and Wright (1983) suggested that UMPK*3 may be an 'American' allele, since it is rare. The UMPK*3 allele has been detected in Eskimos (where its frequency is 0.20), Aleuts (0.232), Siberian Yupik (0.133), Athabaskan Indians (0.131), Cree Indians (0.114) and Venezuelan Mestizos (0.001) (Scott and Wright, 1983; Lucciola *et al.*, 1974; Gallango and Suinaga, 1978). Giblett *et al.* (1974) also reported a low frequency of this allele among persons of European ancestry in Seattle; most likely it was introduced through gene flow from northwestern Amerindians. The UMPK*3 allele has been found in the Warao Indians of Venezuela at a frequency of 8.6% (Gallango and Suinaga, 1978).

The UMPK*3 allele appears to be a promising marker for Amerindian populations.

Data on frequencies from additional populations would be of great use in reaching an understanding of the evolutionary and historical factors that may have affected variation among contemporary Amerindian populations.

Phosphogluconate dehydrogenase (PGD)

Phosphogluconate dehydrogenase (PGD) is one of a series of enzymes involved in red-cell glycolysis. This enzyme catalyzes the oxidative carboxylation of 6-phosphogluconate to ribulose 5-phosphate. The PGD locus is on the short arm of chromosome 1 (1p36.2–p36.13). Fildes and Parr (1963) were the first to describe an inherited variant at this locus and since then the geographical distribution of the two common alleles has been investigated by numerous scientists. For population studies, this locus is not highly informative because the frequency of the commoner of the two alleles, PGD*A, varies only between 75 and 100%. There is no detectable pattern in its distribution among New World populations; PGD*A reaches fixation in North, South and Central America. PGD*C is most common among the Guaymi of Costa Rica, among whom it has attained a frequency of 10%. There are a number of private alleles of PGD, which are limited to specific tribes or populations. These rare genes are found primarily in South American groups and include PGD*MAK, PGD*AY and PGD*CAI among the Makiritare, Aymara and Caingang, respectively.

Sukernik and associates (1978, 1980, 1981) have described the patterns of PGD gene frequency distributions for Siberia. As in the New World, there is little variation at this locus, with frequencies of PGD*C varying between 4 and 6% among the Nganasan, Chukchi and Eskimos.

Acid phosphatase (ACP1)

Acid phosphatase (ACP1) is a heterogenous red-cell enzyme that is most active in a pH range of 4.8–6.1 and takes up phosphorus from the substrate and transfers it to acceptor molecules (Giblett, 1969). Hopkinson et al. (1963) described electrophoretic variation at this locus, and attributed it to the presence of three alleles, initially called P[a], P[b], and P[c]. Additional alleles, such as P[r], were described in Afro-Americans (Karp and Sutton, 1967). According to Roychoudhury and Nei (1988), this locus is believed to be carried on the short arm of chromosome 2 (2p25).

There is considerable variation at the ACP1 locus in Siberian populations. The frequency of the allele ACP1*A ranges from 19% among the Forest Nentsi (Sukernik et al., 1980) to 62% among the Itelmeni (Alexseyeva et al., 1978). The latter frequency of ACP1*A is most likely an overestimate related to the appallingly small number of samples examined ($n = 25$). The 19% frequency of the ACP1*A allele in the Nentsi is an average over four small hunting and fishing settlements in which this allele varies in frequency from 6 to 25%. Given these small settlements and tiny samples, there is a great deal of variation in frequency from group to group and no clear-cut pattern is observed. In most Siberian populations, ACP1*B is more frequent than ACP1*A and ACP1*C is extremely rare, occurring at polymorphic frequencies only among the Nentsi and Evenki (Crawford et al., 1994). A dated summary of the acid phosphatase frequencies for Siberian populations is contained in Szathmary (1981).

Eskimo and Aleut allelic frequencies at the ACP1 locus also vary in part because of the small sizes of the villages, which foster divergence in frequencies by drift and repeated founder effect. For example, Scott and Wright (1983) report gene-frequency data for 25 Central Yupik Eskimo villages with an observed range of frequencies of ACP1*A from 19 to 73%. The average frequencies computed for the ACP1*A allele are 59% among the Inupik, 69% among the Siberian Yupik and 56% among the Central Yupik. The Eskimos of St. Lawrence Island exhibit ACP1*A frequencies of 60% in Savoonga and 72% in Gambell (Crawford et al., 1981b). Eskimo villages on Kodiak Island have lower frequencies of ACP1*A ranging from 32% to 47% (Majumder et al., 1988).

Unlike the case in most of the Siberian and some of the Eskimo populations, in North, Middle and South American Indians the ACP1*B allele is much more common than the A allele. North American populations such as the Ojibwa, Cree and Dogrib Indians exhibit the B allele at frequencies of 50–60%. In Central American populations, the ACP1*B allele is higher, ranging from 75 to 83%. According to Salzano et al. (1986), the B allele ranges in frequency from 63 to 99% among South American Indians. Cavalli- Sforza et al. (1994) have plotted the frequency of ACP1*B in the Americas and display the highest frequencies in South America (90–100%) in two foci, with gradients radiating out.

Esterase D (ESD)

Esterases are a group of isozymes that have been identified electrophoretically by their substrate specificities and inhibitors. ESD (also known as S-formylglutathione hydrolase) was detected by Hopkinson et al. (1973) through the use of fluorogenic substrates. This red-cell enzyme is located on the long arm of chromosome 13 (13q14.11) and is known to be polymorphic in Europeans (Germany) (Bender and Frank, 1974), Africans (Gambia) (Welch, 1974) and Asian Indians (Hopkinson et al., 1973). Variation at this locus in Amerindian populations was first demonstrated by Goedde et al. (1977) in the Shuara of Ecuador and by Neel et al. (1977a,b) in a number of Brazilian tribes.

The esterase D locus is highly polymorphic among New World populations. In human aggregates distributed from Alaska to Chile, the less common allele, ESD*2, ranges in frequency from 4 to 64%. The few studies of Eskimos have revealed a low incidence of ESD*2, with frequencies varying between 4 and 8% in Yupik and Inupik populations of northwestern Alaska and St. Lawrence Island (Scott and Wright, 1978; Crawford et al., 1981b). Scott and Wright (1983) screened a total of 25 Central Yupik Eskimo villages in Alaska and found a range of ESD*2 frequencies of 4–24%, with a mean of 10.7%. Most of the Yupik villages are of small size and none of the sample sizes used to represent these populations is over 100 individuals. Some samples dip down to as low as 19 persons. As a result, it is not surprising to observe this wide range of variation for the ESD*2 allele in Eskimo groups. Mestriner et al. (1980) summarized the available data from ten South American Indian tribes and found that the frequency of ESD*2 varied between 0 and 54%. However, in most of the tribes surveyed the frequency of ESD*2 was between 13 and 34%. Higher frequencies were noted among the eastern and Atlantic coastal groups and lower ones characterized the tribes of the tropical forest.

Few Siberian populations have been screened for esterase D variation. The two Evenki communities, Surinda and Poligus, have intermediate frequencies of ESD*2 (13.6% and 12.1%). Preliminary results for the Udehe and Kets (Sulamai) communities reveal frequencies of ESD*2 of 3% and 27.6%. However, the accuracy of these proportions is in question when the small sample sizes of the latter two populations are considered (Crawford et al., 1994).

Other electrophoretic loci

There are many serum proteins and red-cell enzymes that are highly polymorphic and useful in measuring populational affinities and tracing origins. Roychoudhury and Nei (1988) summarized the available hematological genetic markers and listed 138 known enzyme-encoding, 80 protein-encoding and three immunoglobulin-encoding loci. Add to this list 115 blood-group systems and private antigens and five loci controlling histocompatibility, and a total of 341 protein-coding loci, plus the highly variable DNA polymorphisms, are available for sampling the gene pools of human populations. Granted, 341 genes out of a possible 100 000 or more is still a small sample of the total genetic makeup of an idividual. However, the genetic variation revealed by both mitochondrial and nuclear DNA has allowed us not only to distinguish between populations but also to produce individual DNA 'fingerprints' that identify each of us genetically.

Given the limitations on space in this book, I cannot review more than a small sample of the total number of genetic markers that are available for the characterization of Amerindian, Eskimo and Siberian populations. There are many more promising protein-encoding loci that may reveal some episode in the evolution of Amerindians during their odyssey of expansion across the vast continents of the New World. These include highly polymorphic loci such as glutamate pyruvate transaminase (GPT) with an allele, GPT*1, that varies in frequency between 30 and 75% among New World populations, and glyoxalase (GLO1) and the allele GLO1*1, which ranges in frequency from 0.8 to 67% in native American groups. The frequency of GLO1*1 also varies (from 23 to 51%) in Siberian populations (Spitsyn, 1985). However, our sampling of the Evenki for GLO1*1 (Crawford et al., 1994) reveals a frequency (0.185) outside the range provided by Spitsyn (1985). Ceruloplasmin (CP) is polymorphic among some of the isolated Amerindian groups of South America. In Siberia, the CP*A allele occurs at polymorphic levels among groups speaking Turkic languages (Spitsyn, 1985).

4.6   IMMUNOGLOBULINS (GMs AND KMs)

Gamma globulins (GMs)

Introduction

GM allotypes are structural variants of the heavy gamma chains of IgG immunoglobulins in humans. The most common GM allotypes occur on the IgG1 and IgG3 subclasses of immunoglobulins. The allotypes located on the gamma chains are encoded by tightly linked genes on the long arm of chromosome 14 (14q32.33). Because of this linkage, the allotypes are inherited in units called haplotypes, which vary in their frequencies in human populations. Unlike other genetic markers, GM haplotypes tend to be population-limited (Schanfield, 1980). As a result, the GM system has been utilized to study populational affinities and to estimate genetic admixture.

The complete designation of a GM haplotype tends to be a awkward and lengthy process; I have therefore implemented a shorthand notation in this chapter. Table 14 lists the common GM haplotypes found in New World populations and provides a shorthand designation.

Geographic distribution

New World A number of publications have attempted to apply the observed GM variation in New World and Siberian populations to questions concerning the peopling of the New World. Williams *et al.* (1985), after identifying GM haplotypes of ten Amerindian populational samples and summarizing other data, argued that the data on GM support the three-migration hypothesis. They claim that the GM variation can best be explained by one migration of Paleo-Indians 16 000–40 000

Table 14. *A listing of GM haplotypes and their shorthand notation in Amerindian and Siberian populations*

Shorthand	Haplotype
GM*A G	GM*A,Z G,G5,U,V
GM*X G	GM*A,X,Z G,G5,U,V
GM*A T	GM*A,Z B0,B3,B5,S,T,V
GM*F B	GM*F B0,B1,B3,B4,B5,U,V
GM*A,F B	GM*A,F B0,B1,B3,B4,B5,U,V

years BP, a second wave of Na-Dene hunters 12 000–14 000 years BP, and an Eskimo–Aleut migration 9000 years BP. They base this three-wave hypothesis on the presence of GM*A G and GM*X G in South American Indians; GM*A G, GM*X G and GM*A T (previously called GM*A,ZBO,3,5,S,T) in the Na-Dene Indians; and the presence of GM*A G and GM*A T in the Eskimo–Aleut group. This tripartite migration theory has been challenged by Schanfield *et al.* (1990), who proposed, on the basis of the distribution of GM haplotypes in Siberia and the Americas, four migrations into the New World. They agree with Williams *et al.* on the characterization of the Paleo-Indian and Eskimo–Aleut components, but indicate that the Na-Dene had a high frequency of GM*A G and intermediate frequencies of GM*X G and GM*A T. The second migration of Schanfield *et al.* is judged to be one of a population with a high frequency of GM*A G and low frequencies of GM*X G and GM*A T. Field *et al.* (1988) described the GM haplotype frequencies among the Bella Coola and Haida of British Columbia. Although their results were complicated by relatively high levels of European admixture, they reported that frequencies of GM*A G, GM*X G and GM*A T were intermediate between those hypothesized by Williams *et al.* (1985) for Paleo-Indians and the Na-Dene. Field and her colleagues concluded that these data support the archeologically derived hypothesis that these British Columbian Amerindians are a product of admixture between the descendants of the Na-Dene and the Paleo-Indians. The same data could be used to argue for the occurrence of a fourth migration. The distributions of GM haplotypes support no fewer than two and possibly three Amerindian migrations. This evidence erodes the earlier concept of a single Amerindian move-ment into the New World after the Wisconsin glaciation, followed by a later expansion of the Eskimos and Aleuts.

Siberia The GM haplotype with the highest incidence in most indigenous Siberian populations studied to date is GM*A G (see Table 14). This haplotype reaches its highest frequencies in the region of Chukotka, ranging from 66% among the inland Reindeer Chukchi to 86% among the coastal Chukchi. The Eskimos of New Chaplino exhibit a GM*A G frequency of 79.5% (Sukernik and Osipova, 1982). The frequencies of GM*A G appear to diminish as one moves westward from northeastern Siberia to the Tamyr Peninsula. Similarly, the Evenki of Surinda and Poligus exhibit GM*A G frequencies of 0.488 and 0.418, respectively. The Udehe and Kets have higher frequencies but are represented by small sample sizes (Crawford *et al.*, 1994). The Samoyed groups, as exemplified by the Nganasan, Forest Nentsi, and Yenisey Samoyeds, exhibit lower frequencies of GM*A G than are found in northeastern Siberia. The Nganasan and the Evens, who had mixed with the Yukaghirs, are exceptions in Siberia in that their most common GM haplotype is GM*A T, which appears at frequencies of 0.486 and 0.449, respectively (Posukh *et al.*, 1990). However, the GM*A G haplotype is found among the Nganasan at the Siberian low frequency of 0.353. The Eskimos of New Chaplino appear to exhibit the least amount of genetic variation at the GM loci with only two GM haplotypes, GM*A G (at a frequency of 0.795) and GM*A T (0.205). GM*X G is most frequent among the Samoyed groups and least frequent among the populations of Chukotka. The Yakuti of the Middle Lena River exhibit the highest frequency of GM*A,F B (a southern Asian haplotype) in Siberia. The Evenki harbor the GM*A,F B haplotype at polymorphic levels (0.074) but at a lower frequency than the Yakuts (Crawford *et al.*, 1994).

Upon examination of the haplotype frequencies among contemporary Siberian indigenous populations, can we shed light on the various theories of the peopling of the New World? There appear to be at least three Siberian groupings based on GM frequencies, these are the Samoyeds, Chukotka inhabitants, the Yakuts and Touvinians. The Samoyeds (Nentsi, Yenisey, and Nganasans) have high frequencies of GM*A G and intermediate frequencies of GM*X G and GM*A T. This pattern seems to parallel that observed among the Na-Dene Indians of North America. At this time it is not possible to state with any degree of certainty that a common ancestor of the Samoyeds and the Na-Dene crossed Beringia and peopled North America. The second cluster, which includes the Eskimos and Chukchi, exhibit high frequencies of GM*A G and intermediate frequencies of GM*A T. Most likely, this group gave rise to the Eskimos and Aleuts of the New World. Caution must be exercised when trying to reconstruct the peopling of the Americas based on contemporary Siberian distributions and gene frequencies. Admixture with Europeans has contributed the GM*F B haplotype to the indigenous populations on both sides of the Bering Strait. As in the Americas, indigenous Siberian populations have experienced major demographic upheavals, diminutions and population explosions. The Yukaghirs, considered by some scholars to have shared ancestors with at least one group of migrants into the New World, are now close to extinction. Prior to European contact, the Yukaghirs were numerically one of the largest groups in Siberia. In addition to the ravages of disease epidemics, they were further squeezed by the expansions of the Chukchi, Yakut and Tungus peoples. Little is known about the GM haplotypes and their frequencies among the populations of Kamchatka and the South Eastern Siberian groups. Their expansions into Siberia may have contributed to the peopling of the New World. My cultural anthropological colleague from the University of Illinois, the late Demitri Shimkin, was convinced that the Gilyaks (Nivkhi) were major contibutors to the gene pools of peoples of the New World. However, the mtDNA evidence presented by Torroni *et al.* (1993b) shows that the Nivkhi mtDNA haplotypes appear to be Asian in pattern and are highly distinctive when compared to the New World populations.

The effects of gene flow on the genetic structure of indigenous holarctic populations characterized genetically by both GMs and KMs are shown in Fig. 30. The arrow indicates the gradient of European admixture associated with increases in the frequencies of GM*F B. This principal components analysis reveals considerable genetic diversity among the Siberian groups, with the Nentsi of Central Siberia differing from the Eskimos and Chukchi of Chukotka. The Greenland Inuit and the Ottawa Canadian Amerindians exhibit the greatest effects of European gene flow (Steinberg *et al.*, 1974). The Amerindians show patterns of GM and KM that are similar to, yet distinct from, those of Siberian and Eskimo groups.

South America Salzano *et al.* (1986) have summarized the ranges of frequencies of GM haplotypes among South American Indians. They found that the GM*A G allele occurs at frequencies between 44 and 100%, the range of GM*X G frequencies is 1–54% and that of GM*A T is 0–5% (see Table 16). The GM*A T allele was found among the Jandu Cachoeira subdivision of the Icana River Indians. The size of this sample ($n = 363$) was an adequate representation of the population. The low frequencies of GM*A T in South American Indians contradict the assumption of Williams *et al.* (1985) that no GM*A T alleles were carried by the hypothesized Paleo-Indian founders. The indigenous populations of the New World are characterized by the presence of GM*A G, GM*X G and GM*A T, in decreasing frequencies. These three haplotypes must have been brought by the Siberian migrants and then acted upon by selection, founder effect, and other stochastic processes. Schanfield (1992) points out that the distribution of GM*A T in South American Indians is limited to language families that are coastal or riverine, such as the Caribs. He suggests that GM*A T is recent in South America and could easily represent the second migration.

Callegari-Jacques and her colleagues (1993) summarized the GM haplotype distri-

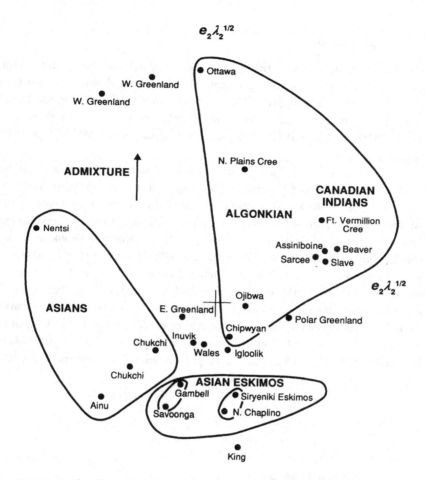

FIGURE 30 The effects of European gene flow on the genetic structure of indigenous holarctic populations, characterized through a principal-components analysis of GM and KM allotypes.

butions in 60 Eskimo, North, Central, and South American Indian populations. In comparisons between North + Central America vs. South American indigenous populations, nearly identical F_{st} and average heterozygosities were found for the GM locus. A North–South cline was observed along the first axis of a correspondence analysis. The two haplotypes that are responsible for this gradient are GM*A G and GM*A T, both haplotypes are thought to have been brought into the New World by Siberian migrants. Whether, GM*A T represents the second migration, as argued by Schanfield (1992), remains to be seen.

KM (Inv) system

The genetically simpler KM (formerly Inv) genetic markers are located on the kappa or light chains of human immunoglobulins. They have been shown to be genetically independent from the heavy (GM, AM) chains of immunoglobulins. The KM locus is located on the short arm of chromosome 2, 2p12. Although at least three alleles (KM*1, KM*2, and KM*3) have been identified, usually only KM*1 is reported in population surveys.

The KM*1 allele varies from 5% among the Nganasan (Sukernik and Osipova, 1982) to approximately 60% among the Dogrib of the North West Territories (Szathmary, 1983). The lowest frequency of KM*1 occurs among the Evenki (0.019) and the Samoyed groups of Siberia; frequency increases among the Eskimos and Chukchi inhabitants of Chukotka (see Table 15). The Na-Dene and Algonkian speakers exhibit frequencies of KM*1 of between 30 and 60%. Schanfield (1980) summarized the frequencies of the KM*1 allele: 20% among Eskimos, 32% among North American Indians and 35% in South American aboriginal populations. These figures can be viewed as averages for the geographic areas but ignore the observed range of variation. The KM*1 frequencies vary among South American tribes from a low (15%) in the Siriono to near fixation (91%) in the Piaroa (Gershowitz and Neel, 1978). However, the majority of South American aboriginal populations vary in frequency from 30 to 60%

4.7   HISTOCOMPATIBILITY SYSTEMS

Human leukocyte antigens (HLA)

The human leukocyte antigens (HLA) are serologically defined antigens located on the surfaces of nucleated human cells. Initially, this major human histocompatibility system was detected by antisera (from multiparous women) containing antibodies to antigens on the white cells of the blood. The serologically defined antigens of the HLA system belong to one of five different loci that are located on the short arm of chromosome 6 (6p21.3). Pedigree studies have revealed that the recombinant fraction between HLA-A and HLA-B loci is about 0.8%, i.e. the map distance between the two loci is about 1/3000 of the total genomic map length and includes at least several hundred genes (Bodmer, 1975). The other three loci are HLA-C, HLA-DQ and HLA-DR. Each individual can express a maximum of four antigens, which are inherited in groups called haplotypes.

These five HLA loci exhibit extraordinary variability. Several years ago, Bodmer calculated that the four loci and 56 alleles known at the time could generate more

Table 15. *Immunoglobulin haplotypes (GMs and KMs) among Siberian and Native American populations*

Population	GM*A G	GM*X G	GM*A T	KM*1
Samoyeds				
Forest Nentsi[1]	50.2	16.3	13.9	9.9
Yenisey Samoyeds[1]	50.6	17.3	4.5	10.3
Nganasans[1]	35.3	15.4	0.8	4.8
Chukchi				
Reindeer[1]	73.1	10.9	15.3	17.7
Coastal[1]	86.0	2.2	9.9	20.0
Evenki				
Surinda[12]	48.8	12.0	31.4	2.6
Poligus[12]	41.8	17.9	33.1	1.1
Eskimos (Siberian)				
New Chaplino[1]	79.5	0	20.5	20.6
Siryeniki[2]	81.5	0	18.1	23.0
Eskimos (Alaskan)				
Savoonga[2]	64.3	1.8	27.2	28.2
Gambell[2]	65.0	1.3	28.1	25.0
King Island[2]	72.7	0	27.3	44.0
Wales[2]	74.6	2.3	17.9	22.7
Eskimos (Canadian)				
Copper Igloolik[3]	78.2	0.5	17.1	26.9
Mckenzie Inuvik[2]	72.8	0	19.8	20.4
Amerindians (North America)				
North Athapaskan				
Dogrib[4]	79.3	6.3	13.5	59.7
Chipewyan[2]	78.3	2.2	15.8	28.0
South Athapaskan				
Navajo[2]	74.1	6.0	17.4	25.2
Apache[5]	75.1	5.2	16.8	25.1

Table 15 (cont.)

Population	GM*A G	GM*X G	GM*A T	KM*1
Algonkian				
Northern Cree[2]	98.1	0 1.5	43.5	
Ojibwa[6]	85.8	7.2	7.0	28.6
Siouan[2]				
Assiniboin[2]	94.3	0	2.8	34.3
Amerindians (Middle America)				
Cuanalan Residents[7]	86.1	8.7	2.4	43.4
Saltillo (Chamizal)[7]	58.6	7.7	1.8	30.2
Cora[7]	74.2	9.7	11.4	19.2
Mayan[7]	64.9	12.3	4.4	nt
Amerindians (South America)				
Yanomama[7]	85.2	14.8	0	38.7
Makiritare[10]	56.3	43.5	0	57.4
Cayapo/Txukahme[11]	70.9	28.8	0.3	49.7

References: [1]Sukernik and Osipova, 1982; [2]Schanfield et al., 1990; [3]McAlpine et al., 1974; [4]Szathmary et al., 1983; [5]Williams et al., 1985; [6]Szathmary et al., 1974; [7]Schanfield et al., 1978; [8]Steinberg et al., 1967; [9]van Loghem, 1971; [10]Gershowitz and Neel, 1978; [11]Salzano et al., 1967b; [12]Crawford et al., 1994.

than 300 million genetically different combinations producing more than 30 million distinguishable phenotypes. Despite this wealth of variability, few investigations of Amerindian populations have utilized the C and D loci. There are few publications on the HLA types of indigenous populations of Siberia.

Layrisse et al. (1976: 137) argued that

'The HLA tissue antigen system is probably the most suitable genetic tool for the analysis of genetic polymorphism in populations and for the determination of the ethnic groups that have substantially contributed to their gene pool.'

This praise of the HLA system is made despite its numerous limitations. In contrast to those of the GM system, the HLA antigens are not population-limited, i.e. no antigen occurs in only a single geographically defined population. Therefore, the presence of a single antigen does not necessarily connote affinity to, or gene flow from, a particular population. In most human populations, the HLA alleles occur at frequencies between 1% and 15%. However, in Amerindian populations there is a small number of detectable antigens, with higher frequencies.

Schanfield (1980) summarized the frequencies of antigens encoded at the A and B loci in Eskimo and North and South American Indian populations. In Eskimo groups A9 and B40 alleles were the most common, their frequencies averaging 65 and 42%, respectively. The antigen A2 is of intermediate frequency and a number of antigens (A1, A3, A10, A28, B5, B7 and B15) occur at lower rates. Schanfield characterized the North American Indian as having high frequencies of A2 (51%), A9 (34%), B40 and B35 (22%). The South American Indians also exhibit a high frequency of A2 and B40. B15, B5 and A9 are intermediate in frequencies. The genes A9, AW24, AW10 and AW16, relatively common in South American tribes, occur at approximately the same frequencies in Asian populations. The South American tribes (Warao, Yanomama and Makiritare) exhibit most of the A and B antigens that are seen in Asian populations, with some exceptions (Layrisse et al., 1976). One exception is the AW31 antigen, which is present among the South American groups but is absent from Asian populations. Similarly, the majority of antigens that are present among Asian groups are absent from New World populations.

Callegari-Jacques et al. (1994) summarized HLA haplotype results in four Amazonian tribal groups: Cinta Larga, Karitiana, Surui, and Kararao. Their HLA allelic and haplotypic frequencies were compared to the average frequencies from 32 South American Indian populations. The most striking finding was the restriction in the number of observed haplotypes among the Karrarao and Karitiana (only 6 and 7, respectively). In contrast, the Surui and Cinta Larga exhibited more than twice as many haplotypes. The Karitiana and Kararao show an absence of a wide array of alleles. It appears that the HLA system, because it contains many alleles at low frequencies, is particularly susceptible to the effects of stochastic processes in small populations, such as those described by Callegari-Jacques et al. (1994). A multivariate analysis of the HLA and 16 other genetic markers is suggestive of a genetic diversity between the Amazonian groups to the north and south of the river. There are two most likely explanations for this diversity: (1) the Amazon river constitutes a geographical barrier to migration; or (2) two different past migrations entered the region from the west.

Black et al. (1980a,b) after surveying the HLA loci in eight Indian populations of Brazil and Chile, concluded that there is limited heterogeneity in this system, probably resulting from reproductive isolation. The most common antigens in these groups are A2, A28, AW24, AW30, AW31, AW32, B5, B15, B40, BW35, BW39, CW1, CW3 and CW4. Surprisingly absent are A9, AW10 and AW16, the alleles most characteristic of Amerindian populations. Most of the antigens found in South American populations also occur in North American populations. Similarly, Hansen *et al.* (1986) described relatively restricted polymorphism of the A and B antigens in the Eskimos of Alaska. They detected the presence of approximately 35% of the A-locus specificities and 37% of the possible B-locus specificities. In comparisons between the Yupik and Inupik of Alaska, the frequencies of the HLA A antigens were significantly different. However, no significant differences in allelic frequencies at the B, C, D and DR loci were detected between the two linguistic groups.

Greenacre and Degos (1977) attempted to characterize variation at three HLA loci in 124 populations by using correspondence analysis (which includes both principal components analysis and multidimensional scaling). This method clearly separates Asians, Eskimos and Amerindians from Europeans, Africans and Oceanic populations along the second axis. More than 40% of the variation is subsumed by the plot of the first and second principal axes. The Amerindian populations are separated from the rest of the world by their frequencies of BW16, BW35, BW21, A2, A9 and A28. From this analysis it is clear that, on the basis of the HLA system, New World populations can be distinguished from all other groups but Asians. Asian populations share many HLA alleles with Amerindians, reflecting their close evolutionary ties.

After principal components analysis (PCA), the first versus second scaled eigenvectors (e_1 vs. e_2) were plotted using allelic frequencies for the HLA-A locus in 25 Amerindian and Eskimo populations. The first eigenvector separated the Tikuna from the other populations, largely because of the AW24 allele among the Tikuna. No distinct clusters became evident, except that the South American populations (Parakana, Kayapo and Aymara) were separated along the second eigenvector. The two eigenvectors accounted for 61.45% of the observed variation (see Fig. 31).

The HLA-B locus provided slightly better discrimination among 24 Amerindian and Eskimo populations. A two-dimensional PCA plot of an R-matrix revealed that the first two eigenvectors ($e_1 = 36.55\%$, $e_2 = 35.54\%$) accounted for 72.09% of the total variation (see Fig. 32). The first eigenvector separated the Venezuelan Mestizo from all other populations. The Lumbee,the Mexican Indians and the Mestizo occupied an intermediate position in the plot. Instrumental in this dispersion were the alleles B12 and B7, both indicative of gene flow from the Old World. The second eigenvector clustered the Eskimo groups separately from the Amerindians. The BW35 and BW22 alleles were involved in the separation of Eskimos from Amerinds.

The use of three HLA loci (A, B, and C) increased the number of available alleles to 21, but decreased the number of populations that were tested for all three loci to 11 (see Fig. 33). The resulting two-dimensional plot provided better

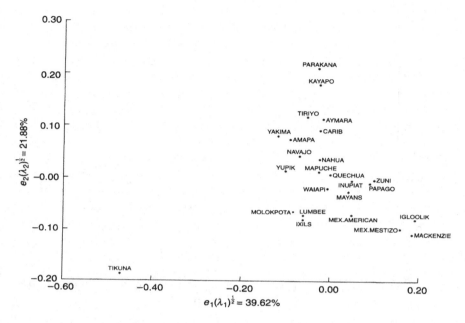

FIGURE 31 Principal-components analysis of 25 Amerindian and Eskimo populations based upon HLA-A locus allelic frequencies.

discrimination between the populations, but accounted for less of the total observed variation (54%). With the exception of the Mapuche, the South American Indian populations formed a relatively tight cluster. However, the Eskimo groups (Inupiat and Yupik) are located in intermediate positions between the South Americans, the Navajo and the Lumbee. Thus, despite earlier criticism, the HLA systems appear to be reasonably useful markers for the calculation of affinities among populations provided multivariate statistics are employed.

Bodmer (1975) and others have argued that the relative homogeneity at HLA loci of Amerindians is due to selective pressures associated with infectious diseases. These pressures would have had to reduce HLA heterogeneity without reducing the diversity of other markers. Given no evidence of relationships of elements in the HLA system to infectious diseases and to more than few genetic maladies, it appears that a more reasonable explanation for the restriction of the HLA variation is the founder effect. Most likely, the Siberian founders brought with them a limited spectrum of related HLA types (Black et al., 1980a). However, Black and Salzano (1981) have demonstrated a significant deficit of Amazonian Indians homozygous for HLA haplotypes. A similar phenomenon was reported by Degos et al. (1974) for the Tuaregs of the Sahara. Given the restricted variation of antigens and haplotypes and village endogamy, one might

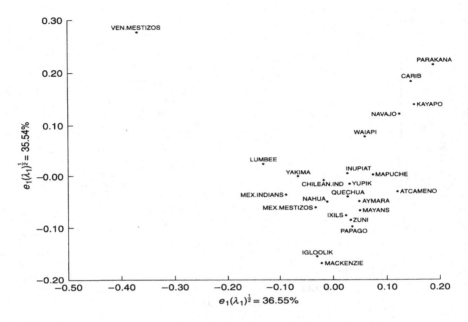

FIGURE 32 Principal-components analysis of 24 Amerindian and Eskimo populations based upon the allelic frequencies of the HLA-Blocus

expect to find extreme homogeneity at these loci. Instead, there were 56% fewer homozygotes than were expected statistically. Migration could explain this deviation from expectation (through the addition of new haplotypes to the community, their descendants would admix with the residents and produce an excess of heterozygotes). However, demographic and familial reconstructions do not provide evidence of sufficient migration. Therefore, in the absence of evidence to the contrary, it is likely that heterozygosity for HLA haplotypes confers selective advantage.

4.8   DNA POLYMORPHISMS

With recent developments in the use of restriction fragment length polymorphisms (RFLP), amplification through polymerase chain reaction (PCR) and nucleotide sequencing, it is now possible to explore the variation of the DNA molecule (Cann, 1985). Most of the research on genetic variation described earlier in this volume has dealt with protein polymorphisms, i.e. variation in gene products. Two forms of human DNA have been investigated, mitochondrial DNA and nuclear DNA.

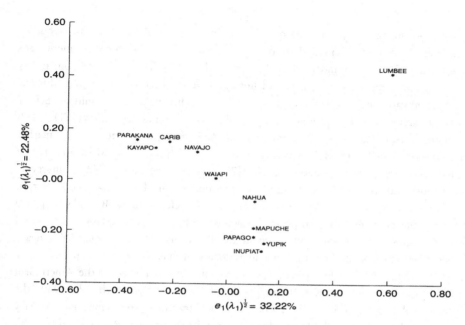

FIGURE 33 Principal-components analysis of 11 Amerindian and Eskimo populations based upon the allelic frequencies of three HLA loci, -A, -B, and -C.

Mitochondrial DNA

Mitochondrial DNA (mtDNA) is a small, circular molecule of 16 569 nucleotide pairs (np) located in mitochondria of the cytoplasm (Anderson *et al.*, 1981). Although mtDNA is maternally inherited, it does recombine with a minute paternal component (Case and Wallace, 1981; Gyllensten *et al.*, 1991). This molecule evolves by the accumulation of mutations in the maternal lineages (Brown *et al.*, 1979) and is believed to fix new mutations more than ten times faster than does nuclear DNA (Wallace *et al.*, 1987). This rapid evolutionary change permits the use of this molecule to study the evolutionary divergence of human populations, such as the Eskimos and Amerindians, who split from Siberian groups a few thousand years ago (Johnson *et al.*, 1983).

Even though Alan Wilson and Rebecca Cann first applied mtDNA to the study of human phylogeny, Douglas Wallace and his research group used mtDNA to examine questions concerning the peopling of the New World. Wallace *et al.* (1985) utilized a series of restriction endonucleases, *Hpa*I, *Bam*HI, *Hae*II, *Msp*I, *Ava*II and *Hinc*II, to identify haplotypes of mtDNA. Their initial restriction analysis of the mtDNA of the Pima and Papago Indians revealed a distinctive marker, *Hinc*II morph-6, in 42% of the people tested. This haplotype is observed at low frequency in the mtDNA of Asians and its presence at high frequencies among Amerindians is explained by the founder effect (Schurr *et al.*, 1990). There is some controversy surrounding the number of maternal lineages that are necessary to account for the observed variation in mtDNA haplotypes of Amerindian populations. Initially, Schurr *et al.* argued for the presence of four different lineages, as follows. (1) *Hinc*II morph-6 (site loss at nt 13 259) and an *Alu*I site gain at nt 13 262) marker (haplotype AM10) found in virtually every mtDNA from Pima, Maya and Ticuna Indians, later termed haplogroup C. (2) Asian-Specific COII-tRNA^{Lys} intergenic deletion (haplotype AM2) was found in Amerindians who lacked the *Hinc*II morph. This was defined by a 9-bp deletion and is now refered to as B haplogroup. (3) *Hae*III site at nt 663 polymorphism (haplotype AM6) first observed by Cann (1982) in Chicanos and Chinese, was noted in the Pima, and Maya, Ticuna. This lineage was later renamed A. (4) The D haplogroup was defined by an *Alu*I site loss at nt 5176. Schurr *et al.* (1990) placed this haplogroup in the middle of the Amerindian tree. Because most of these haplotypes can be derived from each other with the sequential accumulation of mutations, Wallace (personal communication) initially claimed the origin of all Amerindians from a single lineage. Pääbo *et al.* (1988), after an analysis of the mtDNA from a 7000-year-old Indian brain from Florida, proposed the existence of another founding lineage. However, Schurr *et al.* explained these results in terms of the loss of this founding haplotype from the three Amerindian tribes they had studied.

A more extensive analysis by PCR amplification with 14 endonucleases of 321 Amerindians from 17 populations confirmed the presence of the four different lineages or haplogroups A, B, C, and D that account for 96.9% of Amerindian mtDNA variation and are of Asian ancestry (Torroni *et al.*, 1993a; Torroni and Wallace, 1995). These findings supported the hypothesis that the four Amerindian mtDNA haplogroups resulted from four separate demic expansions. Three of the four haplogroups (A, C, and D) observed in the Americas are present in indigenous Siberian populations (Torroni *et al.*, 1993b). None of the Siberian populations (a total of ten tested) exhibited the B haplogroup. The presence of the B haplogroup in Asia and the New World, and its absence from Siberia, is suggestive of its separate expansion into the New World, possibly prior to the peopling of Siberia (circa 20 000 years ago). When dates of divergence were calculated using the mtDNAs of Amerindians and Siberians, they fell between 17 000 and 34 000 years BP (Torroni *et al.*, 1993b). These dates provided additional evidence for the presence of pre-Clovis populations in the New World (see Fig. 34).

Torroni and Wallace (1995) reported that out of 743 Native Americans tested to date, 25 individuals, scattered among eight tribal groups of North, Middle and

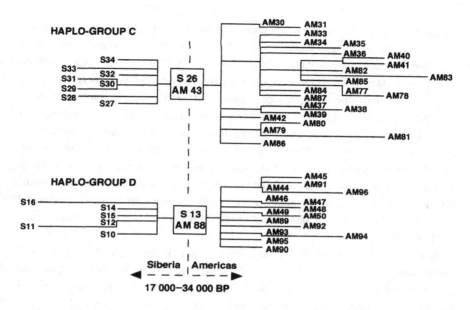

FIGURE 34 The mitochondrial DNA haplotypes shared by the Siberian and Amerindian populations, used to reconstruct the chronology of the peopling of the New World (Torroni *et al.*, 1993b).

South America, displayed some mtDNA variants that differed from the four common haplogroups (A,B,C,D). They suggested that these variants may be the result of: (1) a second mutational event; (2) possible admixture with Europeans or Africans; or (3) additional Asian haplogroups brought into the New World by Siberians. Thus, Torroni and Wallace cautioned researchers against classifying mtDNAs from Old World populations by using only the primary variants found in Amerindian mtDNA. They went on to point out that the 9-bp deletion had occurred independently in different regions of the world. This conclusion was supported by Soodyall *et al.* (1996), who discovered the presence of the so-called Asian-specific deletion in sub-Saharan Africa. From these data it appears that this 9-bp deletion arose independently at least twice, once in Asia and once in Africa. Bailliet *et al.* (1994) have proposed the existence of as many as ten possible mtDNA founder haplotypes in Amerindian populations. However, they do believe that some of these haplotypes may be due to mutations in Amerindian populations and/or admixture with Europeans (Bianchi and Rothhammer, 1995).

Ward and his colleagues (1991) sequenced a 360-nucleotide segment of the mtDNA control region from 63 individuals of the Nuu-Chah-Nulth (or Nootka) from Vancouver Island. They identified 28 mtDNA lineages as defined by 26 variable positions within the control region. Ward *et al.* computed the average sequence divergence between the lineage clusters using a maximum rate of evolution of 33% divergence per million years for the control region. They obtained a range of 41 000–78 000 years, with an average of 60 000 years. These data suggest that the mitochondrial lineages within a single Amerindian tribe diverged approximately 60 000 years ago. Ward *et al.* interpreted these data as evidence that the lineages were established prior to Amerindian entry into the New World, and they concluded that the founding populations of Amerindians contained considerable genetic diversity. These conclusions support my earlier premise that the founders of the Amerindians were not merely a few bands of Asians following animal herds, but that the peopling of the New World was a demic expansion involving numerous populations. I tend to agree with Ward *et al.* that the mtDNA sequence data 'will allow a test of the "three wave" hypothesis' (Ward *et al.*, 1991: 8723).

More recently, Merriwether *et al.* (1995) extensively investigated the geographic distribution of the four founding mtDNA lineage haplogroups in Native populations of the Americas. They observed a north–south increase in the frequency of the B haplogroup, accompanied by a north–south decrease in the frequency of A. Based upon the extensive distribution of the four lineages in the New World, Merriwether and his colleagues concluded that the pattern is consistent with a single migratory wave from Siberia into the Americas, followed by genetic divergence. However, these data can also be interpreted to represent a number of migrations from Siberia reintroducing the same haplogroups.

A recent comparison of mtDNA RFLPs from Mongolians of Ulan Bator with an array of frequencies of the founding lineage haplogroups in New World, Asian, and Siberian populations reveals considerable similarity between the Mongolian and New World populations (Merriwether *et al.*, 1996). In this study the haplogroups were further subtyped into A1, A2, B1, B2, C1, C2, D1, D2, X6, X7, and 'others'. Unlike the Northeastern Siberian populations, this Mongolian sample exhibited all four of the New World primary haplogroups and shared the highest number of haplotypes with Amerindian populations. To date the B haplogroup has not been detected in any of the Siberian populations in closest proximity to the Bering Strait. However, this haplogroup occurs at a frequency of 75% among the Atacameno and 50% among the Pima, but is absent in a number of Amerindian groups (e.g. Makiritare, Dogrib and Haida). The virologic HTLV II data also support a Mongolian origin for the founding population(s) of the New World (Neel *et al.*, 1994). Since Mongolia is land-locked and at some distance from Beringia, the most parsimonious explanation of the data is that the populations of Mongolia and the New World descended from a common ancestral population.

Nuclear DNA markers

Nuclear DNA can roughly be subdivided into coding and non-coding regions. The coding regions of the human genome are involved in the production of a gene product, protein. Of the approximately 3 billion bases that make up the human genome, it has been estimated that only 3–5% are involved in protein synthesis. Because of their physiological functions, these coding regions tend to be more evolutionarily conservative than the non-coding regions.

Coding DNA

Kidd *et al.* (1991) have argued in favor of using nuclear DNA over mitochondrial DNA for evolutionary studies. They point out that: (1) mtDNA cannot be utilized to study selection on specific genes in the nucleus; (2) there is greater genetic variation in nuclear DNA with 1000 times (or more) genes in a nucleus than in a mitochondrion; (3) male migration cannot be addressed in mtDNA since genetic transmission is through mothers.

To date, the majority of investigations of nuclear DNA variation in Amerindians have utilized RFLPs (restriction fragment length polymorphisms). RFLPs are DNA fragments cleaved by restriction enzymes at specific sites; these DNA fragments are separated by electrophoresis on the basis of size. Bowcock *et al.* (1987) argued that RFLPs are important because the number of markers is almost limitless, and the biological function of encoding genes can be derived. They examined the variation of 47 nuclear DNA markers in five human populations and demonstrated that the variation between populations (as measured by F_{st}) was highly significant.

In their study of nuclear DNA variation among Amerindian populations, Kidd and her co-workers (1991) examined a total of 37 RFLP systems in three native American populations. They compared two tribes of the Amazonian basin of Brazil with a Mayan population from the Yucatan. All 31 loci of the 37 RFLP systems utilized in this comparison were known to be polymorphic in individuals of European descent. Thus, all comparisons of rates of heterozygosity in these Amerindian groups were made with those of Europeans. A slight reduction in average genetic heterozygosity in the three Amerindian populations was observed when compared with that of their European counterparts. Some systems, such as the REN locus (digested by *Hind*III restriction enzyme and identified by the hREN probe) had gone to fixation in both Amazonian populations. Out of the 37 systems studied, almost 7% had gone to fixation in their sample of New World population and approximately 10% were fixed in the Amazonian tribes. Kidd *et al.* (1991) interpreted these slight reductions in genetic variability (*vis* à vis European groups) as evidence that Amerindian gene pools never experienced severe bottlenecks. However, an alternative explanation is that a narrow bottleneck was followed by admixture with Europeans and Africans. Kidd *et al.* (1991) also suggested that, with more DNA loci and better estimates of dates of arrival of Siberians into Alaska, it will become possible to determine the mininmum effective sizes of the founding population(s).

Beta-globin haplotypes Recently, Bevilaqua *et al.* (1995) re-examined the question of whether South American Indian populations after Contact underwent massive genetic bottlenecks, which would be evident in the reduction of genetic variability. They examined the beta-globin gene clusters (haplotypes) in five Brazilian tribes (Xavante, Zoro, Gaviao, Surui and Wai-Wai), and compared them with European, Asian and Oceanic groups. The five Brazilian tribes were first compared on the basis of haplotype frequencies, with the Yanomama Indians. The most frequent haplotypes observed among the Brazilian populations were haplotype 2 (+ – – – –), which ranges from 93.3% among the Zoro to 60% among the Xavante Indians, and haplotype 6 (– + + – +), which ranges from 18.3% among the Xavante to 3.4% in the Gaviao. These haplotypes are characterized by the presence or absence of specific restriction sites. These same two haplotypes are the most prevalent in Eurasians (Wainscoat *et al.*, 1986; Long *et al.*, 1990; Chen *et al.*, 1990). The degree of genetic diversity was measured by the Gini–Simpson Index

$$\text{GSI} = 1 - \sum_k p_k^2,$$

where p_k is the frequency of the kth allele in the population. The highest haplotypic diversity was noted among the Xavante (0.582); the lowest was in the Zoro (0.124). Unlike the study by Kidd *et al.* (1991), which found a small reduction in the variability of 30 RFLPs for two Brazilian Indian groups, Bevilaqua *et al.* (1995) observed a considerable reduction in variability (56%) when comparing Brazilian Indians with European populations. However, a comparison between Brazilian Indians and Asians yielded a small reduction of only 8%. The Asians (with whom the Brazilian Indians probably share more recent common ancestors) when compared for heterozygosity levels with the Brazilian tribes have lower reduction of genetic variation and are suggestive of the absence of significant bottleneck effects in the early colonization of South America.

Long *et al.* (1990) have argued that haplotypes 2, 6 and 5 were probably present in the ancestral Amerindians who colonized South America. These haplotypes are separated from each other by at least two steps of mutation and recombination.

Haplotypes 7 (– + + – –), 11 (– – – + +) and 16 (– + – – –) could have been present among the colonizers of South America or have arisen by recombination events in tribal populations.

Non-coding DNA

Approximately 95–97% of the human genome does not code for any protein product and has been characterized as 'junk DNA', possibly providing the frame or scaffolding essential to the structure of the DNA molecules. This non-coding DNA, in the form of tandem repeats, apparently does not play a major role in the physiology of the human organism and has been thought to be selectively neutral (Harding, 1992). However, recent research has prompted the reassessment of the role of these short repetitive DNA segments distributed over much of the human genome.

Several cases of triplet repeat amplification have recently been found to be associated with monogenetic diseases, such as myotonic dystrophy (CTG repeats), spinal and bulbar muscular atrophy (CAG), Fragile X mental retardation (CGG), and Friedreich's ataxia (GAA) (Barinaga, 1996; Campuzano et al., 1996). These diseases vary in incidence among human populations. For example, Huntington's disease (CAG repeat within a gene) has a worldwide average of 5 per 100 000, yet occurs in Malta at a frequency of 11.9 per 100 000 (Gallo et al., 1997). There appears to be a relationship between the amplification of the numbers of repeats, age of onset, and the severity of the disease. Thus, there is some question about the evolutionary neutrality of these repeats.

The tandem repeats of non-coding DNA have been categorized by their array structures and the numbers of repeats. They are termed: (1) microsatellites (or Short Tandem Repeats, STRs) if the repetitive DNA consists of 2–3 nucleotide repeats, e.g. CA or CAT, forming arrays of several to tens of repeats in length; (2) minisatellites (Variable Number Tandem Repeats, VNTRs), with tens of nucleotides repeated from tens to hundreds of times; or (3) satellites, regions of repeats that extend for thousands of kilobases.

Flint et al. (1989) tested the capacity of minisatellites occurring about 400 bp 5′ to the insulin gene on chromosome 1 to discriminate between Oceanic populations. They found that Polynesians have lost a significant amount of genetic diversity, probably through evolutionary bottleneck effects. Thus, these highly variable loci could be utilized to reconstruct human evolutionary history.

Variable number of tandem repeats (VNTRs) VNTRs are stretches of DNA in which a short nucleotide is tandemly repeated 20–100 times. Different VNTR alleles are composed of different numbers of repeats. To date, because of their high mutation rates

Table 16. *Listing of DNA loci and the probes required to identify them*

Locus	Probe	Repeat length (base pairs)
D7S104	CRI-PAT-pS194	50
D11S129	CRI-PAT-pR365-1	28
D18S17	CRI-PAT-pL159-1	48
D20S15	CRI-PAT-pL355-8	33–35
D21S112	CRI-PAT-pL427-4	26–30

and exceptional individual variability, VNTRs have primarily been applied to forensic studies.

Much as in the Flint *et al.* (1989) study, we observed fewer VNTR alleles in isolated Siberian reindeer-herding populations than in large, less isolated groups (McComb *et al.*, 1995). VNTRs are pieces of non-coding DNA of varying sizes and numbers of repeats. Table 16 lists the five VNTR loci, the probes and the repeat lengths (in base pairs) utilized in the analysis of the genetic structure of Siberian populations (see Fig. 35 for the location of the Siberian groups). Table 17 compares the number of binned (because of experimental error) DNA fragments that are grouped into bins of ±2% for the Evenki, Keto (fishermen of Central Siberia), Navajo, Asians, Afro-Americans, Americans of European origin, and Mexican Americans. The more reproductively isolated populations, such as the Evenki, Keto and Navajo, have the fewest 'alleles' whereas the larger, panmictic samples exhibit almost double the number of alleles. These results remind me of the many more immunoglobulin (GM) phenotypes shown by the triracial hybrid populations of the Caribbean than are found among Europeans or Amerindians (Schanfield *et al.*, 1984). The distributions of eight groups (Navajo, Mexican, four Siberian and African and European American) are compared for the bimodally distributed (has a large gap of 600 kb) locus D11S129. This locus is the best discriminator between Siberian, Amerindian and Old World populations (see Fig. 36). The Siberian and Amerindian distributions can be discriminated by the more frequent presence of the larger fragments (2688–3034 bp) and the less frequent small fragments (1350–2023 bp).

A comparison of VNTR DNA fragment frequencies for five loci (D18S17, D11S129, D20S15, D21S112 and D7S104) in Siberian, Amerindian and various other regional populations reveals similarities between Siberian and Amerindian groups (see Fig. 37). A total of 45 'allelic' frequencies were used to measure affinities. The first two eigenvectors explain 73.4% of the total variation. The Keto, a small population of fishermen from Central Siberia who reside in a mixed Keto-Russian village, show some

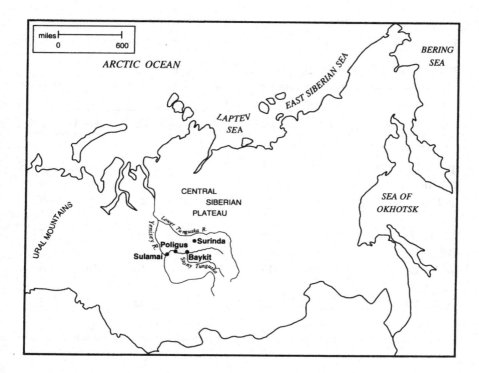

FIGURE 35 A map of Siberia locating the Evenki populations that were involved in the study of VNTR variation in Amerindian and Siberian indigenous groups.

affinity to the European American sample. This is most likely due to admixture between the Russians and Keto in the village. Similarly, the Altai community (Mindur-Sokkon) because its gene pool is a melange of Turkic–Mongol–Chinese, displays some affinity to the European group. The Mexican sample shows some affinity not only with the European sample but also with Afro-Americans. Thus, on the bases of the frequencies of VNTR alleles, the Siberian populations tend to cluster with the New World populations (McComb et al., 1995). The addition of an Asian (Chinese–Japanese) mixed sample to this analysis shows that the Asian population clusters with the Keto and Evenki, while the Navajo show less affinity to the Siberian groups but are closer to them than are the Caucasian or Afro-Americans.

Y-chromosomal markers

The non-recombining (Y-specific) portion of the human Y chromosome has been of great interest to anthropological geneticists in reconstructing human phylogeny. Much

Table 17. *Number of alleles (DNA fragments) at five loci in six populations.*

The loci are indicated by 'D'; the probe number is preceded by 'p'

Locus	No. of bins	Population					
		Navajo	Mex-Amer.	Afro-Amer.	Cauc.	Evenki	Keto
D18S17 (pL159-1)	18	8	13	9	10	7	6
D7S104 (pS194)	29	9	18	27	23	9	9
D20S15 (pL355-8)	20	9	10	15	9	13	7
D11S129 (pR365-1)	22	9	13	17	16	10	6
D21S112 (pL427-4)	35	19	31	31	32	7	8
TOTAL	124	54	85	99	90	46	36

like the mtDNA, but a male mirror image, the Y-specific portion evolves through the accumulation of mutations. Only a small portion of the Y chromosome (the pseudoautosomal region) recombines with the X chromosome. Markers on the Y-specific region provide some indication of male migration and admixture. The initial research was somewhat disappointing because of the paucity of variation in the Y chromosome. This prompted Kenneth Kidd to joke:

> 'There is a long-standing presumption that the Y chromosome is a dud that contains only junk DNA with the sole exception of that trivial bit of DNA that turns on male sex development. What has been learned though is that there are only three genes found in the Y-specific region'

(quoted by Gibbons in Science 251: 378).

However, Kidd's joking lament proved to be an exaggeration with the information contained within the so-called junk DNA. Y-specific polymorphisms have successfully been used to construct informative haplotypes that are specific to geographic regions and to historical and migration events. In addition, Y-chromosome-specific deletions and transitions have been discovered that apparently have arisen once in human evolution and serve as markers for phylogenetic reconstruction (Karafet *et al.*, 1997).

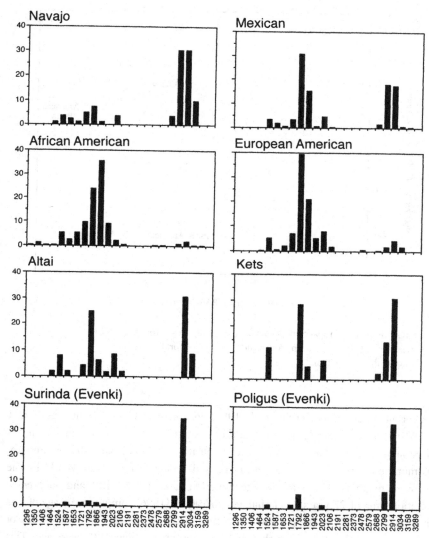

FIGURE 36 A comparison of distributions of VNTR alleles in Siberian indigenous populations with larger synthetic world populations (after McComb et al., 1995).

However, a statistical test of Greenberg's (1987) tripartite model based on the distribution of the frequencies of the 1T Y chromosome haplotype is far from convincing. Karaphet et al. (1997) found no statistically significant frequencies between the Eskimos, Na-Dene, and Amerinds using sample sizes as low as 10 for the Eskimos plus Bonferroni protection of alpha.

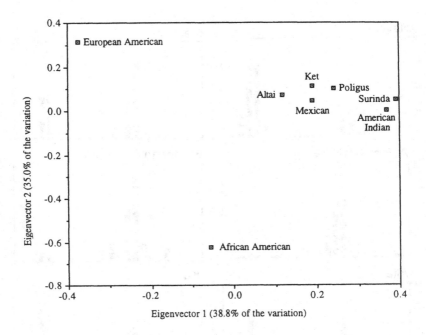

FIGURE 37 A principal-components analysis using alleles from five VNTRs representing several Siberian, Amerindian, and Old World populations (McComb et al., 1996).

A recent report by Underhill et al. (1996) documents a C → T point mutation at the DYS19 microsatellite locus. To date this muatation has been found only in Inuits and Navajos of North America and other populations of South and Middle America. This mutation may have occurred in Siberia and brought to the New World by the first Asian migrants. Given an average mutation rate of 1.5×10^{-4} and an average generation time of 27 years, Underhill and colleagues (1996) computed the age of this transition as 30 000 years ago. However, they acknowledge that mutation rates of smaller magnitude provide dates that are more recent, i.e. only 2147 years ago. This polymorphism may shed some light on the colonization of the Americas, particularly if it is present in some Siberian populations and not in others (Santos et al., 1995, 1996).

Lin et al. (1994) investigated the variation in Asian, European and African–American populations for Y- and X-associated polymorphisms by using the 47z (DXYS5) probe. Although both the X1 and X2 alleles were detected in most of the populations, Y1 and Y2 were polymorphic in only the Japanese, Koreans, and the Hakas and Folo of Taiwan. To date, these X- and Y-associated markers have not been tested in Amerindian populations. If polymorphic, these molecular markers would be useful in studying sex-specific migration.

4.9   CONCLUSION

Based on the information presented in this chapter on the genetic variation observed among contemporary Amerindians, we can conclude that the native peoples of the New World constitute a distinct entity with considerable genetic variation. Construction of genetically based 'trees' reveals that New World populations form distinct clusters, and that Amerindians and Eskimos descended from several populations of Asians who crossed Beringia perhaps at different times.

Given the evolutionary history of the native peoples of the Americas, it is not surprising to observe the vast variation in gene frequencies between populations and in genotypes of individuals within populations. The founding populations must have been small, probably made up of extended lineages, and were not randomly constituted subsets of the ancestral groups. Thus, in these groups gene pools were highly subject to stochastic processes and past effects of selection might not be discernible to us. There are few clear-cut clinal distributions over populations of the New World; where they do exist, population movements and unique historical events are the more plausible explanations. In South America, the differences in gene frequencies between small groups are particularly evident, with a number of loci displaying fixation, little variation, or a substantial incidence of familial variants, i.e. genes that are rare elsewhere.

Based on standard blood markers, Salzano (1968b) demonstrated the broad range of intra-tribal variation among the South American Indians (see Fig. 38). He was careful to eliminate samples that were a result of admixture and/or laboratory error. The frequency ranges for P, M, RH*R2, FY*A and JK*A are considerable and approximate those among all the world's human populations. Some of the systems were monomorphic, such as K, Lutheran and Lewis. The observed intratribal variation was not as great as the intertribal variation among the South American groups, but for the seven Caingang and three Xavante groups that Salzano described, the variability is surprisingly extensive.

The genetic diversity in Amerindian populations as revealed by the DNA markers is greater than that measured by standard markers. When isoelectric focusing was first used to assess genetic variation, heterozygosities at a given locus reached a level of between 0.30 and 0.58 in the highly admixed Black Carib populations (Dykes *et al.*, 1983). In contrast, short tandem repeats (STRs) attain heterozygosities of 0.7–0.9 even in relatively homogeneous European populations. Thus, DNA markers appear to

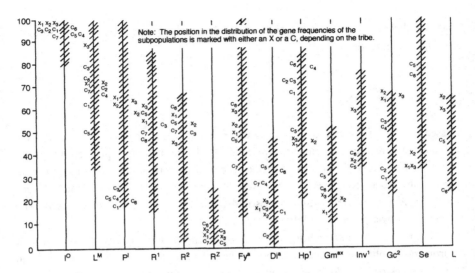

FIGURE 38 A visual presentation of the variation observed among South American Indian tribes using 14 different blood-group and protein alleles (Salzano, 1968b).

be much more informative for anthropological genetics. However, when the genetic distances among populations, based on standard markers, were correlated with distances based on DNA markers, the correlations were relatively high and the same affinities and historical patterns were revealed (McComb and Crawford, unpublished data, 1996). Yet on the individual level, heterozygosities based on standard markers cannot predict the level of genetic variation measured in VNTR loci. Apparently VNTRs, with high mutation rates, reveal more current evolutionary history, whereas the variation in gene products (more conservative evolutionarily) is the result of the action of natural selection in more ancient times. We can learn much about the evolution of our species from studying our genetic heritage i.e. genetic variability.

5 Population structure of Native Americans

5.1 INTRODUCTION

There are numerous definitions of population structure. Some of these refer to the relationships between the elements within populations, such as genes, genotypes, phenotypes, and groups of individuals (Workman and Jorde, 1980). Schull and Mac-Cluer (1968) include within 'population structure' all attributes or parameters of a population (such as geographic, cultural, demographic and social parameters) in time and space. Yet others view population structure as the correction of ideal populations (with properties such as panmixis, infinite size, equal genetic contributions of phenotypes) with real populational characteristics (Cavalli-Sforza and Bodmer, 1971). This latter approach comes from the seminal work of Sewall Wright (1921, 1943, 1945, 1978), who developed much of the theory concerning effects of non-random mating and finite population size. In this chapter, the concepts of population structure will be presented in two parts. First, the effects of small size and non-random mating will be examined within individuals or subdivisions of populations (intrapopulation comparisons). Second, the relationships between populations sharing similar environmental (e.g. geography), temporal (time subdivided by periods) and cultural factors (e.g. languages) will be examined (interpopulational studies). This categorization is imperfect because the lines drawn between populations and subdivisions of a population are often ambiguous. Given sufficient temporal depth, all human populations can be viewed as subdivisions of a single founding population. In fact, the concept of 'population structure' may be defined as the totality of factors that interfere with random mating, i.e. population subdivision. Estimates of admixture or gene flow will be considered in the second section of this chapter. Obviously, the effects of admixture on the native populations of the New World were evolutionarily critical and this feature of population structure is considered in some detail.

5.2   INTRAPOPULATION SUBDIVISON

Most human populations exhibit some subdivision, either geographic (spatial), linguistic, economic or social. Amerindians are subdivided into various social and spatial entities such as linguistic groups, nations, tribes, villages, clans, lineages and other aggregates. Often these social and spatial subdivisions are hierarchical and serve as barriers to reproduction of varying degrees of effectiveness. The evolutionary effects of subdivision can be assessed by sampling the subdivisions genetically and comparing the gene frequencies by an assortment of analytical methods ranging from genetic distances to heterogeneity Chi-square to Wright's F_{st}. All of these approaches have been successfully utilized in studies of Amerindian populations.

Genetic heterogeneity (χ^2)

Workman and Niswander (1970) examined the population structure of the Papago Indian reservation, which is subdivided by political districts that are roughly analagous to their ancestral defense villages. The observed variations in genic proportions are subjected to a contingency χ^2 analysis; significant differences among these Papago political districts were reported. The observed differences between the districts were probably due to non-systematic forces. Workman and Niswander identified the most likely causes of genetic heterogeneity as: (1) founder effect; (2) isolation by distance, i.e. as the distance between individuals or groups increases, their degree of genetic similarity decreases. They applied an isolation-by-distance model to the Papago districts and found a correlation of $r = 0.494$ between genetic and geographic distances, and that approximately 24% of the total genetic variation could be attributed to geographic distances between the groups. Comparisons of the degree of heterogeneity among the Papago subdivisions with that observed among Yanomama Indian villages reveals that marked differences in genetic heterogeneity exist between these tribes. The Yanomama villages exhibit greater genetic heterogeneity than do the Papago subdivisions. Given the fission–fusion process (Neel and Ward, 1972) (fission by lineage, sometimes followed by fusion of several lineages) in the creation of new Yanomama villages, this heterogeneity is not surprising. Smouse (1982) points out that increasing social tension (owing to kin groups competing for dominance) can bring about hostilities, which are often followed by village fission. Small splinter groups that experience hard times may fuse with another village or village fragment. The evolutionary effect of random fission is the genetic divergence of daughter villages. I found analogous population subdivision processes among the Mennonite communities of Kansas and Nebraska (Crawford et al., 1989). Original coalescence (believers came together under a religious leader) was followed by fission (lineages splitting off from a congregation because of doctrinal or personal disagreements).

In search of the elusive force of natural selection, Rothhammer *et al.* (1990) applied a χ^2 heterogeneity analysis to detect the possible relation between altitude and the gene frequencies of resident populations. This between-niche heterogeneity produced no significant results, indicating that there had been no systematic changes in gene frequencies among their samples of Ayamara Indians of Chile. However, utilizing a parent–offspring stochastic backward matrix for the Ayamara villages, an expected Wahlund variance (a measure of deviation from Hardy–Weinberg expectations, based upon the variance of subpopulation gene frequencies around the mean) of 0.0106 was generated for the villages. Together with the calculation of the Wahlund variance (Wahlund, 1928) based on genetic data, they concluded that 'the bulk of genetic variation encountered among the Ayamara is the result of chance, despite the fact that this population is exposed to contrasting, rigorous environmental conditions' (Rothhammer *et al.*, 1990: 200).

Wright's *F*-statistics

In 1921, Sewall Wright formulated a fixation index that can be used as a measure of deviation from panmixia. He defined it as,

$$F = 1 - H_o/H_e, \tag{3}$$

where H_o is the observed number of heterozygotes in a population and H_e is the expected number based upon Hardy–Weinberg equilibrium conditions. Wright (1965) further expanded this *F*-statistic into a hierarchical model in which the population is characterized by the following parameters in terms of the total population (t), its subdivision (s) and the individuals (i). The *F*-statistics are as follows:

(1) F_{is} (local inbreeding) is the correlation between uniting gametes relative to the gametes of the subdivision averaged over all subdivisions. It measures the effects of non-random mating within population subdivisions. F_{is} can be computed using diallelic loci as follows:

$$F_{is} = (H_t/2pq) - (p^2/pq)/1 - (H_t/2pq), \tag{4}$$

where H_t is the observed proportion of heterozygotes in the total population, and p and q are the mean allelic frequencies of the array.

(2) F_{st} is the correlation between randomly selected gametes relative to the entire population. For diallelic loci this is

$$F_{st} = \text{variance}_p / \bar{p}\bar{q}. \tag{5}$$

Table 18. R_{st} values for Circumpolar populations

Ethnic grouping	No. of subdivisions	R_{st} value
Circumpolar (all groups)	47	0.085
Circumpolar (plus Touvinians and Tophalars)	40	0.122
Circumpolar (without Touvinians and Tophalars)	35	0.075
All Eskimos	19	0.079
Inupik-speaking Eskimos	12	0.054
Siberian tribes	18	0.047

Source: Crawford and Bach Enciso (1982).

This F-statistic is usually utilized as a measure of genetic microdifferentiation of subdivisions of finite size as a result of stochastic processes. It is often used as a measure of the degree of among-group variation.

(3) F_{it} is the correlation between uniting gametes relative to the total population, given as

$$F_{it} = H_t / 2\bar{p}\bar{q}. \qquad (6)$$

This F-statistic measures the combined effects of non-random mating and finite size of populations. The relation between the three F-statistics can roughly be summarized as

$$F_{it} = F_{st} + F_{is}(1 - F_{st}). \qquad (7)$$

Of the three F-statistics, F_{st} has been widely utilized as a measure of genetic microdifferentiation among subdivisions of populations. However, Jorde (1980) cautioned against the unqualified use of this statistic across populations. He warned that variability in size of the subdivisions, differences in the technology of the groups compared, and sampling biases may obscure the actual level of genetic differentiation. Similarly, Nei (1973, 1977) was also critical of the use of F-statistics, because of such underlying assumptions as an infinite number of subdivisions, and proposed a corrected measure, G_{st}. Jorde (1980) has demonstrated that a strong positive relation exists between the number of subdivisions and the magnitude of F_{st}. Crawford and Bach Enciso (1982) confirmed this finding for circumpolar populations.

Harpending and Jenkins (1973) demonstrated that the R_{st} statistic computed by the R-matrix method is equivalent to Wright's F_{st}. The circumpolar populations described by Crawford and Bach Enciso (1982) have similar subsistence patterns and are of relatively small size. Based upon 47 populations, an R_{st} value of 0.085 was observed (see Table 18). Ward (1973) described the F_{st} value of 0.06 for the Yanomama villages as one of the highest values recorded for human tribal populations. Siberian populations, a much more heterogeneous collection of subdivisions of the tribe, have an F_{st} value for 18 tribes of 0.047. These data suggest that, despite the narrow environmental range that exists in the Arctic and Subarctic regions, the level of genetic microdifferentiation is almost equivalent to those observed among the major human races (Wright, 1978).

Gershowitz and Neel (1978) computed F_{st} values for 32 American Indian subdivisions. They observed genetic microdifferentiation among the South American tribes ($F_{st} = 0.094$) comparable to that noted for circumpolar groups. However, F_{st} analyses for the Yanomama (37 subdivisions) and Makiritare (7 subdivisions) produced lower values of 0.0633 and 0.0358, respectively. Similarly, Salzano (1975) measured an F_{st} value of 0.057 among 29 South American Indian tribal subdivisions.

The only measure of F-statistics in Central America that I could locate in the literature was based upon a migration matrix for six Guatemalan highland villages. Cavalli-Sforza and Bodmer (1971), computed an F_{st} of 0.013 for the subdivisions of these Maya villages. Comparisons of F_{st} values between populations is questionable, but migration-based versus genetic-based estimates of F_{st} are of interest when comparing predicted and actual differentiation of subdivisions of the same population.

Workman and his colleagues (Workman and Niswander, 1970; Workman et al., 1973) compared the F_{st} values for ten Papago Indian reservation subdivisions based upon genetics versus migration. They observed that the F_{st} values based on migration from 1900 to 1959 were lower than those based upon genetics. It is difficult to conclude whether these differences (0.0077 versus 0.0198) are statistically significant because the distributions for the F-statistics had not been compiled and statistical measures of significance were not computed.

Roychoudhury (1977) argued that Nei's (1973) measure of genetic microdifferentiation of subpopulations has advantages over Wright's F_{st}. Using Nei's method, he partitioned the total genetic differentiation between three Amerindian tribes into two components, within and between subpopulations. Roychoudhury concluded that the amount of genetic differentiation (G_{st}) in the subdivided tribes – Papago, Makiritare and Yanomama – varies between 2 and 7%. Thus, only a small proportion of the total variation can be attributed to subpopulation differences, whereas the remaining 93 to 98% of the diversity is within subpopulations.

Analysis of clans

Because of the high degree of genetic variability displayed by DNA markers it is possible to correctly classify individuals into their respective social clans (extended kin groups) on the basis of their DNA fingerprints. Through linear discriminant function

Table 19. *Discriminant function analysis of lineages from a village of Mindur-Sokkon, Altai*

	Clans		
	Irkit	Todosh	Kipchak
Percentage correctly classified	0.533	0.846	0.800

Source: After McComb (1996).

analysis, each individual from the Altai village of Mendur-Sokkon was entered as an unknown and on the basis of three VNTR loci (D7S104, D11S129, and D18S17) was classified into one of three patrilocal clans (McComb, 1996). If this assignment of individuals was due entirely to chance then only 33% of the individuals would be correctly classified into the appropriate clan. However, within Mendur-Sokkon individuals were assigned into the appropriate clan at 72% accuracy. This accuracy of assignment into tribal groups increased to more than 90% over classification into clans (see Table 19). McComb (1996) computer-simulated the distributions of 500 random classifications and showed that the actual discriminant function ranks in the upper 6% of all of the functions. What these data indicate is that extended families share certain VNTR distributions that can be detected by discriminant function analysis. Thus, within a population the genes are not distributed randomly, but there is genetic structure that can be detected by DNA markers.

Non-random mating

One of the most extensive studies of non-random mating in North American Indians was done by Spuhler and Kluckhohn (1953) on the Ramah Navajo. This group of Navajo was isolated geographically from the larger concentrations of Navajos of northwestern New Mexico and in 1948 totaled 614 persons. The pathway coefficients computed from genealogies revealed considerable variation in inbreeding for individuals (0.001–0.098). The mean coefficient of consanguinity (F) for the population was a modest 0.0066, indicating that despite the geographic isolation and small population size the Ramah Navajo did not practice an extensive system of non-random mating. The continuation of this study by one of Spuhler's former students, Kenneth Morgan, revealed an increase in inbreeding from 0.0066 to 0.0092 from 1950 to 1964 (Morgan, 1968). Spuhler (1989) provided an update to the original 1953 article. He summarized the results of several other studies and placed his results within the theoretical context of the latter time.

Short (1972) estimated inbreeding in another southwestern Amerindian group, the Papago. Their present-day reservation is located near Tucson, Arizona, bounded by the Gila and Santa Cruz Rivers, and Sonora, Mexico (Short, 1972). In pre-Contact times the Papagos used to live in small, seasonally nomadic groups that moved between the mountains and summer fields. Short utilized an extensive Population Register, which contained information on 22 525 individuals, subdivided by their patrilineages. Inbreeding coefficients were calculated for each on-reservation sibship with the total mean coefficient of inbreeding for the population of $F = 0.000\,885$. Inbreeding coefficients by district varied between 0.000 122 and 0.002 513. These inbreeding coefficients are much lower, by a factor of 10, when compared with two neighboring southwestern tribes, the Ramah Navajo ($F = 0.006\,60$) and the Hopi ($F = 0.007\,97$) as reported by Woolf and Dukepoo (1969). Based on isonymy, Short argues that the largest proportion of inbreeding is due to random factors (77%), and 23% of inbreeding can be attributed to assortative mating.

Few studies of inbreeding, based upon pedigree data, and other forms of nonrandom mating exist for Amerindian populations because until recently there were few reliable marriage records. However, the file cabinets of cultural anthropologists are bulging with genealogies that can be used to reconstruct the breeding patterns in Amerindian populations. Recently, Markow and Martin (1993) combined the genealogical data collected by Leslie Spier with censuses, birth rolls, and their own data to reconstruct the genealogies, over eight generations, of the Havasupai tribe of northern Arizona. They attempted to relate levels of inbreeding to developmental stability and dermatoglyphic asymmetry. They found individual inbreeding coefficients (F) reaching a high level of 1.3% in the period from 1981 to 1987. The mean population inbreeding coefficient of 0.01 calculated for the Havasupai (1%) was among the highest reported in New World populations. Markow and Martin claim to show evidence for the disruption of developmental stability in the form of increased asymmetry in the inbred Havasupai.

5.3   INTERPOPULATION SUBDIVISION

Genetic distances

Genetic similarities and differences between a population's subdivisions due to differentiation can be measured by comparison of gene frequencies. It is difficult to visualize the affinities between populations on the basis of matrices consisting of many alleles and populations. As a result, a number of genetic distances have been developed for comparing patterns of among-group variation through the use of summary statistics. Many of the distance measures are based upon squared differences between gene frequencies from a series of populations, followed by a transformation of these differences (Jorde, 1985). Some of the early distance measures do not account for correlations between the genes, whereas others provide mathematical transformations to make the variances independent of the gene frequencies. Most of these genetic distances provide similar results, as attested by high correlations between the measures. Some distance measures, such as Nei's, are particularly useful because they permit the estimation of standard errors and confidence levels (Nei *et al.*, 1985). Other measures are less useful, such as Hedrick's gene identity measure based upon genotypes rather than gene frequencies (Hedrick, 1971). Nei (1973) criticized Hedrick's measure on the grounds that 'genotypic frequencies are quadratic functions of gene frequencies in diploid organisms and (are) affected strongly by (the) mating system.' He also warned that this distance is not linearly related to evolutionary time even in the simplest case. Instead of reviewing all of the different measures of genetic distances and their various assumptions, I refer you to Jorde's review article on genetic distances based on standard blood markers (Jorde, 1985). Suffice it to say that most of the distance measures provide similar results and the choice of method is one of availability of software.

Because the microsatellite loci may evolve by a stepwise mutation process and alleles may mutate by possible slippage of a small number of repeats, Goldstein *et al.* (1995) challenged the applicability of standard genetic distance measures to microsatellite loci. Instead Goldstein and colleagues developed a distance measure linear with time and based on the stepwise mutation model corrected by the number of allelic repeats. Certainly, such a distance measure would be more appropriate for VNTRs and STRs than the measures derived for standard diallelic systems.

The literature is replete with studies of Amerindian populations that utilize various measures of genetic distance in order to characterize population affinities. Many of these articles are almost 'formula-driven', i.e. they report gene frequencies for a particular tribe or group of tribes, followed by a 'mandatory' genetic distance analysis. However, some of these studies do provide useful information about the evolutionary relationships among Native American populations.

The Tlaxcaltecan population affinities were first examined by the use of genetic distances. Distances between pairs of populations can be graphically illustrated a number of different ways. One of the simplest methods was to construct a three-dimensional model with populations represented by spheres and the distances between them by rods of the appropriate length. These archaic methods of representing population affinities utilized the square root of a D^2 value in order to circumvent a Euclidean constraint that the sum of any two sides of a triangle must exceed the length of the third (Lees and Crawford, 1976). Figure 39 shows such a

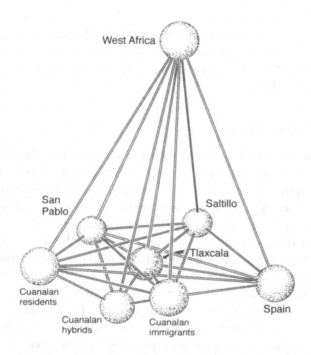

FIGURE 39 A three-dimensional model used to represent genetic distances and affinities between populations.

reconstruction of population relationships between Tlaxcaltecan, West African and Spanish populations.

Rothhammer (1990) employs Nei's measure of genetic distances on South American Indian populations grouped by languages in order to resolve some controversies surrounding their origins and affinities. These weighted linguistic groupings revealed the close affinity of the Arawak to the Ge-speakers, thus grouping these Arawaks with a prehistoric population that may have peopled eastern Brazil.

Geographic subdivisions

The role of geography in the distribution of genes and gene frequencies in human aggregates has been approached by means of two different conceptual models. One approach compares genetic distances between populations with their geographic distances. This approach yields correlations between geographic and genetic distances. The second approach (based upon Malecot's isolation-by-distance model; see below) assumes a population continuity along a planar surface in which the densities in each subdivision are constant in space and time (Malecot, 1948). This model is based on a conditional probability: that the probabilty of two alleles being the same by descent is a function of distance. Thus, the differentiation of the populations is a function of the migration patterns. In New World populations, structure has been studied by using both analytical approaches.

Geography and genetic distances

Workman and Niswander (1970) were among the first to examine the relationship between genetic and geographic distances in Amerindian populations among the Papago districts. They correlated the geographic distances (measured as the shortest distance by road or trail used by the Papago) with a genetic distance G, based upon Sanghvi's (1953) method. The overall correlation for all districts was 0.494, which means that approximately 25% of the total genetic variation can be attributed to geography. However, in some subdivisions genetic distances were much more affected by geography than were those of other districts. For example, district 10 had a correlation between geographic and genetic distances of 0.911, whereas district 9 had the lowest correlation, 0.25. Thus, a considerable proportion of the observed genetic variation can be ascribed to the geographic distances. Workman and Niswander hypothesized that three factors may be responsible for the observed correlation patterns in the Papago: (1) the geographic distances may reflect original genetic differences or similarities of the founders; (2) isolation by distance, i.e. an inverse relationship between frequency of matings between districts and geographic distances; (3) probability of migration to certain districts appears to be inversely proportional to the geographic distances. They further conclude that the Papago districts have differentiated from each other as a result of isolation by distance, founding effect and random genetic drift.

A number of studies have focused upon the relationship between geographic distances and genetics of South American Indian populations. The correlations derived from these comparisons vary from almost zero (0.007) for nine South American populations (Murillo *et al.*, 1977) to a high 0.716 for Chilean Indians (Chakraborty *et al.*, 1976). The larger comparisons of 22 South American Indian populations (Blanco and Chakraborty, 1975) revealed an intermediate relationship ($r = 0.47$) between genetics and geography. Similarly, the Yanomama exhibit intermediate correlations between geographic and genetic distances (Neel *et al.*, 1974). A rotation of matrices (genetic distance on geographic distance) to maximum congruence by the MATFIT computer program for all Black Carib populations produced a correlation between coordinates of 0.36 (Devor *et al.*, 1984). However, the coastal villages of Central American Black Caribs show a correlation between geographic and genetic distances of 0.89. This extremely high correlation is due to the fact that all of these Carib villages

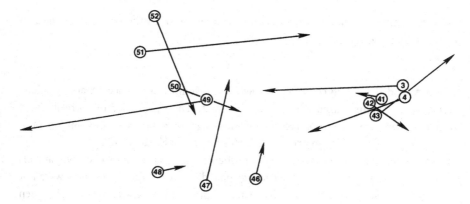

FIGURE 40 A plot of matrix-fitting (by MATFIT) of genetic distances to geographic distances for 12 Arctic Eskimo populations.

are coastal and that roads between them are infrequent. Therefore, the geography of migration between villages is one-dimensional, conforming to the shape of the coastline. The lower correlations for all Black Carib populations are due to long-distance migration (from St. Vincent to Honduras) and patterns of gene flow.

There is considerable variation as to the relationship between geography and genetics in circumpolar populations (Crawford and Bach Enciso, 1982; Crawford, 1984). The correlation between genetic and geographic distances is 0.686 for Siberian populations. Thus, 47% of the observed genetic variation can be explained by geography in Siberia. This is a fairly high correlation when the effects of European intrusion into Siberia are considered. A number of Siberian tribes, such as the Yukaghirs, were either decimated or squeezed out of their traditional territories and incorporated into other gene pools. Other populations, such as the Yakut and the Chukchi, have expanded their ranges in a most dramatic fashion. The correlation of geography and genetics is slightly lower among the Inupik Eskimos (0.562) and for all Eskimos (0.457). Figure 40 is a plot of matrix-fitting of genetic distance to geographical distance for 12 Arctic Eskimo populations using MATFIT. The Inupik-speaking Eskimos expanded in historic times from the Norton Sound region, along the northern slope into Canada and Greenland. Their distribution is coastal and linear and therefore conducive to high correlations between geographic and genetic distances. Yet the observed intermediate correlations may reflect the relative recency of their numeric expansion with insufficient time for greater genetic differentiation.

One major problem in the use of the MATFIT measure of congruence is its failure to provide any measure of significance of the correlations. However, Mantel's test (Mantel, 1967) compares matrices and tests the significance of correlations. Given two distance matrices, A and B, an association is tested between the elements of the two matrices by using the statistic

$$Z_{AB} = A_{ij}B_{ij} \tag{8}$$

where A_{ij} and B_{ij} are the elements of row i and column j of matrices A and B, which results in an unnormalized correlation coefficient. The significance of correlations is tested by comparing the observed correlations against a sampling distribution of Z based on a randomized B matrix B_R (Crawford and Duggirala, 1992).

Although high correlations have been shown by MATFIT in some comparisons between matrices of genetic and dermatoglyphic distances, these correlations were not significant (Enciso, 1983). On the other hand, all product-moment correlations between blood genetic distances and geography proved to be highly significant in the Tlaxcaltecan population studies.

The relationship between geography and genetics is not entirely due to social factors and geographical distances affecting migration. Geological and ecological events have undoubtedly contributed to the reproductive isolation of human populations during various periods of time. All too often ignored are the Pleistocene and post-Pleistocene ecological conditions, such as the existence of gigantic inland seas or lakes in Siberia and in Amazonia, that helped sculpt the genetics and distributions of human populations. This ecological approach has been successfully utilized by Rogers et al. (1991) with regard to possible peopling of the New World. Yet the effect of an inland sea during the Pleistocene on the contemporary human population distributions and migration patterns is yet to be fully examined.

Table 20 summarizes a recent analysis of the relationship between genetics, geography, and languages based on Mantel statistics in Siberian indigenous populations (Crawford et al., 1997b). The correlations between genetics and geography, and genetics and language, are both significant. These correlations are not surprising when the geographic expanses of Siberia are considered and the possible roles that geographic barriers may have played in the evolution of languages and genetics during the Pleistocene. The significant association between language and genetics disappears in the partial correlations when geography is kept constant (Dow et al., 1987). Yet the correlation between geography and genetics remains strong when language is kept constant. Using multiple correlation, the relationship between genetics and the combined effects of geography and language is high and significant (see Table 20). A total of 30.6% of variation in the genetics of Siberian populations is explained by the joint effects of language and geography. Judging from the moderate correlation between genetics and linguistics and the insignificance of the relationship between genetics and linguistics, if geography is kept constant, it appears that most of the genetic differentiation in Siberia is geographically patterned.

Table 20. *Correlations of (1) genetic (GENE), geographic (GEOG), and linguistics (LING) distance matrices; (2) partial correlations between two matrices influencing the third matrix; (3) multiple correlation obtained through multiple regression of genetic distance matrix against both geographic and linguistic distance matrices among 13 Siberian indigenous groups and Alaskan Eskimos*

Distances compared (1) or test of relationship (2, 3)	Correlation (r)	p^a
(1) Correlations		
GENE•GEOG	0.546	0.001
GENE•LING	0.351	0.001
GEOG•LING	0.500	0.001
(2) Partial correlations		
GENE•GEOG (LING)	0.456	0.001
GENE•LING (GEOG)	0.108	0.152
(3) Multiple correlations		
GENE•GEOG, LING	0.553	0.001

[a] Mantel test probabilities.
Source: After Crawford *et al.* (1997b).

Spatial autocorrelation

Robert Sokal and his colleagues first applied the techniques of spatial autocorrelation to human population structure. This technique is based upon the geographic principle that 'Everything is related to everything else, but near things are more related than distant things' (Tobler, 1970). This technique analyzes the spatial relationship of allelic frequencies in populations through correlation by distance intervals (Sokal and Oden, 1978; Sokal, 1988; Sokal and Friedlaender, 1982). Spatial autocorrelation can be viewed as the correlation of the frequency of one allele with the values of the frequencies of the same allele at all points of a two-dimensional surface. Spatial correlograms are constructed for various distance intervals. From this technique, patterns due to selection or gene flow may be extracted.

Crawford *et al.* (1997) computed Moran's standardized *I* statistic for 61 contemporary Siberian indigenous populations, characterized by 10 genetic loci and 19 alleles. Moran's *I* is a product-moment coefficient which has been standardized to Z-scores. This autocorrelation statistic was calculated for different distance classes and graphed as correlograms. These graphs enable the general trend of declining similarity among localities to be investigated across increasing distances. Seven distance classes were

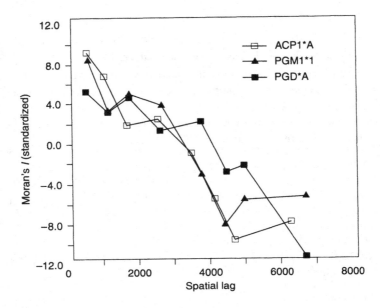

FIGURE 41 Moran's *I* correlogram for Siberian indigenous populations using three loci and eight spatial lags. This correlogram displays a monotonic decline in the level of spatial autocorrelation.

defined with upper limits of 1000, 2000, 3000, 4000, 5000, 6000, and 7000 kilometers. Distances between the Siberian populations range from 12 to 6742 km, with a median distance of 2815 km.

All of Moran's correlograms for the 19 alleles display large values of spatial correlation at the initial spatial lag, indicating that populations that are spatially close tend to have similar allelic frequencies. Figure 41 is a plot of three alleles, ACP1*A, PGM1*1 and PGD*A, displaying an essentially monotonic decline in the level of spatial autocorrelation, from strongly positive values at small spatial lags to strongly negative values at large spatial lags. This pattern for the loci GM, ACP, PGM, and PGD, is consistent with an isolation-by-distance model operating on a continental scale of up to 7000 km.

In contrast to Fig. 41, the following figure (Fig. 42) displays some reduction of similarity as a function of distance, but also includes considerable deviation from the Malecot model and shows statistical 'noise'. In particular, GM*A T (GM. zabst) declines in similarity for the first five spatial lags, but then the correlation increases dramatically in the sixth and seventh lags. Thus, there appears to be some additional genetic structure in Siberian populations other than isolation by distance in spatial lags of

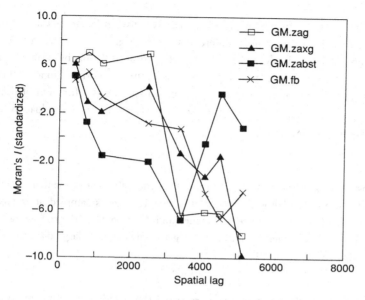

FIGURE 42 A plot of Moran's *I* correlations versus eight spatial lags for four GM allotypes in Siberian indigenous populations.

3500–4500 km. Most likely this structure can be explained by the historic population expansions and movements associated with European contact.

Sokal *et al.* (1986) applied their spatial autocorrelation techniques to 50 Yanomama villages from southern Venezuela and northern Brazil. They utilized 15 allelic frequencies, representing blood-group and protein polymorphisms. As observed in the Arctic populations, there was a marked decline in genetic similarity with geographic distance. However, few moderate clinal patterns were discerned. This method of analysis provided the following insights into Yanomama population structure: (1) the villages displayed considerable genetic heterogeneity; (2) significant spatial patterns were revealed for most of the allelic frequencies; (3) the isolation-by-distance model of Malecot is fully supported in Yanomama populations; (4) most of the population structure is hierarchical in nature. Their findings support earlier analyses about the role of both stochastic processes and social factors in determining village allelic frequencies.

Malecot's isolation-by-distance model

The concept of 'isolation by distance' was first introduced by Wright (1943), who observed a tendency of populations to exchange genes with their nearest neighbors.

This eventually results in greater genetic similarity between geographically proximal populations and increasing genetic differences between groups that are further and further apart. Malecot (1948, 1950, 1959) developed an isolation by distance model in which it is assumed that, in a population distributed uniformly along an infinite line, the probability of migration from one point to another is a function of distance between the two points. Thus, a mean kinship coefficient, θ_d, for individuals who are at distance d can be expressed as

$$\theta_d = a\, e^{-bd} \tag{9}$$

where a is a measure of local kinship and b is the rate of exponential decline of kinship. This coefficient (θ) is the probability that two genes sampled at random at distance d are identical by descent. Lalouel and Morton (1973) provided a scalar correction factor, L, in order to eliminate a negative kinship over large distances. Their correction changes equation (9) to:

$$\theta_d = (1 - L)\, a\, e^{-bd} + L. \tag{10}$$

A number of population studies have reported the a and b values for the Malecot isolation-by-distance model. Both parameters are usually estimated by non-linear regression techniques (Jorde, 1980). This approach of Malecot has been broadly criticized and defended. Felsenstein (1975) states that Malecot's model is internally inconsistent and dismisses this model as 'biologically irrelevant.' Lalouel (1977) has defended this model and has shown some misinterpretations and errors in Felsenstein's attack. In my opinion, this method has a limited use in that it permits the examination of a large number of subdivisions with a standard method. This method of kinship bioassay has often been applied inappropriately in instances of non-continuous distribution of populations.

Morton and his colleagues have estimated the a and b parameters for several South American Indian populations. Roisenberg and Morton (1970) utilized a bioassay of kinship by phenotypic pairs in 12 South and Central Amerindian countries. However, these initial values of a and b (0.025 and 0.003) were shown to be underestimates of kinship. Roisenberg and Morton later corrected the a parameter and raised its value to 0.038. Usually the measure of local kinship, a, in non-Western horticultural populations tends to be greater than 0.03 as it does among Amerindians. Similarly, Lalouel and Morton (1973) report a value of a of 0.053 for eleven Makiritare villages on the basis of 11 polymorphic systems. For the other Central and South American groups, the rate of exponential decline of kinship, b, was considerably lower than its value of 0.04 km^{-1} among the Makiritare. These data indicate that there is a clear-cut exponential decline in kinship with distance among the South American Indians.

Genetics and languages

Languages have been viewed as possible barriers to reproduction and an influence on the patterns of genetic variation. There has been considerable disagreement over the exact nature of the relationship between linguistic and genetic variation, with the results of studies often dependent on the method of comparison employed. Sokal (1988) has demonstrated a clear-cut relationship between genetic distances and linguistic distances in European populations. However, attempts to correlate genetic distances with some linguistic measures of similarity in the New World have often failed. Murillo *et al.* (1977) found no correlation when analyzing distances among nine South American Indian tribes. In contrast, Salzano *et al.* (1977) found a negative correlation of 0.27 between the linguistics and genetics of the Ge-speaking Indians of Brazil. Spuhler (1972), in his analysis of the relationship between North American Indian populations and linguistic affinities, found an *r* value of −0.33. In his later *magnum opus* on the study of genetic and cultural interrelationships among North American Indian groups, Spuhler (1979) demonstrated slight, but significant, associations of biology, language and culture. Using stepwise discriminant functions, genetic distances classify tribes into their present culture areas in only 58.5% of trials, and languages into their families in 64.7%. Spielman *et al.* (1974) demonstrated that the linguistic diversity among the Yanomama corresponds closely to the patterns of genetic microdifferentiation. Spielman and his colleagues tentatively concluded that Yanomama dialects have been diverging for about 1000 years. A study by Zavala *et al.* (1982) of the relationship between genetic distances, kinship coefficients, geography and languages among 23 Mexican Indian tribes indicated a poor fit between genetics and languages. They attributed this discrepancy to small sample sizes and genetic admixture.

Studies using some form of genetic distance and historical connections between languages often show an association between genetics and linguistics. For example, in our study of circumpolar genetic variation and population structure, the genetic characteristics of populations from Siberia, Alaska and Greenland clearly clustered by linguistic affiliation (Crawford *et al.*, 1981b). In particular, the Inupik-speaking Eskimos, despite their extremely broad geographic distribution (from Norton Sound, Alaska, to Greenland), formed a tight cluster. Similarly, the Samoyed-speaking populations of Siberia, the Paleo-Asiatic-speaking groups and the Turkic speakers all formed their own genetic clusters (Crawford and Bach Enciso, 1982). Rothhammer (1990) showed that the linguistic classification of South American Indians corresponds to genetic distances based upon markers expressed in blood.

Barbujani *et al.* (1989) utilized Womble's method for detecting biological boundaries by averaging absolute values of the derivatives of functions describing biological variation in space at various locations. They applied this method to 50 Yanomama populations characterized by 15 alleles. The boundaries separate the Yanomama into dialectic/linguistic clusters. Barbujani et al. stated that, except for geographic continuity among villages, there is no reason, according to the boundary map, to unite the Yanomama localities. The group of dialects has expanded over the past hundred years and its members have diverged. The regions of rapid genetic change are in concordance with observed linguistic differences. Womble's (1951) method appears to be a useful technique for examining abrupt changes of biological variables and their linguistic consequences.

Hulse (1957) compared the blood-group frequencies on reservations of Yakima, Okanagon and Swinomish Indians and concluded that languages are stronger barriers to gene flow than is geography. He based these conclusions on the similarities in the frequencies of MN and RH alleles on these reservations. Both the Swinomish and the Okanagon speak a Salishan dialect, whereas the Yakima belong to the Sahaptin linguistic stock. Although the Okanagon and Yakima are closer geographically, they apparently are most different genetically. Unfortunately, Hulse failed to statistically test any of these purported differences and instead relied on opinion and conjecture.

5.4   DISPLAY TECHNIQUES

In studies that simultaneously consider the genetic relationships among many populations it is difficult to visualize and interpret multiple population affinities. As a result, a number of different methods for graphic display, based upon genetic matrices, have been developed. These displays can be subdivided roughly into dendrograms and topologies.

Dendrograms

The construction of dendrograms and phylogenetic trees provide not only a graphic display of the genetic data but also information about the fission of populations and the time of divergence. A number of methods utilizing different assumptions have been developed for the creation of evolutionary trees. Some methods construct rooted trees whereas others provide unrooted ones. Although the statistical error in tree building can be high, Cavalli-Sforza et al. (1988) have introduced a bootstrap method for comparing trees obtained by resampling. Cavalli-Sforza and Edwards (1967) compared a number of geographic regions of the world, each containing a wide assortment of populations, and observed that Amerindian populations clustered with Asian groups. The most ambitious study of North American Indian phylogenetic relationships, one that included construction of evolutionary trees, was carried out by Spuhler (1979).

He measured population affinities, as based upon a few blood-group systems, for 53 North American Indian tribes. The trees reflected the evolutionary histories of these tribes, and hence included the effects of migration, gene flow, genetic drift and possible selection. With some exceptions, groups that shared a common history tended to cluster together. Considering the devastating depopulation of North America attributed to epidemic disease, and the effects of European admixture, it is amazing that elements of the early genetic structure persist. In Spuhler's tree for the Arctic culture area, there are a number of clusters that make little evolutionary or historic sense. For example, the Commander Island Aleuts cluster with West Greenland Eskimos. Based on history and geography, the West Greenland population should cluster with the East Greenland and Thule Eskimos whereas the Commander Island Aleuts should cluster with the Western Aleuts. In contrast, our analysis of the population structure of the circumpolar populations revealed that all of the Inupik-speaking Eskimos formed a tight cluster (Crawford et al., 1981b). Most likely, the anomalous affinities shown by Spuhler's tree are consequences of European admixture with all of these populations and the heterogeneous origins of the Commander Island inhabitants. The population of this island nearly reached extinction but was repopulated by Russians, Eskimos and Aleuts from other islands. The dendrogram by Ferrell et al. (1981) also indicated the clustering of the Central, Eastern and West Greenland Eskimos, and the distinctiveness of the Aleuts. Yet in their dendrogram, the Blackfoot Indians cluster with East Greenland Eskimos.

Figure 43 is a dendrogram based upon application of a hierarchical cluster technique utilizing unweighted, squared Euclidean distances for 62 Central American populations (several groups from the Southwest are included). The available data on frequencies of alleles common to these populations covered five loci and included 12 blood-group alleles. A minimum sample size of 50 individuals per population was applied in selection of populations. This plot shows a deep divergence of the populations with high African admixture from those predominantly Amerindian. The second branch separates the Amerindian groups with considerable Spanish admixture from those with less. Regional proximity and historical relationships are evident on a finer level. For example, San Pablo, the Papago, and Hueyapan are together in a cluster. This clustering reflects the northern origins of the Nahua speakers and the geographic proximity of the Hueyapan and San Pablo populations in the State of Puebla and Tlaxcala.

Figure 44 shows the relationships among Amerindian groups when the Black Carib populations are removed from the analysis. In this case, regional and historic relationship take precedent in the observed affinities. These dendrograms reduce the complexity of gene-frequency data, by using 72 intermediate nodes or points of bifurcation, to a total of 145 data points. A genetic map, based on an R-matrix, reduces the intricacy of the data to only 72 points. In both dendrogram and map,

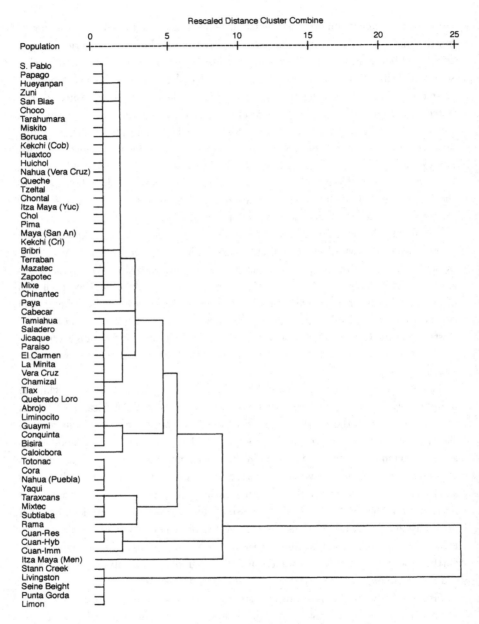

FIGURE 43 A dendrogram based on a hierarchical cluster technique with unweighted, squared Euclidean distances for 62 Central American populations.

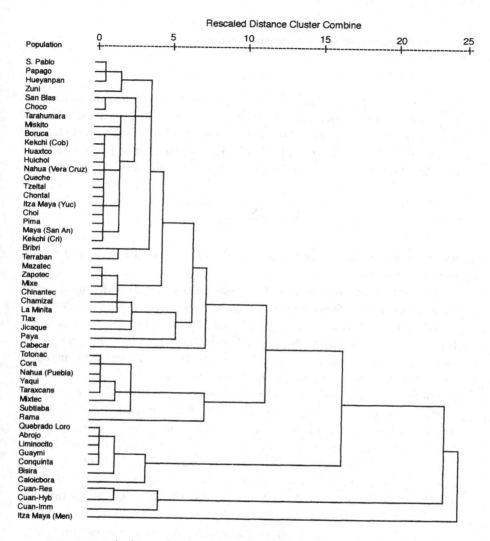

FIGURE 44 A dendrogram based on a hierarchical cluster technique with
unweighted, squared Euclidean distances for the same populations as in Fig. 43,
but minus the Black Carib groups.

the net relations of the populations to each other are shown at a glance. Generally,
the map summary is less distorted than the dendrogram. The dendrogram corresponds closely to both the genetic map and the synthetic maps as to the unique
position of the Black Caribs.

Topological approaches to genetic variation

R-matrix

Various forms of principal components analyses have been utilized to reduce gene-frequency matrices to eigenvectors (positions of objects on component axes) for purposes of graphic representation. The variance–covariance matrix approach of Harpending and Jenkins (1973), known as R-matrices, have been applied frequently to Siberian and New World populations. The coefficient of kinship between two groups i and j is:

$$r_{ij} = 1/K \sum_{i=l}^{k} \frac{(P_i - \bar{P})(P_j - \bar{P})}{\bar{P}(1 - \bar{P})} \tag{11}$$

where $\bar{P}$ is the weighted mean frequency of the allele in the study array, and K is the number of subdivisions.

The R-matrix method was applied to 14 genetic loci and 21 alleles in four populations of the Norton Sound region. The genetic loci tested are ABO, RH, MNS, FY, PI, GC, HP, ACP, ESD, PGM, BF, PGD, GM and KM. This geographic area is a linguistic boundary between Siberian Yupik and Inupik. The two communities on St. Lawrence Island, Savoonga and Gambell, are both Yupik-speaking; on the mainland, Wales and the transplanted King Islanders are Inupik speakers. Figure 45 provides a plot of the first and second scaled eigenvectors, which account for 87% of the observed variation. The first eigenvector, containing 56% of the variation, separates the two language groups. The second eigenvector separates the two Inupik-speaking communities from each other. Figure 46 is a plot of the alleles responsible for the dispersion along the two eigenvectors, with GC, acid phosphatase, Ms, Ns and HP contributing to the separation of the language groups. The second axis differentiates Wales from King Island on the basis of CDe, cDE, Ms and NS. The GMs and KMs contribute to this dispersal.

Figure 47 is a gene map of 40 Siberian populations that belong to four different ethnic groups: Eskimo, Chukchi, Samoyed and Turkic. The Chukchi and Eskimos are Paleo-Asiatic speakers; the Touvinians and Tophalars are Turkic speakers. The Nentsi, Nganasan, and Yenisey Samoyeds are all Samoyed speakers. The liguistic affiliation of each population is marked by a symbol in Fig. 47. The genetic uniqueness of the Turkic groups from the Altai causes all of the other Siberian populations to cluster. This genetic distinctiveness is due to the high incidence of the NS blood group haplotype. When compared with the incidence of NS in other Siberian populations, this high frequency in the Altai region seems anomalous. In addition, the two Tuvan

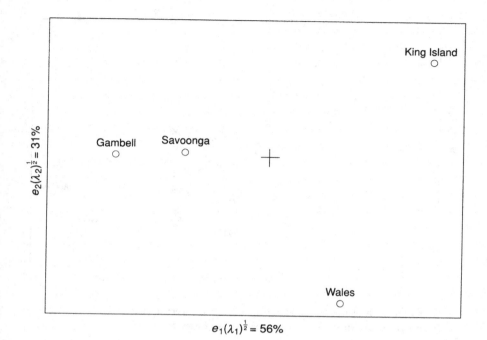

FIGURE 45 An R-matrix reduced spatial plot of four Norton Sound Eskimo populations using 14 genetic loci and 21 alleles (Crawford *et al.*, 1981b).

communities exhibit a CDE frequency of 59%, compared with the frequencies of 1–2% in Eskimos and zero among the Forest Nentsi.

When the Altaic (Turkic) populations are excluded from analysis, the gene maps become clearer and the relationships among Siberian populations begin to emerge (see Fig. 48). The first two scaled eigenvectors account for 62% of the total variation. The various subdivided populations form distinct clusters, in relationships suggestive of Malecot's isolation-by-distance model.

Attempts at documenting the action of evolutionary forces on the genetic structure of human populations have usually focused upon a single process at a time, the process in question being a function of the model used. The gene-frequency distributions were attributed to either selection, migration or stochastic processes. However, the regression of mean per-locus heterozygosity (H) on r_{ii} (distance from the centroid of the distribution) permits the assessment of the relative contribution of systematic versus non-systematic evolutionary pressures on the observed genetic variation in human populations. For example, from the regression of $\dot{H}$ on r_{ii} it

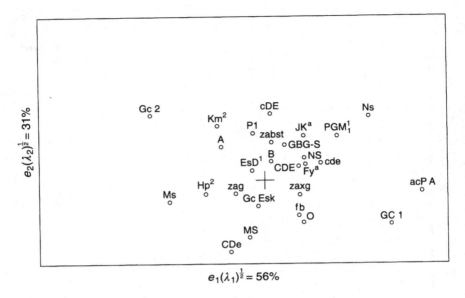

FIGURE 46 A plot of the alleles responsible for the dispersion of the populations (shown in Fig. 45) along the first two eigenvectors.

appears that Anguyema (population 16 in Fig. 49) had experienced the least amount of genetic admixture and had differentiated genetically from the other groups, probably through genetic drift or the founder effect. This population, located at the base of the Amguyema River, is more isolated geographically than the other Chukchi communities.

A similar type of analysis, utilizing the R-matrix and linguistic and geographic subdivisions, was performed for an assortment of Mexican populations (see Fig. 50). The Tlaxcaltecan populations were compared to various Mexican Indian and Mestizo groups described in the literature. In addition to the Tlaxcaltecan populations, Hueyepan was sampled by the research team from the University of Kansas in 1978. Hueyepan is a municipio located in the northeastern region of the State of Puebla, northeast of Tezuitlan. The first eigenvector separates Cuanalan and its subdivisions, Tlaxcala and Saltillo, from the other Mexican Indian groups. This axis apparently separates the admixed or Mestizo groups from the less hybridized populations. Groups dispersed along the second eigenvector represent a geographical gradient and reflect historical relatedness. For example, San Pablo, the Nahua of Puebla and Hueyapan are all Nahua groups that arose in the North and cluster in this gene map.

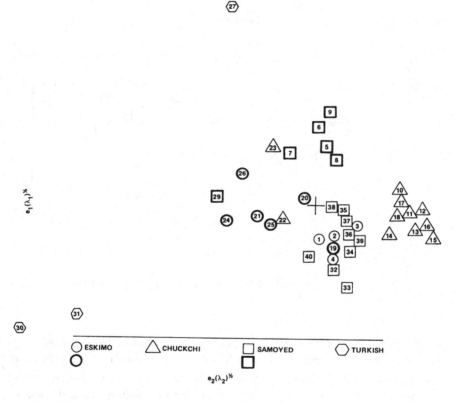

FIGURE 47 An R-matrix plot of the relationship between 40 Siberian populations based on 11 blood-group alleles.

Synthetic gene maps

Cavalli-Sforza and his co-workers have produced synthetic gene maps and initially applied them to investigate the spread of farming technology in Neolithic Europe (Menozzi *et al.*, 1978; Ammerman and Cavalli-Sforza, 1984). This approach is based upon the reduction of gene frequencies to synthetic variables by principal components analysis (PCA). Map surfaces are constructed upon a regular system of lattice intersections derived both from observed data and from application of an algorithm. This

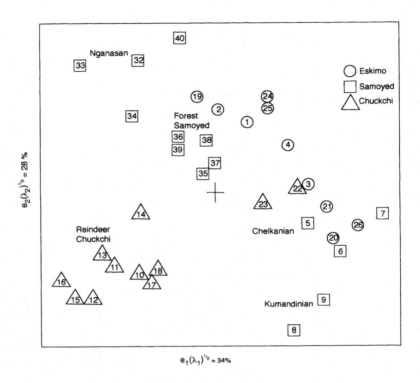

FIGURE 48 A plot of the relationship between the Siberian populations shown in Fig. 47, excluding the Altai region.

technique has been applied to the analysis of spatial variation of blood-groups, serum and red-cell protein frequencies in North and South American Native populations (Suarez *et al.*, 1985; O'Rourke and Suarez, 1986). Suarez *et al.* (1985) characterized the population structure of North and Central America, utilizing 11 alleles and 63 least admixed Indian populations. Judging from the location map that they published, these populations were not distributed evenly over the continent: large lacunae existed along the west coastal and southeastern regions of the United States. A north–south gradient in allelic frequencies was observed, the ABO and Diego blood-group loci dominating the first component (which accounts for nearly 30% of the original variation). The dominant feature of the map is a well-defined projection from west-central Canada through the Plains, which terminates in the Southwest United States (see Fig. 51). Suarez and O'Rourke interpret this pattern to represent the migration of the Southern Athapaskans (Apache and Navajo) to New Mexico approximately 600–800 years ago. The relatively low frequency of the ABO•O allele and high frequency of the A allele in Eskimo populations is reflected in the synthetic gene maps by their

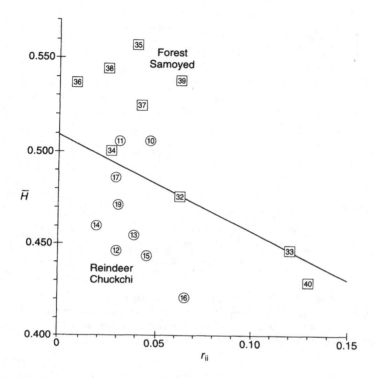

FIGURE 49 A regression of mean per-locus heterozygosity on R$_{ii}$ for 18 Siberian populations.

dichotomy in the first component. The Kutchin Amerindians of the north coast of western Canada are set apart from the surrounding populations on the basis of high PC loadings for the RH•R2 haplotype and Duffy alleles (see Fig. 54). A comparison of the patterns generated by the full data set and the reduced set indicates considerable similarity and the authors tend to support Spuhler's (1979) observation that Indian–Indian hybridization is a more significant determinant of population affinities than is admixture with Europeans.

O'Rourke and Suarez's (1986) synthetic gene-frequency maps of South American Indian populations show a pattern that is different from the North American one. In contrast to the north–south clines of North America, the South American populations exhibit local pockets of differentiation, which correspond to similar patterns generated by bioassays and other analytical methods. This pattern has probably been brought about by the ecological and cultural isolation of tribal populations of South America. An apparent dichotomy exists between the Andean Highland and the Amazonian Lowland populations. However, with the application of the transangulation method of

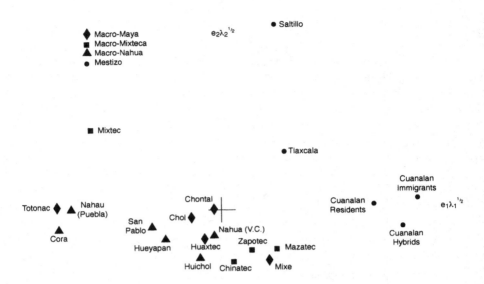

FIGURE 50 A plot of the first versus the second scaled eigenvectors of a principal components analysis of gene frequencies from Mexican indigenous and Mestizo populations. The linguistic groups are indicated by symbols in the left-hand corner of the figure.

fitting surfaces by O'Rourke (unpublished data), the local, patchy pattern disappears in the South American populations. Figure 52 is a synthetic gene map, based on the same data set, which demonstrates that the methodology employed in fitting surfaces plays a major part in observed pattern.

In order to better understand the effects of admixture on synthetic gene maps, O'Rourke and I (unpublished data) examined the patterns for Central America, using a larger sample of populations (see Fig. 53). We included in our sample a number of populations with estimated proportions of European and African admixture. Figure 53 shows a plot of the second principal component (PC) of 63 Central American populations. The high loadings indicated by the first PC are a result of admixture along the Atlantic coast of Central America. The eastern coast of Mexico contains populations that are triracial hybrids with 20–40% African and 28–35% European admixture (Lisker and Babinsky, 1986). The high PC loadings are indicated as dark areas and the low loadings are represented by light sections of the map. The most notable observation is that Belize, Guatemala and the zone surrounding the coast of Mexico City have high loadings; these reflect admixture. The second PC has highest loadings apparently associated with the degree of 'Indianness.' The white zone in the center of Central American may be an artifact of the insufficiency of data points and failure to converge. In order to determine the effects of African admixture on these

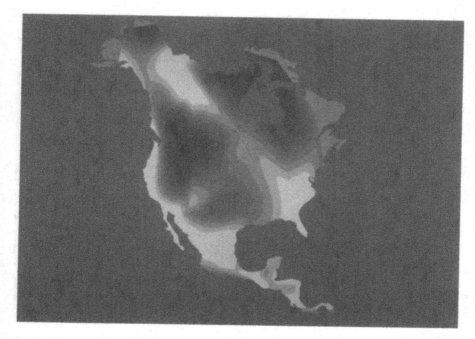

FIGURE 51 A synthetic gene map obtained by the first PC for unadmixed populations of North America (provided by D.H. O'Rourke, 1995).

synthetic maps, the Carib populations were excluded from the analysis and the high PC loadings disappeared from the plots. The results of this removal of the African groups provided a pattern that was almost identical to the plot of the first PC of the total Central American sample. Apparently, removal of the predominantly African populations from the Caribbean coast leaves a pattern of gene frequencies reflecting the undistorted Amerindian relationship. O'Rourke also removed the admixed populations from the original data set and examined the underlying structure for North American Indian populations.

5.5   ADMIXTURE AND GENE FLOW

Genetic admixture (gene flow) is a systematic evolutionary force that increases genetic variability within populations while decreasing genetic differences between groups. New mutations arising in one population may be exposed through gene flow to new genetic backgrounds in different environments. Thus, admixture and gene flow between human populations increase the adaptive potential of the species by providing new combinations of genes which are in turn tested by natural selection.

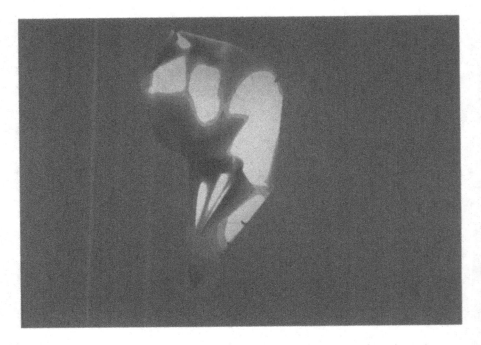

FIGURE 52 Synthetic gene map of South American Native populations based on the first principal component. An angular transformation method was used for fitting the surface (analysis by D.H. O'Rourke, 1995).

Methods of estimating genetic admixture, based upon allelic frequencies, can be traced to the original formulations of Bernstein (1931). Let a gene A, in the parental populations P1 and P2, have frequencies of q_1 and q_2. The frequency of this allele in the hybrid population H is q_H, which is the average of the frequencies in migrant plus recipient populations, or

$$q_H = mq_1 + (1 - m)q_2. \tag{12}$$

Bernstein (1931) demonstrated that if the frequencies of A are known in both the parental and hybrid populations, then an estimate of m, the proportion of migrant genes is

$$m = (q_H - q_2)/(q_1 - q_2). \tag{13}$$

This method of estimating admixture was first applied by Ottensooser (1944) and da Silva (1949) to tropical Brazilian populations.

Bernstein's original model required no assumptions as to the nature and magnitude of the gene flow. Whether the hybrid was a result of a single massive admixture or

FIGURE 53 Synthetic gene map of Central America based on allelic frequencies
from 63 populations. This plot represents the second principal component (PC).

the consequence of many generations of gene flow of varying proportions was not
considered. For example, in the formation of the Afro-American gene pool, gene flow
from the population of European origin into that of African origin occurred at various
magnitudes over many generations. Yet admixture estimates have been restricted to
the computation of the proportion of the total gene pool of the hybrid H derived
from P_1. This simple model of gene flow is diagrammatically represented in Fig. 55.
This model assumes that the ancestral populations' gene frequencies, P_1 and P_2, are
known and that natural selection, mutation and genetic drift are not operating to
significant effect on the hybrid or parental populations. It is also assumed that the
migrants comprise a random sample of their population. Tlaxcaltecan Indian studies
revealed that in fact the migrants differ genetically from the sedentes and that, at
least in Mexico, the migrants carry a greater African component (Crawford *et al.*,
1976b).

Szathmary and Reed (1978) presented a useful method for estimating genetic
admixture when the magnitude of gene flow into the population has been small.
This computation method was utilized for a hybrid European/North American Indian
population and based upon a gene counting method. It provides the upper limit of

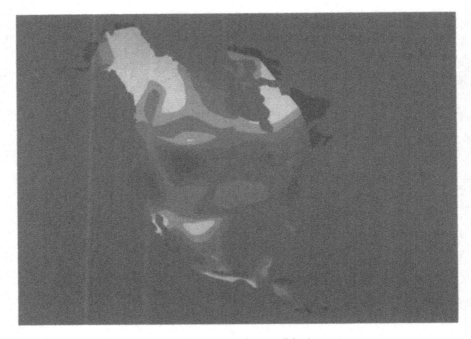

FIGURE 54 Synthetic gene frequency map of FY*A allelic frequency in 74 indigenous populations from North America (O'Rourke and Lichy, 1989).

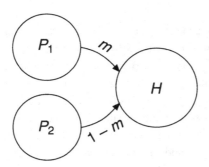

FIGURE 55 A diagrammatic representation of dihybrid model of gene flow with P_1 and P_2 representing the gene frequencies of the parental populations and H the frequencies of the hybrid group.

admixture in situations where the European genes were unequally distributed within the population.

Some researchers have combined information from a number of loci for the computation of a composite m (Elston, 1971). For example, Krieger *et al.* (1965) computed *m* for a large triracial hybrid population in northeastern Brazil, producing a maximum likelihood solution. Roberts and Hiorns (1962, 1965) proposed a least-squares measure of admixture, which assumes n parental populations with known gene frequencies and no selection or drift. A multiple regression method for the computation of *m*, proposed by Crawford *et al.* (1976b), gave an estimation of gene flow similar to those estimated by Chakraborty (1975) and Cavalli-Sforza and Bodmer (1971). These methods were tested on a Tlaxcaltecan Mestizo data set from Mexico. Minor differences in admixture estimates were observed, with the maximum likelihood method apparently overestimating the Amerindian contribution to the Mestizo gene pool. Korey (1978) tested the relative efficiencies of these computation procedures utilizing genetic data generated by Monte Carlo simulation. He found that the maximum-likelihood procedure provides the most accurate results. Chakraborty (1986) reviewed and evaluated the different methods and models of estimating admixture in human populations.

A number of studies have attempted to measure admixture between Amerindians and Europeans or, in some cases, Africans. With some exceptions, the majority of these studies utilized Bernstein's method and one or two selected loci. For example, Matson (1970) based his estimates of European gene flow on the frequency of the ABO•A allele. These estimates of *m* based upon the ABO blood group systems appear not to be reliable. However, the use of alleles that occur at a high frequency in one ancestral population (P_1) but are absent in another (P_2) may provide more reasonable approximations of gene flow. An example of such genetic markers are the GM haplotypes, such as GM•F B, that occur in European populations, or GM•ABS, which is found in African populations.

5.6   HYBRIDIZATION IN THE NEW WORLD

Continental appraisal

One means of providing continent-sized appraisals of the amount of outside admixture into Amerindian groups comes from the almost 500 tribal blood group studies compiled by Post *et al.* 1968) (see Table 21). They used blood-group frequencies to classify the tribal samples into three groups: unmixed; less than 5% outside admixture; more than 5% outside admixture. Table 21 shows the breakdown of the numbers and percentages of tribal samples into these three categories. It is likely that outside admixture is somewhat greater than these percentages indicate, since some field observers eliminated from their samples persons who look like products of hybridization.

Clearly, European and African admixture with North American Indians has pro-

Table 21. *Estimates of admixture comparing samples from North and South American Indian populations*

| | Unmixed | Admixed | | Total |
		Less than 5%	More than 5%	
North America	9	23	62	94 samples
	(9.5%)	(24.4%)	(65%)	
Latin America	121	170	107	398 samples
	(31.4%)	(42.7%)	(27%)	

Source: After Post *et al.* (1968).

ceeded at a more rapid pace than it has with Latin American Indians. Only 9.5% of the North American samples are judged as unmixed, as against 31.4% of Latin American samples. At the other extreme, 65% of North American samples indicated over 5% outside admixture, compared with only 27% of Latin American samples. In the context of outside admixture, then, Latin American Indians are now about where North American Indians were in the 1880s. The accuracy of these computations is a function of the reliability of the estimated frequencies of alleles used to estimate m. It must be understood that the selection of Indian populations to be studied serologically, and then the selection of a sample within the population, may yield a non-representative sample. Thus, this summary of admixture is only an approximation of the real situation.

In North America, the southwestern area houses the most intact and generally least mixed Indian populations. Of the nine populations considered unmixed by Post *et al.* (1968), six are from the Southwest. These groups are: Apache, Navaho, Maricopa, Mohave, Pima and Papago. The allelic frequencies support this conclusion.

M.T. Newman (unpublished notes) summarized in greater detail the proportion of admixture in the Central and South American populations (see Table 22). The analysis of admixture by Newman is somewhat dated, there being fewer unmixed populations now in existence. This summary was made during the early 1970s and represents the best estimate for that time. He characterized the Yupa as wholly unmixed. This group of tribes was forced into an inhospitable, desolate region on the Venezuelan–Colombian border that was of no interest to the settlers. The Lowland Amazonian tribes that were relatively free of admixture during the 1960s have been subjected to an onslaught from settlers and prospectors who are stripping the region of forest and hybridizing with the natives. In southern South America, Argentina, Uruguay and Chile, where European settlement is most dense, only remnants exist of the former Indian populations.

Table 22. *Admixture in Amerindian populations of Central and South America*

Area or group	Unmixed	<5% Admixture	>5% Admixture	Totals
Yupa Refuge	100%	—	—	13
Lowland S. Amer.	67	5	28	93
Central Amer. (less Maya)	56	5	38	16
W. Coast S. Amer.	38	16	47	32
Gran Chaco	35	15	50	20
Maya (Mexico & Guatemala)	33	44	22	27
Andean Highlands	32	21	47	19
Southern S. Amer.	19	44	37	16
Mexico (less Maya)	11	31	57	35

Source: M.T. Newman (unpublished notes).

There is tremendous variability in the degree of admixture with Europeans or Africans among native populations of the Americas. Much of this variation is a result of historical and demographic events such as the extreme depopulation following the introduction of Old World diseases; variation also results from the different sizes and cultures of the Native populations. The gene pools of Black Carib populations of St. Vincent Island and Dominica are an almost equal amalgam of African and Carib/Arawak Indians. There are no unmixed populations of Amerindians remaining in the Caribbean. The coastal groups of Central America have experienced considerable gene flow from the African slaves brought to work on the fruit plantations. By contrast, the Amazonian tribes, because of their residence in tropical, inhospitable environments, managed to avoid much genetic admixture until recently. In order to illustrate the relationship between historical, biological and cultural events in the creation of the hybrid and Mestizo populations, I shall review the processes involved in the formation of a number of populations. I have worked with three of these groups, representing the Arctic (Eskimos), Mexico (Tlaxcaltecans) and the Caribbean (Black Caribs). In addition, I shall provide examples from North and South America.

Eskimos of St. Lawrence Island

Prior to European contact, the population size of St. Lawrence Island (located in the Bering Strait between Siberia and Alaska) has been estimated as between 1500 and 4000 persons (Foote, 1965; Burgess, 1974). The population was subdivided into five main groups, each consisting of up to 35 coastal villages. Most likely, the settlements consisted of extended families that were exogamous. Thus, the breeding units were a series of settlements exchanging mates.

Toward the end of the nineteenth century, the St. Lawrence Island population underwent a significant bottleneck effect. This population was reduced from approximately 4000 persons to a minimum of 222 in 1917 (see Fig. 15, chapter 2). Unusual climatic conditions, disease epidemics, and severe famines were responsible for the dramatic decimation of the St. Lawrence Island population. One result of this diminution was the increased recruitment of mates from the Siberian mainland. The genetic effects of this selective recruitment of wives from Chaplino is apparent in genetic distance comparisons between St. Lawrence Island Eskimos and those from Chaplino, Siryeniki and Noonyamo. Chaplino shows much closer genetic affinities to Savoonga and Gambell than to Siryeniki and Noonyamo (Crawford and Bach Enciso, 1982). In addition, some children from the Alaskan mainland were adopted by the islanders.

In 1917, the surviving population of Gambell underwent fission, when a reindeer-herding settlement was founded in Savoonga. This new population consisted of young herders and their nuclear families. As Byard (1981) and Byard and Crawford (1991) demonstrated, because the founders of Savoonga were young and actively breeding families, the effective size (N_e) of the offshoot was larger than that of the founding population. As a result, there was greater opportunity for the action of stochastic processes in Gambell than in Savoonga. This was documented genetically in a comparison of the two communities (Byard and Crawford, 1991). From this intrapopulational examination of Savoonga and Gambell, it is apparent that unique historical events play an important role in the genetic makeup of the population of St. Lawrence Island.

Because of this unique history, the two settlements of St. Lawrence Island appear to be intermediate in populational affinities between Siberia and Alaska. However, this intermediate position was not attained by systematic pressure of low magnitude between St. Lawrence Island, Siberia and Alaska. The gene pool of the Island was sculpted by a genetic bottleneck, adoption of children from Alaska and recruitment of wives from Siberia. How did all of this effect the flow of European genes to the Island?

Byard et al. (1984b) showed, using GM*F B frequencies, that Gambell had a greater amount of European admixture (0.081 ± 0.026) than did Savoonga (0.043 ± 0.014). The GM*F B-based estimates of admixture appear to be more reliable than those based upon GM*A G because of the lower standard errors. Ferrell et al. (1981) derived a slightly lower level of admixture for St. Lawrence Island using the maximum-likelihood approach of Szathmary and Reed (1978). They claimed 0.073 European admixture for Gambell and 0.024 for Savoonga. Byard et al. (1984b) estimated an m value for Wales, Alaska, of 0.074 ± 0.028, whereas the King Islanders of Alaska contained no European genes and thus had an m value of 0.0. At the other extreme, the Augpilagtok Eskimo gene pool of Greenland exhibits 0.40 admixture with Europeans (Steinberg et al., 1961, 1974).

Arctic whaling destroyed geographic isolation and placed St. Lawrence Island villagers in contact with Europeans. This contact introduced both European diseases and European genes into the Eskimo population. After the severe genetic bottleneck at the turn of this century and the subdivision of St. Lawrence into two settlements, the Eskimo populations experienced less gene flow. Then during World War II, there was some additional European gene flow into Gambell because of the presence of an Air Force base. Although there are no significant differences between these settlements in levels of heterozygosity, a comparison between admixed and non-admixed individuals reveals the expected higher heterozygosities in persons with genealogical histories of admixture.

North American natives

Amerindian admixture with Europeans and Africans varies widely from those regions that have experienced the most intensive contact of longest duration with settlers to groups that have remained geographically and or socially isolated. The Atlantic seaboard and several regions of the southeastern United States have the highest admixture rates, which approach 50% (Pollitzer et al., 1967; Szathmary and Auger, 1983). A small amount of admixture with Africans has been detected in several North American Indian populations. Pollitzer et al. (1967) estimate that 6% of the genes carried by the Catawba (South Carolina) are of African origin. Some African marker genes have been observed in a number of other North American tribes, but at low incidence.

Szathmary and Reed (1972) estimated the degree of European gene flow into two Ontario Ojibwa communities with over 300 years of contact. Based upon the frequencies of alleles in several blood-group systems, one Ojibwa community had an average estimate of 29% European ancestry, but the second virtually none at all. This example is indicative of the vast variation in admixture within the same tribe, probably as a result of unique historical events.

Given full and complete genealogies, which people rarely remember, individual familial histories can be traced, and admixture estimates developed from pooled familial data. Intensive study of the Ramah Navajo in New Mexico produced some genealogies that go back eight generations; taken together, these familial records show that this population has absorbed large numbers from other tribes over the past few centuries, and not inconsequential numbers of genes of European origin. Much of the European ancestry dates to the seventeenth and eighteenth centuries, when it was introduced through the children of Navajo women who were enslaved by the Spanish but eventually escaped and returned to the tribe. Spuhler and Kluckhohn (1953) used some of these genealogies to measure the intensity of inbreeding among the Ramah Navajo. The mean level of inbreeding for the population was 0.0066, a moderate amount when compared with some other human populations.

Caribbean Black Caribs

In the estimation of genetic admixture of the Black Caribs, a simple dihybrid model was initially utilized. It assumed that the Carib Amerindians hybridized with Africans to produce the hybrid gene pool of St. Vincent Island. Yet prehistorical reconstruction revealed that the Amerindian component was actually a hybrid of Caribs and Arawaks. The African ancestral population of the Black Caribs had originated from many different ethnic groups of West Africa and in itself should be viewed as a hybrid population. From 1517 to 1646, the African component was introduced into the population of St. Vincent Island in a number of ways: (1) by runaway slaves from European-held islands; (2) by Carib raids on various settlements on adjoining islands; (3) possibly by slaves surviving wrecks of slave ships. Therefore, the likelihood of the African component coming from a single population of origin is very low. In addition, analyses of European marker genes revealed some gene flow from the Creole populations of St. Vincent and from coastal populations of Central America into the Black Carib gene pool. The presence of albumin Mexico alleles in Central American Black Caribs, in conjunction with its absence from the St. Vincent Island population, documents the magnitude of Highland Amerindian gene flow into the Black Carib enclave.

The Black Carib gene pool experienced a series of hybridizations that underlie the contemporary genetic variation of these people. In fact, this complex model of Black Carib admixture still fails to take into account the numerous population subdivisions that occurred during the colonization of the coast of Central America. The selective forces associated with diseases such as malaria are also ignored by this model of genetic admixture. Clearly, such models only grossly approximate the evolutionary bases for the observed genetic variation among and within human populations, while ignoring a number of other evolutionary factors.

Estimates of admixture for Black Carib populations of St. Vincent Island and the settlements on the coast of Central America pose some interesting contradictions. Although about 41% of the gene pool of Sandy Bay (St. Vincent) is of African origin, most of the Central American Black Carib populations exhibit African contributions of 70–75% (see Table 23). The Black Caribs of Central America originated on St. Vincent Island, from where approximately 2000 of them were deported in 1797 (Gonzalez, 1988). This small founding population gave issue to the 100 000 or so Garifuna presently residing in Central America in 54 villages dispersed from Belize City to La Fe, Nicaragua. Those Black Caribs who avoided the British round-up of 1797 gave rise to the contemporary Black Caribs of St. Vincent. The disparity in the

Table 23. *Estimation of percentage admixture in Black Carib populations of St. Vincent Island and Central America*

Population	African	European	Amerindian
St. Vincent			
Sandy Bay	41	17	42
Owia	58	10	32
Central America			
Livingston	70	1	29
Stann Creek	75	3	22
Punta Gorda	71	5	24

Source: Schanfield *et al.* (1984); Crawford *et al.* (1981).

African contribution to the Black Caribs of St. Vincent (41%) compared with that of Livingston, Guatemala (70%) and Punta Gorda, Belize (75%) has raised some questions about the selectivity of the British deportation. Did the British deport those of most 'African' phenotypes and leave the most Indian-looking Black Caribs? There is no historical evidence to support this contention. Historical accounts from the late eighteenth and early nineteenth centuries described Carib phenotypes in Honduras as being primarily Indian. E.G. Squier, writing under the pseudonym of Samuel Bard while reporting on a Honduran village near Trujillo, wrote:

'Most are pure Indians, not large, but muscular,with a ruddy skin, and long straight hair. These were called the Red or Yellow Caribs.
Another portion are very dark with curly hair, and betraying unmistakably a large infusion of Negro blood, and are called Black Caribs.'

Thomas Young similarly noted great variability among the Black Caribs. He concluded that 'some are coal black, others again nearly as yellow as saffron' (Young, 1842: 123). During my first fieldwork among the Black Caribs of Livingston, I failed to observe such extreme variation. Most of the Black Caribs were phenotypically African, and this was borne out genetically.

Given an almost equal proportion of African and Amerindian genes in Sandy Bay (the 58% African in Owia reflects gene flow from Creoles residing in that village) it is not clear why the Black Caribs of Central America are predominantly African genetically (see Table 23). My initial explanation was that they had admixed with Creoles who were already on the coast of Central America. However, this is not supported by the low levels of European genes present in villages such as Livingston. The Creoles of Central America contain 17–33% European admixture; if there were massive gene flow between the Creoles and Black Caribs there should be more than 1% European admixture in Livingston. Derek Roberts suggested that this disparity may be due to selection (Roberts, 1984). The Black Caribs on the coast of Honduras, Belize and Guatemala were exposed to severe selection by both falciparum and vivax malaria. Individuals who carried hemoglobins S and C and the Duffy FY3FY3 genotype would have had a higher probability of surviving malarial infections. The probability is higher that individuals carrying the abnormal hemoglobins and the Duffy null phenotypes would be more African. Thus, selection may indirectly be operating in favor of those individuals with the greatest amount of African ancestry.

Tlaxcaltecans of Mexico

Because of their unique history, the people of the Valley of Tlaxcala in Central Mexico managed to avoid many of the devasting effects of Spanish colonization. The Tlaxcaltecans had been bitter enemies of the Aztecs, who lived in the adjoining valley. The Tlaxcaltecans established a military alliance with Cortez (known as the Segura de la Frontera) against their traditional enemies. Together, the Tlaxcaltecans and the Spanish brought about the fall of the Aztec empire. In return for their services, the Spanish crown granted the Tlaxcaltecans a number of privileges, which included some degree of self-administration of Tlaxcala from a few administrative centers within the Valley. Knowledge of this history led to the development of a research project to study admixture in the Valley of Tlaxcala in one of the administrative centers and in enclaves of Tlaxcaltecans that had been transplanted to various regions of Mexico.

Admixture estimates in Tlaxcaltecan communities were made using two different approaches. One approach was based on Bernstein's single-locus comparison of GM haplotypic frequencies (Schanfield, 1976; Schanfield et al., 1978). The second method employed several analytical methods and a wide array of allelic frequencies in estimating admixture (Crawford et al., 1976b). Unfortunately, no GM typing was possible for the populations of San Pablo del Monte and the City of Tlaxcala, both in the Valley of Tlaxcala. These two communities were selected because of their contrasting levels of admixture. The town of San Pablo is located on the slopes of the volcano La Malinche and was highly endogamous, according to our cultural anthropological colleagues. However, Lisker et al. (1988) have raised some questions concerning our assumption that San Pablo was primarily Indian with little, if any, European admixture. He argued on the bases of the frequencies of several alleles, such as ABO*A, that this community has experienced some European gene flow. If San Pablo did in fact experience some European admixture then the magnitude of

Table 24. *Admixture estimates (percentages) for populations from Tlaxcala and their transplants in Cuanalan and Saltillo*

Population	Parental groups		
	Indian	Spanish	African
San Pablo[a]	100	0	0
Tlaxcala[a]	70	22	8
Cuanalan			
Residents	96	4	0
Hybrids	83	14	3
Immigrants	66	29	5
Saltillo			
Chamizal (Residents)	52	45	3
La Minita (Immigrants)	58	36	6

Source: Crawford and Devor (1981).

[a] Maximum likelihood estimates; the remainder of the estimates are based on Bernstein's method using the GM haplotypes.

the gene flow into the City of Tlaxcala has been underestimated slightly. San Pablo was used as the Indian parental population in the studies of admixture in Mestizo populations.

Table 24 summarizes admixture estimates for various subdivisions of Tlaxcaltecan populations. Estimates for San Pablo del Monte and the City of Tlaxcala are based upon maximum-likelihood procedures. The City of Tlaxcala gene pool contains approximately 70% Indian, 22% European and 8% African genes. The origin of the African genes in the Tlaxcala gene pool is the subject of some debate. Historically, there is no evidence of gene flow from the coastal areas of Mexico, which do contain populations that have undergone considerable admixture with Africans. In addition, in a biracial hybrid model for Tlaxcala, the European component is estimated to be 30%; thus, the African component in the triracial model appears to come at a 'cost' to the European admixture while the Amerindian proportion remains unaffected. I have suggested that one source of these African genes may be the Moorish component of the Spanish army (Crawford, 1978). Therefore, it is possible that we are observing, 400 years later, genes that came to Spain by way of northern Africa.

There is considerable variation in admixture among the Tlaxcaltecan Indians, as indicated by Table 24. Even if San Pablo is not considered, the core of the transplanted enclave of Tlaxcaltecans, the residents of Cuanalan have experienced little European admixture. In the last two decades, Cuanalan has served as a stopover for migrants coming to Mexico City, and with this population movement comes gene flow. These migrants to Mexico City are not a random sample of Indians: they have sustained considerably more African and European admixture (Crawford, 1976). The Tlaxcaltecans were relocated to Saltillo in 1591, when 400 families from the State of Tlaxcala were moved north. Intermixture with the Spanish garrison located near contemporary Saltillo resulted in a gene pool that has almost 48% non-Indian genes.

This example of the Tlaxcaltecans and their transplanted enclaves indicates how unique historical events play major roles in generating observed patterns of admixture. It also illustrates the complexities of admixture. The relatively high African contribution to northern Mexico was a result of mining operations that brought many people of African origin from coastal regions. In contrast, the African admixture in the City of Tlaxcala may have originated in the Moorish component of the Spanish army.

Lisker et al. (1990) provided estimates of admixture in four Mexican cities: Leon, Merida, Oaxaca and Saltillo. They found that all of the samples from the cities could be characterized as triracial hybrids, bearing alleles of African, Indian and European origin. Oaxaca had the lowest proportion of African and European admixture (2% and 31%, respectively); Leon had the highest amount of admixture (8% and 40%, respectively). These data indicate that Mexico's population is rapidly becoming a triracial hybrid as population movement into the urban centers accelerates.

South American Indians

South American Indian populations vary a great deal in amount of admixture. The coastal regions are primarily trihybrid, but along the tributaries of the Amazon there exist a few populations with little, if any, gene flow from either Europeans or Africans (Long and Smouse, 1983). The latter groups are rapidly disappearing as Brazilian settlers and gold miners cut deeply into the jungle in search of a livelihood and wealth in the form of gold. Neel et al. (1977a) attempted to evaluate the degree of admixture of the Macushi and Wapishana with Neo-Brazilians. Earlier studies had indicated that there was no admixture with either Europeans or Africans, except for two individuals who had an ABO blood type other than O. More recently the picture has changed: on the basis of GM haplotypes, Neel et al. have estimated that 2.1% and 2.5% of the genes in the Wapishana gene pool are of African and European origin, respectively. The Macushi tribe exhibits 1–2% European genes and no evidence of Neo-Brazilian

African admixture. What this study shows is that members of even the recently isolated South Amerindian populations are now interbreeding with non-Indian settlers and miners.

Santos *et al.* (1983) studied a population of triracial hybrids residing in Manaus, Brazil. This community was founded in 1669 and was constructed as a garrison designed to protect Portugese interests and prevent invasions by the Netherlands and Spain. Preliminary analyses utilizing the frequencies of genetic markers expressed in the blood revealed that the gene pool is of approximately 61% Portugese, 27% Amerindian and 12% African origin. Thus, in some regions of Brazil, there are settlements in whose gene pools the Native American contribution is surprisingly small.

5.7   CONCLUSION

Effects of admixture

The evolutionary effects of admixture on the Native peoples of the New World include increasing genetic heterozygosity within populations and fostering greater genetic similarity among populations. The varying proportion of admixture with Europeans and Africans may also obscure some of the underlying Amerindian population structure that evolved over thousands of years. However, the association between latitude and genetic heterozygosity in Amerindian populations may be a result of successions of north–south migrations of Siberian founders.

Heterozygosity

The increase in genetic heterozygosity due to admixture has been documented for the St. Lawrence Eskimos (Byard *et al.*, 1984b). Despite varying degrees of admixture, significant populational differences in heterozygosity could not be demonstrated in four of the Eskimo groups (Savoonga, Gambell, Wales and King Island). However, a comparison of individuals from Savoonga and Gambell who had a family history of admixture revealed a significant increase in heterozygosity (based on ten loci) among the admixed grouped individuals. For example, in Gambell the admixed group had a heterozygosity (H) level of 39%, compared with 28% for those with no history of admixture.

The Black Caribs of Central America exhibit an exceptionally high level of genetic variation, resulting from the triracial, multiple ethnic origin of the founding populations. Out of 29 genetic loci tested among the Black Caribs, 26 were polymorphic and only three were monomorphic (malate dehydrogenase, lactate dehydrogenase and phosphoglucomutase-2). The mean per-locus heterozygosity (D) for St. Vincent Island Black Caribs was 46%. The D for Central American Black Caribs was almost identical, 45%, and reflects the complex origins of the population. Because some GM phenotypes are usually limited to geographically disparate populations, hybridized groups exhibit many new combinations of alleles. For example, whereas European populations exhibit 10–12 GM phenotypes, the Black Caribs of St. Vincent had 42 and the Black Caribs of Belize 28 distinct GM phenotypes. This high genetic variation has been maintained through a number of social and demographic mechanisms. Households practicing serial polygyny instead of permanent pair bonding have resulted in the reshuffling of genes at an accelerated rate each generation. There has been a high migration rate; 12% of matings in small villages are exogamous and 3% are interethnic, whereas 20–30% of the children born in large communities have parents of differing ethnicities. There has been a constant gene flow from the surrounding Mayan populations into the coastal Black Carib villages. Finally, high fertility rates have characterized both large and small communities. Firschein (1984) described an average rate of 5.84 children for mothers over 60 years of age in the towns, and Brennan (1983) found in rural villages 10.9 live births per woman 45 or more years of age.

The Black Carib studies provide some insight into the evolution and dynamics of expanding, colonizing, human populations much like those of early hominids that peopled the world. The intimate relationship between the ethnohistory of a group, in the form of unique historical events such as the fission and fusion of the community, and the genetic structure and variability are clearly revealed in admixed populations such as the Black Caribs. Finally, the evolutionary success of the Garifuna and other African-derived groups can only be understood on the bases of the interaction of both the cultural and genetic characteristics of the populations. The tremendous genetic variation, coupled with the presence of alleles (those specifying hemoglobins S, C, G-6-PD deficiency, and the Duffy FY blood type) that are adaptive in the presence of malaria contributed to the population explosion of the Black Caribs of Central America.

Beals and Kelso (1975) claimed the existence of a relationship between sociocultural complexity and genetic heterozygosity. Suarez et al. (1985) retested this hypothesis on 82 Amerindian populations. They found that, when differences in latitude are controlled, there is no significant relationship between sociocultural complexity and heterozygosity. However, their work (O'Rourke et al., 1985) does indicate a significant association of heterozygosity with climatic variability. It is highly unlikely that climate affects heterozygosity in humans; it is more likely that the cline is a result of north-south migration of the Amerindian founders.

Population structure

The elaborate models used in asssessing the genetic structure of Amerindian populations reveal underlying patterns. Despite the social and genetic upheavals associated with conquest and colonization, traces of prehistoric migrants are evident. Affinities among contemporary groups are suggestive of the operation of Malecot's isolation-by-distance model. Whether this relationship between genetics and geography is pre-Columbian or whether it resulted from the migratory patterns after the Europeans arrived is unclear. From historical demographic studies in other continents, it appears that 'new' structures form relatively rapidly. Thus, some of these observed patterns may be post-Contact. On the other hand, the genetic affinities of groups such as the Inupik speakers are related to their expansion approximately 1000 years ago. The close genetic affinities of the Amerindian groups in the southwestern United States with Nahua speakers of Mexico most likely stem from prehistoric migrations the evidence for which resides in linguistic similarities as well as in the genes.

6 Morphological variation

6.1 INTRODUCTION

The history of physical (also known as biological) anthropology is intimately bound to the study of morphological traits and their uses in attempts to classify human races. Early physical anthropologists, such as F. Boas and A. Hooton, devoted much of their research energy to the collection of anthropometric data (body measurements) and anthroposcopic data (results of subjective grading of morphological characteristics that cannot be concisely measured) for the characterization of human races. Both the anthropometric and anthroposcopic traits have two distinct disadvantages in the study of human populations: (1) technical error (interobserver variation), which can be reduced to some degree in measurements and observations; and (2) ontogenetic changes in specific measurements, such as girth, height and weight. It is this ontological impermanence of measurements under changing environmental conditions that render anthropometrics unsuitable for tracing long-term genetic relationships between the races. These problems associated with morphological traits are the major reasons that there was a 'stampede' by scientists to use the genetically based blood groups for evolutionary studies when methods for rapidly identifying these traits became available. The blood phenotypes remain unchanged throughout the life of the individual. Although there can be laboratory error associated with blood typing, this problem can be overcome by using independent determinations from two laboratories (Osborne, 1958). By the 1950s and 1960s, many American biological anthropologists had stopped utilizing anthropometric measurements in their studies of human population relationships.

The realization of the ecosensitivity (environmental responsiveness) of morphological traits shifted the use of these measurements from studies of race to those of processes of genetic–environmental interaction (Kaplan, 1954). This change in paradigm (from typology to process) can be observed in the investigations of adaptation to high altitude by Baker and his students (Baker and Little, 1976) and the proliferation of growth studies by researchers such as Garn (1958), Greulich and Pyle (1959) and Krogman (1972), among others.

At the turn of this century, visionary researchers, including Franz Boas (1911, 1912), were cognizant of the ecosensitivity of metric traits and utilized them effectively to develop the 'migrant model' for the understanding of environmental–genetic interaction. M.T. Newman (personal communication) characterized the crux of the migrant models as follows: 'same people, different environment or different people, same environment.' This comparative approach to the study of sedentes and migrants has been widely used to tease out possible selective factors affecting the human phenotype. The migration model has been employed by biological anthropologists since early in this century (in research on Jewish and Japanese migrants: see Boas (1911) and Shapiro (1939), respectively) to the 1980s (studies of Samoan migrants to Hawaii and mainland United States (Baker et al., 1986). This model was first applied to a comparison of sedente Mexican families in Mexico with migrant families in Texas (Goldstein, 1943). This initial research by Goldstein was followed by Lasker (1952), who demonstrated that young migrants manifested a plastic response, while adult migrants closely resembled the sedentes. Lasker (1995) provides a brief review of this methodological approach to the study of human plasticity.

Biological anthropologists once stressed morphological characteristics of living persons or the crania of the deceased in classifying humans into races. The resulting racial classifications were based upon typological principles (i.e. that individuals of a given race could be identified by morphological traits that remained unchanged throughout their lives).

What morphological characteristics do the American Indians share? How can a bicontinental series of populations, each population with considerable variation, and exposed to a different environment, be characterized? Stewart (1973) has described Amerindians as evincing the following traits:

1. straight black hair;
2. dark brown skins ('swarthy' according to Stewart);
3. a tendency to the Mongolian eyefold;
4. prominent cheek bones.

Other biological anthropologists have included such traits as darkly colored eyes, relatively glabrous skin (sparse beards and body hair), relatively long trunks and short legs, and special dental characteristics such as shovel-shaped incisors. Obviously, not all Native Americans share all of these traits and there is considerable variation within populations and from population to population.

6.2   RACIAL CLASSIFICATIONS

Racial classifications of Native Americans have been created with two sets of assumptions in mind (Stewart and Newman, 1951). Some scholars viewed Native Americans as a single entity belonging to the Mongoloid race (Hrdlička, 1912; Birdsell, 1951). They considered the Amerindians as 'undifferentiated Mongoloids' or as intermediate between Europeans and Mongoloids. In 1839, d'Orbigny noted the morphological heterogeneity of the Amerindians and von Eickstedt (1934) and Imbelloni (1958) classified South Americans into races. For example, Imbelloni's racial map of South America subdivided the Amerindians into five races: Fuegids, Lagids, Pampids, Amazonids and Andids. These classifications were based upon typological reasoning and the so-called racial groups were founded on the bases of stature, cephalic index, facial shape, nose shape, eye form and geographical distribution (Newman, 1951).

Neumann of Indiana University was one of the last prominent racial classifiers. His entire professional career was devoted to discerning cranial varieties or races of the Indians of North America. Although his system was typological, his personal assessment of Indian skulls was more of an art than a science. He first developed a scheme of ten varieties, or local races, of Amerindians. Eventually Neumann (1952) modified his scheme to accomodate two migrations across the Bering Strait. The migrants he characterized as Paleoamerind, who represented the Big Game Hunters and were short and lightly built with long heads and delicate facial features. The second migrants he called Cenoamerinds, the last of the Indians to cross the Bering Strait. The Cenoamerinds spoke Athabascan languages (at that time thought to be related to Sino-Tibetan). These people were supposedly of medium stature and build, with round heads and large rugged faces. Judging from the criteria used to identify members of these races, Neumann continued to employ typological reasoning.

Long (1966) re-analyzed Neumann's Amerindian cranial series using multiple discriminant functions. By utilizing measurements of the crania and by statistically determining the fit of specific crania into culturally or archeologically defined groups, he tested hypotheses concerning the existence of various racial groupings. This multivariate method disposed of most of Neuman's cranial varieties or races and raised doubt about the hypothesized origin of these subgroups through large-scale migration.

In 1953, Newman published an article that linked the observed variation in body morphology of Amerindians with ecological rules. He plotted the mean anthropometric measurements on a two-continent map and demonstrated that most of these traits were distributed clinally. He interpreted these morphological gradients as examples of the action of Bergmann's and Allen's rules of bodily size and proportions in humans. According to Bergmann's rule, the larger of two otherwise identical bodies has a smaller surface area in proportion to its mass because volume and mass increase as the cube of the radius while surface area increases as the square. In other words, larger and chunkier bodies are better able to retain heat, whereas smaller and linear bodies are better able to dissipate it. Similarly, according to Allen's rule, individuals with longer limbs dissipate and lose more heat in cold environments than those with short limbs. These rules appear to hold for the New World but not for sub-Saharan Africa. Newman's research suggested the existence of a non-genetic environmental component in human morphological clines much like that observed among other widely distributed mammalian species in the New World, such as the puma. This research contributed much to the demise of the use of morphological traits in typologically based racial classifications. The important question was not how many races exist in a region, but rather what processes brought normal human variation into existence.

6.3   ANTHROPOMETRICS

Anthropometric traits are under polygenic control and the phenotypes result from genetic–environmental interaction. Some of these traits, especially the more linear ones, have a larger genetic component (i.e. a greater proportion of the total variance is genetic) than those measurements that reflect circumference and girth in most human populations (Devor et al., 1986). Body circumferences and skinfold measures tend to exhibit lower heritabilities or transmissibilities and are more greatly affected by nutrition and disease.

Geographic variation

North American Indian populations

Biological anthropologists of the late nineteenth and early twentieth centuries compiled large data sets on the various Amerindian tribes of North America. Franz Boas spearheaded the collection of anthropometric measurements of nearly 15 000 Indians according to a standard protocol (Jantz et al., 1992). Until recently, most of these data had not been analyzed in any systematic fashion. These compilations of anthropometric data 'provide us with time depth in studies of population structure' (Relethford, 1988: 123).

Jantz et al. (1992) described the analysis of the mass of anthropometric data collected by Boas and his collaborators near the turn of the century. Twelve anthropometric measurements were made on both males and females from the major culture areas of North America. Measurements of 6458 persons (putatively full-blooded Indians) from 64 tribes were utilized in this analysis. Although the results may be affected by interobserver error, because there were 50 different observers, Boas tried to minimize this effect by providing a standard protocol and either training some of the observers himself or supervising the training. Jantz and his colleagues utilized a principal-coordinates method of reducing the data to two-dimensional plots. These plots indicate that geography plays an important role in the patterning of anthropometric variation, supporting the earlier suggestion by Newman (1953) of the relationship between climate and body morphology in American Indians. However, in the study by Jantz et al. the cranial and facial dimensions exhibited considerably more inter-tribal variation than did the body measurements. They argued that facial height responds to climate, and that this, rather than interobserver error, is the primary explanation for the intertribal variation. These findings also support the earlier conclusions by O'Rourke et al. (1985) that a strong correlation exists between genetic variation (measured as differences in frequencies of genetic markers) and climate. In addition, Jantz and his colleagues noted that body morphology predicts culture area better than it predicts language. These findings contrast with Spuhler's (1979) conclusions that genetic markers are better predictors of language than of culture area. This difference may be due to greater ecosensitivity of body morphology, which might then better reflect the geographic distributions of populations.

Summaries of the mean values of anthropometric measurements taken from various North American tribal groups are contained in Ubelaker and Jantz (1986) and Johnston and Schell (1979). Ubelaker and Jantz included means for 12 anthropometric traits and indices from 28 Native American populations subdivided by culture area and linguistic phylum. These heterogeneous measurements fail to provide a clear cut north–south gradient in stature, but confirm Boas' earlier observations (Boas, 1903) that the Plains and Mississippi Indians were the largest and the southeastern and western Indians were smallest. These metric comparisons transect different temporal periods and the effects of the secular increase in stature and nutritional changes stemming from contact tend to confound the observed patterns. However, Johnston and Schell, restricting themselves to larger samples that are representative of North, Central and South America, found the northern North American Indians to be the largest, followed by the Eskimos, South American Indians and Mesoamericans. Their analysis revealed that body weight was the variable most closely related to geographic location.

Siberian populations

A special issue of *Human Biology* has been devoted to the analyses of the mass of anthropometric data assembled by Boas from 1888 to 1903 (Jantz, 1995). Besides the measurement of 15 000 Native Americans, Boas and his group also measured 2000 indigenous Siberians. The availability of the Siberian data permitted 'construction of a bridge' between the historical populations measured during the Jesup North Pacific Expedition, such as the Evenki, and their contemporary counterparts (Comuzzie *et al.*, 1995). Apparently, the differences between the two populations reflect increasing levels of admixture with other populations, such as the Russian settlers and other aboriginal Siberian groups relocated during Soviet collectivization in the 1930s. Based on these anthropometric data, Ousley (1995) found a close affinity between the Nivkhs of the Lower Amur region and Sakhalin Island with Northwestern Amerindians. Yet these affinities are challenged by mtDNA findings by Torroni *et al.* (1993b). When affinities measured by anthropometrics are contradicted by molecular evidence, which is correct? Are the plastic or ecosensitive components of the anthropometric phenotypes responsible for these similarities? Apparently, the Nivkhs and Northwest coastal Amerindians share similar environmental conditions that may result in their morphological convergence.

Most of the indigenous Siberian populations have been studied anthropometrically (Alexseev, 1979). The data collection goes back to Yarho's (1947) investigations in the Altai, Debets' (1947, 1951) measurements of indigenous populations of Kamchatka and Chukotka, and Levin's (1958, 1963) study of an assortment of populations in eastern and central Siberia. Alexseev (1979) summarized most of the available anthropometric and anthroposcopic data on Siberian populations. Interestingly, the mean stature in Siberian populations appears to be lower than that published for northern Amerindian groups. Considering the wealth of the Siberian anthropometric data, it is unfortunate that Alexseev failed to analyze it using multivariate techniques. He continued to employ typological methods to group local races (Alexseev, 1979).

Latin American Indian populations

The mean stature in Central American Indian populations shows considerable variation (Faulhauber, 1970). In Falhauber's extensive compilation of the mean anthropometric values for the populations of Mexico, stature varies from 155.1 cm for Mayan males (samples of at least 50 individuals) to 169.6 cm for northern Yaqui males. Papago and Pima males are taller than any of the Mexican Indians, with mean statures of 170.9 and 171.8 cm, respectively. The Mayan females are also the shortest Mexican Indian group, with a mean height of 141.5 cm; the tallest females, with a mean height of 155 cm, come from northern Mexico. It is interesting to note that the mean stature recorded by Starr (1902) for Tlaxcaltecan Indian males was almost identical to that recorded by my research team more than 65 years later (Lees and Crawford, 1976). The mean statures for males and for females recorded by us in the village of San Pablo del Monte are 160.2 and 147.7 cm, respectively, comparable to Starr's (1902) reported mean values of 160.3 cm for males and 148.4 cm for females. Despite the secular trend of increasing stature observed in most of the world, the stature of Tlaxcaltecan Indians has not changed significantly in 65 years.

Tlaxcaltecans Eveleth and Tanner (1976) concluded that there is no secular trend among Amerindian populations because environmental conditions have not improved. By contrast, the admixed or Mestizo populations are considerably taller than are those of the Indian villages. For example, the male residents of the City of Tlaxcala have a mean stature of 163.4 cm. This difference in stature is most likely due to a combination of factors, including admixture with the taller Spaniards, higher income, and better nutrition. The greatest amount of variation in male stature is found in Saltillo (an urban center in northern Mexico represented by a heterogeneous sample that included both affluent residents and poor, shanty-town residents). Among females there was less variation in Saltillo in height than in the City of Tlaxcala, whose population had the highest variances (Lees and Crawford, 1976). Most likely, the high variance in Tlaxcala is a function of the genetic heterogeneity of females marrying into the patrilocal community.

The Tlaxcaltecan populations offered an opportunity to measure evolutionary change, both genetic and morphological, over time. They were transplanted from the Valley of Tlaxcala to the Valley of Mexico (Cuanalan) and to northern Mexico (Saltillo). Utilizing demographic and historical records, we were able to subdivide these enclaves by migrant/resident status and sociocultural factors (such as residence in shanty towns versus town residence). The original research design mandated a comparison of the Tlaxcaltecan enclaves living under different environmental conditions and for differing temporal intervals (Crawford, 1976). This design was viewed as practical in light of our prior understanding that the Tlaxcaltecans were reproductively isolated from neighboring populations. This isolation was due in part to the hatred of the Tlaxcaltecan so-called 'traitors' who, according to other Mexican ethnic groups, sold Mexico out to the Spaniards. However, old animosities diminish in time and, during the past two generations, the barriers to intermarriage seem to have vanished. Thus, gene flow during the past two generations complicated our original experimental design.

The computation of genetic distances and canonical variates for all the anthropometric traits and populations provides some indication of the genetic structure of Tlaxcaltecan populations (Lees and Crawford, 1976). Irrespective of sex, the subdivisions of Cuanalan showed the smallest genetic distances (Mahalanobis D^2) and the highest correlations. These subdivisions of Cuanalan shared similar environments and originated from the same founding population of Tlaxcaltecans. The primary differences between the three subdivisions of Cuanalan involved degrees of admixture with migrants. Three-dimensional scattergrams of male group means on the first three canonical variates reveal that Saltillo and San Pablo del Monte differed most (see Fig.

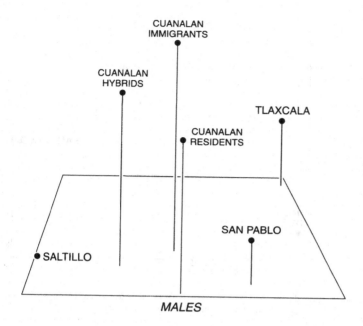

FIGURE 56 A three-dimensional plot of the first three canonical variates of the anthropometric variation observed among male Tlaxcaltecans.

56). These three canonical variates explained 96% of the variation. However, the females of San Pablo differed almost equally from Cuanalan immigrants and from females of Saltillo (see Fig. 57). The morphological differences between Saltillo and San Pablo was not surprising when the genetic makeup of Saltillo was considered. This large industrial center has primarily grown through migration from the surrounding regions of northern Mexico; this migration comprises large Chicimec and triracial gene flows. Our sample of the Saltillo population was drawn from a barrio that purported to be of Tlaxcaltecan origin, and from a shanty town. Unfortunately, the sample sizes for the anthropometric data set were insufficiently large to permit any further subdivision.

Black Caribs An anthropometric comparison of the Black Caribs of St. Vincent Island and Livingston, Guatemala, was made for purposes of assessing their divergence during 180 years of reproductive separation (Lin, 1984). Although the males of Livingston were taller than their counterparts on St. Vincent Island, the reverse was true for the females. Mean statures of males and females in Livingston were 170.3 and 155.6 cm, respectively, compared with 168.6 cm and 157.1 cm in St. Vincent. In comparison with Central Amerindian stature, the average stature of males in Livingston corresponded to

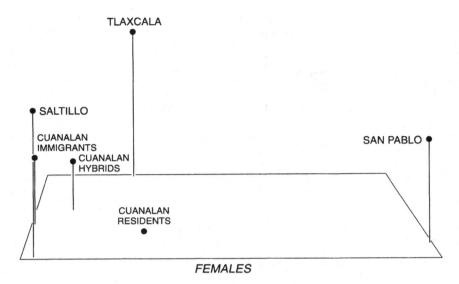

FIGURE 57 A three-dimensional plot of the first three canonical variates of the anthropometric variation observed among females of four Tlaxcaltecan populations.

the maximum stature of Central Amerindian males, and that of males in St. Vincent approximated the average stature of Central American Indians. The average abdominal circumference of females from Livingston was significantly larger than that of males. The males of Livingston were thinner than those of St. Vincent. Factor analysis revealed that the anthropometric variables were more highly correlated among males than among females. In stepwise discriminant analysis the best discriminating variables among the four groups (Livingston males and females and St. Vincent males and females) were thigh length, upper arm length, trochanteric height, and tricepital skin folds. These variables reflect a more linear phenotype with longer arms and legs in the Livingston population when compared with that of St. Vincent. These differences suggest a greater proportion of African ancestry in Livingston than on St. Vincent Island. The results of genetic analysis (indicating that 70–80% of the genes in the gene pool of Livingston are African, compared with approximately 50% in that of St. Vincent) are in accord with this suggestion (Crawford and Comuzzie, 1989).

South American populations

Variation in bodily morphology based upon anthropometric measurements has been described by a number of researchers (Spielman, 1973b; Marcellino et al., 1978). Juan

Comas (1971) compiled the means, standard deviations and the coefficients of variation of an assortment of anthropometric measurements made of members of South American Indian tribes. Steggerda (1950) listed measurements from 88 tribes, of which only 15 tribes were represented by samples larger than 50 persons. More recently, Salzano and Callegari-Jacques (1988) summarized, in their well-crafted volume, the average statures in 43 South American Indian groups and discussed the interpopulational differences in anthropometric traits.

The purported relationship between morphology and climate in South American Indians was recently examined by Stinson (1990). On the bases of four anthropometric measurements (height, weight, sitting height and relative sitting height) on members of 62 South American groups, she reported the existence of a gradient in stature from north to south. Correlation coefficients calculated between height and precipitation indicated a significant negative relationship. She was unable to demonstrate the existence of any significant relationship between morphology and temperature. An analysis of variance revealed no significant differences in mean male height among five linguistic stocks.

Anthropometrics and population structure

Considering the polygenic nature and ecosensitivity of anthropometric traits, we can ask: how accurate are they as measures of population genetic affinities or as measures of population structure? The Tlaxcaltecan research of the late 1960s and early 1970s suggested that the morphologically based distances were similar to those measured by Mendelian genetic markers. However, measures of fit between matrices were unavailable at the time.

In the late 1970s and 1980s, Relethford extended the use of anthropometric variation for the study of population structure, using Hooton, Dupertuis and Dawson's Irish survey, conducted in the 1930s (Hooton and Dupertius, 1951, 1955). Relethford and his colleagues applied to this data set a series of genetic models that included isolation by distance (Relethford et al., 1981) and bioassay of kinship (Relethford, 1980). The use of quantitative traits for the study of human population structure has flourished and a special issue of the journal *Human Biology*, entitled *Quantitative Traits and Population Structure* (Feb. 1990), was devoted to this topic.

A number of studies have explored the relationship between morphological and genetic variation in Amerindian populations. Most of this research is limited to South American Indian groups. Rothhammer and Spielman (1973b) correlated anthropometrically based distances with genetic distances among six Aymara Indian populations of Chile and derived a correlation of 0.33. Similarly, Spielman (1973a) compared distributions of genetic markers and anthropometric measurements among 19 Yanomama villages and found a relatively low correlation of 0.19. Neel et al. (1974) demonstrated a slightly higher Kendall's rank-order correlation (0.30) between matrices of anthropometric and genetic distances among seven Yanomama villages. Although there is a similarity between genetic and morphological affinities, the relationship is not intimate. Most likely, the sharing of similar nutritional habits can result in similar anthropometric patterns in populations that may differ genetically. Spielman (1973a) showed that a close relationship exists between anthropometrics and geography among the 19 Yanomama villages ($p = 0.80$). It is interesting to note the existence of this high correlation despite the constant fission and fusion of village populations and movements into new locales. Spielman and Smouse (1976) used discriminant function analysis to determine how well anthropometric measurements can be used to statistically place individuals into villages. Only 36% of the individuals were thus assigned to their villages of origin.

Konigsberg and Ousley (1995), in an ingenious bit of research, demonstrated that anthropometric measurements can appropriately be used to address questions of population affinities and phylogeny. They demonstrated that the additive genetic variance–covariance matrix (G) is proportional to the phenotypic variance–covariance matrix (P). Thus, where the heritability (h^2) is some constant of proportionality, then Mahalanobis' distance will be proportional to genetic distance.

Genetic–environmental interaction

Anthropometric traits, with their ecosensitivity, are useful tools that can be utilized to explore genetic–environmental interactions in Amerindian populations. The observed north–south clinal variation in size appears to bear a 'made in America' label. In other words, this gradient in Amerindian populations probably resulted from environmental factors (such as nutrition and climatic variation) and possibly the migratory patterns generated by Siberian migrants after their journey into the New World. Newman (1960a), dissatisfied with climatic temperature as an explanation of clines in body size, correlated average adult male body weights with coldest mean monthly temperature for 60 New World samples and showed a steep regression line but with high standard deviations. These results suggested that factors other than temperature were affecting body weights, and that almost 50% of the total variance was due to these other factors. Newman tried to account for the observed clines in stature by invoking nutritional variation, arguing that there was a decrease in bodily size after the introduction of maize agriculture and the consequent reduction in reliance on animal protein. Therefore, he argued that the gradients reflect the spread of maize agriculture. However, this scenario cannot account for the observed north–south genetic clines, described by O'Rourke et al. (1985).

6.4   DERMATOGLYPHICS

Introduction

The dermal ridges of fingers, palms, and soles have been used to measure affinities of and differences among human populations. Dermatoglyphic traits are transmitted polygenically but environmental factors intervene during a brief period of embryogenesis. The dermal ridges on the hands and feet are formed during the first 12–16 weeks of development. After formation, the number and pattern of ridges remain unchanged throughout the life of the individual (Cummins and Midlo, 1943). Figure 58 illustrates some of the morphological characteristics of the fingers and palm that are commonly used as dermal phenotypes.

Although these dermal traits have been used widely to characterize populations, there are a number of methodological problems associated with their genetic analysis. The commonly used phenotypes were defined in an arbitrary fashion (Crawford, 1977). The number of ridges per finger depended entirely on the presence or absence of triradii (see Fig. 59). By definition, individual fingers that contained arches had no triradii and were scored as having no ridges even though ridges could be observed. Thus, the traditional definition of ridge counts resulted in the loss of dermal information and skewed the distribution of ridge counts. This latter problem affects the use of multivariate statistics that assume normality of distribution. Most investigators who utilize multivariate statistics lament the problem but then proceed to ignore it by stating that multivariate techniques are sufficiently robust to withstand such deviations from normality (Friedlaender, 1975). There are, in fact, several solutions to the problem. One can normalize the remainder of the distribution; however, the loss of information persists. I recomended that redefinition of ridge counts be accomplished by ignoring triradii and counting all ridges intersected by a perpendicular line drawn from the crease to the fingernail (Crawford, 1977). The problem with such a drastic methodological switch is that much of the ridge-count data described in the literature would become obsolete and would have to be re-analyzed for comparative purposes. One solution is to use digital-pattern frequencies instead of ridge counts for the study of human population structure (Lin et al., 1983; Enciso, 1983). Bach Enciso has demonstrated empirically that distance matrices based upon digital-pattern frequencies are more highly correlated with genetic data than with traditional ridge counts in Tlaxcaltecan populations.

Intrapopulation variation

Multivariate statistical analyses have revealed the existence of underlying intercorrelations of human dermatoglyphic traits within populations. Knussman (1967), utilizing

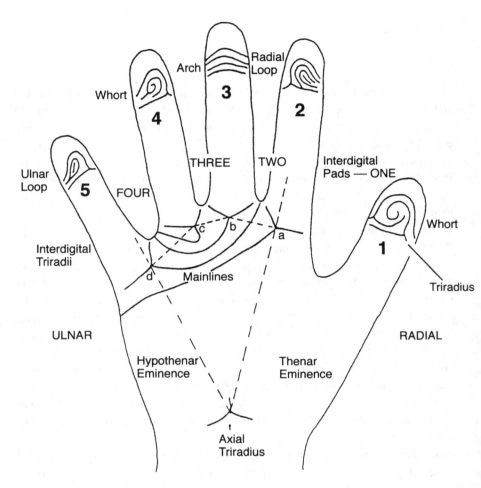

FIGURE 58 Illustration of the dermatoglyphic traits of the palm and fingers
used to describe human populations (after Holt, 1968).

factor analysis, and Roberts and Coope (1975), using principal components analysis,
observed that several dermal traits were highly intercorrelated. Roberts and Coope,
on the bases of 20 ulnar/radial finger-ridge counts, proposed the application of 'field
theory' (previously applied by Butler (1939) to dental complexes) to dermatoglyphics.
This theory is predicated on the concept that the activities of gene complexes are
affected by the environment resulting in a range of expression. Thus every trait does
not require an independent gene for its expression. Crawford *et al.* (1976a) proposed
that the intercorrelated entities or 'fields' correspond to the observed patterns of

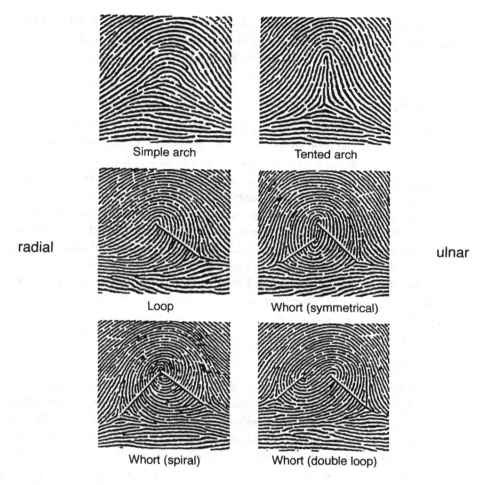

Simple arch   Tented arch

radial   Loop   Whort (symmetrical)   ulnar

Whort (spiral)   Whort (double loop)

FIGURE 59 Triradii and the numbers of ridges on human fingertips (after Holt, 1968).

innervation and vascular supply. Lin *et al.* (1979), after observing the same fields in several populations, proposed that these entities are characteristic of *Homo sapiens*.

Crawford *et al.* (1976a), utilizing principal components analysis (PCA) and factor analysis, demonstrated the existence of these fields in Tlaxcaltecan Indian populations. Digit 1 (the thumb) constitutes a separate field, the second and third digits act in concert to form a second field, and the fourth and fifth digits make up the last field. We hypothesized that the innervation of the skin of the palm and digits corresponds to these fields. To date, this structure, which underlies the observed digital variation, has been observed in all populations studied. Thus, the designation of this structure as 'universal' (Lin et al., 1979) appears to be correct.

Interpopulational comparisons

Comparison of dermal traits in various populations has been complicated by the use of different traits in different studies, sometimes combined with idiosyncratic methods of scoring. Most of the studies of New World populations have been descriptive, producing lists of dermal features with means and sometimes standard deviations. Rarely have analytical methodologies, multivariate statistics, or measures of population structure been applied with the aim of elucidating evolution and microdifferentiation of Amerindian groups. There are, however, some exceptions in which the relationships between dermal traits, geography and genetics have been examined (Rothhammer et al., 1973; Neel et al., 1974). Descriptive statistics for an assortment of dermal phenotypes have been compiled and included in this volume with the hope that they will be utilized by the next generation of dermatoglyphic specialists who apply more sophisticated analytical procedures.

Eskimo Populations

There is more known about the dermatoglyphic variation among Eskimo populations than among any other groups in the New World. For example, there is information on the frequency distributions of finger patterns in a total of 12 Eskimo populations distributed from Norton Sound, Alaska, to Greenland. Table 25 provides percentages of finger patterns for both hands, data on individual fingers being available for only six populations. In these Eskimo populations loops (L) are most frequent, followed by whorls (W) and, finally, arches (A). Ulnar loops (UL) occur many times more often than do radial loops (RL). An interpopulational comparison reveals a wide range of variation in the incidence of various finger patterns. For example, in males, the frequency of whorls ranges from 61.7% at Point Hope to 27.6% at Wainwright, whereas in females, the frequency of whorls range from 62.7% in Scoresbysund to 21.8% in Wainwright. It is difficult to pinpoint the reasons for such variation, particularly since Wainwright and Point Hope are both populated by Inupik Eskimos and are geographically in close proximity. Small sample sizes and differential gene flow from Europeans could underlie some of the observed variation. However, most of these Eskimo settlements are of small size (for example, Wales has a total population of 125 persons; Savoonga has approximately 250 inhabitants) and the samples collected constitute a reasonable proportion of the total community.

Table 25. *Frequency distributions of finger patterns of both hands in Eskimo populations subdivided by sex*

M, males, F, females.

Population	Sex	N	A (Arch)	L (Loop)	UL (Ulnar Loop)	RL (Radial)	W (Whorl)
Kodiak Isl.[1]	M	56	8.3	49.4	46.9	2.5	42.3
	F	40	4.0	55.6	53.8	1.8	40.5
Savoonga[2]	M	81	1.5	51.1	48.6	2.5	47.4
	F	83	2.7	59.0	56.8	2.2	38.4
Gambell[2]	M	41	0.5	57.3	52.7	4.6	42.2
	F	39	3.3	62.2	59.9	2.3	34.5
Wales[2]	M	31	1.6	63.6	59.4	4.2	34.7
	F	28	2.5	58.3	55.4	2.9	39.2
Point Hope[3]	M	53	2.7	35.7	33.4	2.3	61.7
	F	69	5.1	48.2	44.0	4.2	46.7
Point Barrow[3]	M	58	4.7	62.3	58.8	3.5	33.0
	F	71	3.4	60.5	56.6	3.9	36.1
Wainwright[3]	M	91	2.7	69.8	64.2	5.6	27.6
	F	90	6.3	71.9	68.3	3.6	21.8
Anaktuvuk Pass[3]	M	35	0.9	64.8	63.1	1.7	34.3
	F	36	1.7	67.5	63.6	3.9	30.8
Baffin Isl.[4]	M	234	1.8	58.6	55.0	3.6	39.6
	F	218	4.1	66.9	64.3	2.6	29.0
Southampton Isl.[5]	M	28	2.2	57.8	55.3	2.5	40.0
	F	34	6.5	48.1	46.0	2.1	45.4
Angmagssalik[6]	M	100	2.4	41.3	—	—	56.3
	F	100	2.5	46.3	—	—	51.2
Scoresbysund[7]	M	25	2.0	44.4	40.8	3.6	53.6
	F	25	4.0	33.2	32.0	1.2	62.8
TOTAL[a]	M	733	2.6	55.9	52.6	3.3	41.5
	F	733	4.0	57.4	54.6	2.8	38.7

Sources: [1]Meier (1966); [2]Crawford and Duggirala (1992); [3]Meier (1978); [4]Auer (1950); [5]Popham (1953); [6]Ducros (1978); [7]Ducros and Ducros (1972).

[a]Only samples where both sexes were studied separately, and where a distinction was made between ulnar and radial loops.

Table 26. *Total ridge counts in eight Eskimo populations*

Population	Sex	N	Total ridge count		References
			Mean	SD	
Kodiak Isl.	M	56	120.8	52.6	Meier (1966)
	F	40	122.3	38.4	
Point Hope	M	53	157.7	52.6	Meier (1978)
	F	69	136.5	58.6	
Point Barrow	M	58	128.7	48.2	Meier (1978)
	F	71	130.7	50.5	
Wainwright	M	91	113.5	46.8	Meier (1978)
	F	90	104.1	44.2	
Anaktuvuk Pass	M	35	133.8	40.6	Meier (1978)
	F	36	124.6	37.8	
Julianehaab	M	63	146.8	75.0	Cummins and Fabricius-Hansen (1946)
	F	63	152.8	67.3	
Angmagssalik	M	100	169.6	43.7	Ducros (1978)
	F	100	161.3	36.2	
Scoresbysund	M	25	182.5	38.8	Ducros and Ducros (1972)
	F	25	172.4	51.3	

Table 26 presents the range of variability for total ridge counts in eight Eskimo populations. Ridge counts by finger are available from only two populations, namely Angmagssalik and Scoresbysund (both in Greenland). Interestingly, males do not exhibit higher total ridge counts in all of the communities. For example, females from Kodiak, Point Barrow, and the vicinity of Julianehaab exhibit higher mean total ridge counts than do the males. Males' mean total ridge counts vary from 113.5 in Wainwright to 182.5 in Scoresbysund. Those of females vary from 104.1 in Wainwright to 172.4 in Scoresbysund. The Greenland Eskimos have by far the highest number of ridges, whereas other Inupik groups such as those of Point Barrow and Wainwright exhibit the lowest.

Data have been published on frequencies of palmar patterns in seven Eskimo populations. However, in only four of these populations have the data been subdivided by sex (Murad, 1975). Females in these populations have more hypothenar (29.9%) and IV interdigital (57.7%) patterns than males (26.8% and 54.1%, respectively). Males have a higher proportion of thenar/I (12%), II (2.2%) and III (27%) interdigital patterns

Table 27. A compilation of a–b ridge counts in Eskimo populations

Population	Sex	N	Left hand		Right hand		Both hands
			Mean	SD	Mean	SD	Mean
Savoonga[1]	M	81	39.0	6.2	39.0	6.2	78.0
	F	83	38.5	6.0	38.4	5.8	76.9
Gambell[1]	M	41	39.7	6.2	39.6	5.6	79.3
	F	39	41.2	7.3	42.7	8.4	83.9
Wales[1]	M	31	41.3	3.6	40.9	5.1	82.2
	F	28	42.8	7.2	41.6	6.8	84.4
Point Hope[2]	M	102	39.5	6.4	38.7	8.2	78.2
	F	108	39.0	6.3	38.4	4.8	77.4
Point Barrow[2]	M	100	37.8	6.6	37.6	6.4	75.4
	F	117	38.2	5.5	37.4	5.5	75.6
Wainwright[2]	M	114	38.5	5.9	37.5	6.3	76.0
	F	122	38.5	5.5	37.0	5.5	75.5
Anaktuvuk Pass[2]	M	34	39.0	5.8	40.1	7.0	79.1
	F	35	38.7	4.4	36.6	3.7	75.3
TOTAL	M	503	39.1	—	39.1	—	78.3
	F	532	39.6	—	38.9	—	78.4

Sources: [1]Crawford and Duggirala (1992); [2]Murad (1975).

than do females. The interpopulational differences are considerable, with 52.9% of the males of Anaktuvuk Pass having hypothenar loops in contrast to 17.3% in Gambell (Murad, 1975; Cummins and Midlo, 1931). The interdigital patterns are more common on the right hand (38.2% of males and 34% of females) than on the left. These palmar patterns may be useful in studying the population structure of Amerindians and Eskimos.

The variation in the a–b palmar interdigital ridge counts is shown in Table 27. These traits have been described for a total of seven Eskimo populations. There is little fluctuating asymmetry between the left and right palms and an insignificant difference in the frequency of the trait between the sexes. A summation of the total a–b ridge count (right plus left hands) over the seven populations shows a small range of variation, from a low mean of 75.3 in females of Anaktuvuk to a high of 84.4 among the females of Wales, Alaska.

The relationship among the populations of St. Lawrence Island, King Island and Wales, Alaska, have been examined by using dermatoglyphics (Crawford and Duggirala,

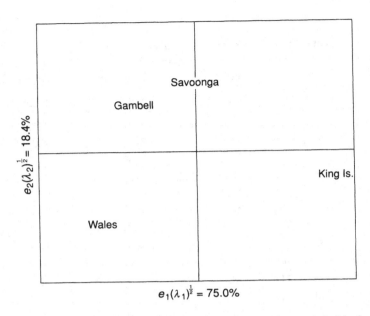

FIGURE 60 The relationship among Eskimo populations of St. Lawrence Island, King Island, and Wales, Alaska, based on 37 dermal traits.

1992). The R-matrix method of Harpending and Jenkins (1973) was applied to the frequencies of fingertip patterns for these four populations. Each population, subdivided by sex, was characterized by using 37 traits with frequencies of whorls, ulnar and radial loops, and arches compiled by finger. As expected, the populations of Gambell and Savoonga on St. Lawrence Island showed considerable affinity, but those of King Island and Wales were distinctly different from each other and from the Yupik-speaking St. Lawrence Islanders (see Fig. 60). The genetic map, a principal component (PC) 1 versus PC2 plot, reveals that the Yupik and Inupik populations are separated along the second scaled eigenvector. The first eigenvector separates King Island Eskimos from the other three settlements. The most informative traits are radial and ulnar loops, which separate the population of Wales from those of Gambell and Savoonga. The population of King Island is distinguished from the other three by the high incidence of whorls on the third finger of the right hand.

I also computed Mahalanobis' D, utilizing all available dermatoglyphic traits (i.e. mean ridge counts and pattern frequency by finger (see Table 28)). As expected, the populations of Gambell and Savoonga are least genetically distant and their distributions are not statistically significantly different. The ethnohistorical reconstructions (see chapter 2) indicate that Savoonga was formed through fission of Gambell's population; thus, these two villages should show little genetic divergence. However, it is

Table 28. *Mahalanobis' D² and F-statistic comparisons of four Eskimo villages, subdivided by sex*

Populations	Savoonga	Gambell	Wales	King Island
Males				
Savoonga	—	3.31 (1.82)	5.15 (2.27)	6.93 (2.63)
Gambell	1.21	—	4.53 (2.13)	7.71 (2.78)
Wales	1.55*	1.08	—	10.13 (3.18)
King Island	1.55*	1.44	1.71**	—
Females				
Savoonga	—	2.52 (1.59)	3.09 (1.76)	4.15 (2.04)
Gambell	0.88	—	4.99 (2.23)	5.77 (2.40)
Wales	0.85	1.07	—	5.75 (2.39)
King Island	1.11	1.21	1.04	—

Males: d.f. = 52, 119; females, d.f. = 53, 121 (the first of the two degrees of freedom indicates the number of variables entered into computation). *F*-matrix lies below the diagonal; Mahalanobis' D^2 matrix lies above the diagonal; *D* is in parentheses; *, significant at the 5% level; **, significant at the 1% level.

interesting to note that for females none of the distributions is significantly different, nor are the distances as great as they are for the males. The overall similarities between the four villages may reflect the importation of female mates from Alaska and Siberia as a result of dramatic diminution of population. These genetic distances suggest that genetic analyses based on dermatoglyphics can provide results similar to those obtained by the use of frequencies of genetic markers.

Amerindian populations

North America There is considerable variation in the frequencies of fingerpattern types in North American Indian populations. Both males and females of Apache, Choctaw Navajo and Seneca populations exhibit relatively high unweighted averages of whorls (males, 42.6%, females, 44.6%). These averages are slightly higher than those observed in Eskimo groups (see Table 29). The North American Indians also exhibit relatively high incidences of ulnar loops (47.8% in males and 43.9% in females), but these proportions are lower than those in Eskimo groups (52.6% in males and 54.6% in females).

Table 29. *Frequency distribution of finger patterns (%) for both hands (combined) in North American Indian populations*

Populations	Sex	N	A	L	UL	RL	W
Micmac[1]	T	150	10.4	60.0	55.9	4.1	29.7
Arapahoe[2]	T	50	4.6	47.8	44.2	3.6	47.6
Cherokee[3]	M	343	5.5	62.6	—	—	31.9
	F	327	5.6	63.1	—	—	31.3
Mohawk[3]	M	108	1.9	47.2	—	—	50.9
	F	92	1.7	54.0	—	—	44.2
Commanche[4]	T	67	6.3	50.4	48.5	1.9	43.3
Seneca[5]	M	27	5.6	51.8	48.5	3.3	42.6
	F	55	10.5	51.8	48.7	3.1	37.6
Seminole[6]	T	146	2.2	55.2	42.7	—	—
Apache[7]	M	44	5.5	45.5	44.6	0.9	49.1
	F	50	3.6	45.5	43.0	2.4	51.0
Choctaw[7]	M	53	13.0	56.5	52.3	4.2	30.6
	F	51	14.9	46.0	42.5	3.5	38.0
Navajo[7]	M	48	3.3	48.7	45.6	3.1	47.9
	F	54	5.4	42.8	41.5	1.3	51.9
Pueblo[8]	T	131	6.8	40.1	37.0	3.1	53.1
Hopi[8]	T	59	2.9	53.0	51.5	1.5	44.1
Alalakaket[9]	T	70	0.6	29.8	28.4	1.4	69.6

Sources: [1]Chiasson (1960); [2]Downey (1927); [3]Rife (1972); [4]Cummins and Goldstein, 1932; [5]Doeblin *et al.* (1968); [6]Rife (1968); [7]Flickinger, 1975; [8]Cummins (1941); [9]Meier (unpublished).

Central America The dermatoglyphic patterns of Central American Indian populations have been reviewed by a number of researchers (Newman, 1970; Coope and Roberts, 1971). Some of these reviews emphasize the dermal differences between the Mayan and non-Mayan populations (Garruto *et al.*, 1979); others have focused upon variation within Mayan groups (Newman, 1960b). Here I summarize the variation of such selected dermatoglyphic traits as finger patterns, total ridge count, palmar patterns and a–b ridge counts (see Table 30).

Table 30 provides the frequencies of finger patterns of both hands in various Mayan and non-Mayan populations. Both male and female Mayans have a higher incidence of

Table 30. *The frequencies of total (left plus right hand) finger patterns in various Mayan and non-Mayan populations*

A, *frequency of arches; L, incidence of loops; UL, ulnar loops; RL, radial loops; W, incidence of whorls; M, males; F, females; T, both sexes.*

Populations	Sex	N	A	L	UL	RL	W
Mayan							
Tzotzil-tzeltal[1]	M	90	6.4	58.6	55.3	3.3	34.9
Chamala[2]	M	100	3.4	52.8	48.1	4.7	43.8
Hacienda Acu[3]	M	25	6.4	42.4	40.8	1.6	51.2
Chichen Itza[4]	T	127	7.6	59.2	57.0	2.2	33.2
Amatenango[5]	M	49	1.2	46.4	42.5	3.9	52.4
Huixtan[5]	M	50	2.5	48.7	46.6	2.1	48.9
Finca Tzeltal[5]	M	47	2.8	57.0	53.2	3.8	40.2
Patzun[6]	M	72	6.9	58.3	55.4	2.9	34.8
	F	32	14.0	49.9	48.7	1.2	36.1
Soloma[6]	M	90	2.4	48.8	46.3	2.5	48.7
	F	22	2.0	54.9	53.2	1.6	43.1
Santa Clara[6]	M	68	5.6	54.9	51.6	3.2	39.5
Solola[6]	M	82	6.3	53.7	50.0	3.7	40.0
Chiapas (Tzotzil)[7]	M	230	7.2	57.0	—	—	35.8
Maya totals	M	—	4.39	52.2	49.0	3.2	43.4
	F	—	8.0	52.4	51.0	1.4	39.6
Non-Mayan							
Tarahumara[3]	M	26	2.7	45.5	43.5	2.0	51.9
Aztec[8]	T	78	3.2	57.3	55.0	2.3	39.7
Tarascan[9]	M	116	4.2	61.6	58.0	3.6	34.2
Mixtec[10]	M	78	3.3	56.5	52.9	3.6	40.3
Zapotec[10]	M	50	3.2	60.3	56.7	3.6	36.6
Zapotec[1]	M	104	6.6	59.1	56.3	2.8	34.2
Nahua Vera Cruz[1]	M	37	9.5	47.4	45.7	1.7	43.3
Tarasco[1]	F	40	11.3	51.5	50.0	1.5	37.3
Nahua (Otomi)[1]	M	34	6.5	58.3	56.2	2.1	35.3
Tlaxcala[11]	M	55	8.0	62.9	58.5	4.4	29.1
	F	82	10.7	58.3	53.9	4.4	31.0
Non-Mayan total	M	500	5.5	56.5	53.5	3.0	38.1
	F	122	11.0	54.9	52.0	3.0	34.2

Sources: [1]Zavala et al. (1971); [2]Leche (1936a); [3]Leche (1934); [4]Cummins and Steggerda (1936); [5]Leche et al. (1944); [6]Newman (1960b); [7]Kalmus et al. (1964); [8]Leche (1936b); [9]Leche (1936c); [10]Leche (1936d); [11]Lin et al. (1979).

ulnar loops when compared with either the North American Indians or the Eskimos. The ranges of the Mayan and non-Mayan frequencies overlap, but the non-Maya exhibit higher incidences of ulnar loops. Comparisons of the samples are of limited value because of their small sizes and the preferences of the investigators for certain traits while ignoring others.

Little information is available concerning the mean a–b ridge counts among Central American Indian populations. To date, only one Mayan population, the Tzotzil-Tzeltal, has been scored for the a–b ridges (Zavala et al., 1971). Their ridge-counts are available only by total for both hands, and exhibited a mean of 81.9 (s.d. 10.9). The ridge counts in non-Mayan populations range from 61.9 (Zavala et al., 1971) to 89.4 (s.d. 10.2) for 149 males of Cuanalan (Crawford et al., 1976a). The Tlaxcaltecan population subdivisions vary from 82.2 mean a–b ridges for females from San Pablo del Monte to 89.4 for males of Cuanalan. The least admixed population, San Pablo, appeared to have a lower mean number of a–b ridges in the Mestizo groups. In addition, the admixed Tlaxcaltecans possessed higher a–b counts than did any of the North American Indian groups studied to date.

The Tlaxcaltecan populations provided an opportunity to investigate affinities between genetically characterized groups of known history. Crawford et al. (1976) utilized multivariate techniques to examine intra- and interpopulational variation of four Tlaxcaltecan groups. The intrapopulation variation revealed the existence of an underlying structure much like that described by Roberts and Coope (1975). The interpopulation affinities were established by means of Mahalanobis' genetic distances. San Pablo differed from the other three populations, reflecting its Amerindian origin and very small amount of admixture. The two transplanted populations more closely resembled that of the City of Tlaxcala and differed significantly from each other. Considering only males, Tlaxcala is closer to San Pablo and Saltillo than is Cuanalan. Using female groupings, San Pablo is closer to Cuanalan than to Tlaxcala, even though that community is located in the same valley as is Tlaxcala and only a short distance away. Discriminant function analysis indicated that dermatoglyphics assign individuals correctly to their populations at a relatively low frequency (see Table 31). The 'most correct' classifications resulted from assignment of almost 52% of Cuanalan males and San Pablo and Tlaxcala females to their appropriate populations of origin (Crawford et al., 1976a). On the other hand, 'correct' classification of between 37% and 52% of the individuals is considerably better than random placement into the categories, which should result in 25% correct classification. These correct classifications are complicated by the varying degrees of admixture of Tlaxcaltecans with Spanish and African parental ,populations.

In a follow-up to the earlier dermatoglyphic study of Tlaxcala (Crawford et al.,

Table 31. *Number and percentage of cases classified correctly into four groups based upon discriminant function analysis*

	San Pablo	Tlaxcala	Cuanalan	Saltillo	Total	Percentage of correct classification
Males						
San Pablo	33	14	14	8	69	47.83
Tlaxcala	12	19	11	10	52	36.54
Cuanalan	27	16	77	29	149	51.68
Saltillo	14	16	11	42	83	50.60
Total	—	—	—	—	353	48.44
Females						
San Pablo	61	22	16	19	118	51.69
Tlaxcala	16	35	3	14	68	51.47
Cuanalan	47	39	74	39	199	37.19
Saltillo	22	37	32	81	172	17.09
Total	—	—	—	—	557	45.06

1976a), Enciso (1983) re-analyzed the Tlaxcaltecan data set. She applied R-matrix analysis to digital pattern frequencies among male and female Tlaxcaltecans and compared her results to published results from studies of other groups. Maps based on genetic and dermatoglyphic data from the four populations explained approximately 80–90% of the total variation in the first two axes. The populations of San Pablo and Saltillo are the most distant and are separated along the first eigenvector. In analyses of genetic markers alone, this separation along the first eigenvector is caused primarily by the MN system. Dispersion along the second eigenvector is most influenced by the RH system, with the cde gene complex affecting the plot for male and CDe influencing that for females. In an R-matrix plot based on digital pattern frequencies, the dispersion along the first axis is a result of the frequency of whorls versus ulnar loops. San Pablans are always distinct along the first axis because of their high frequency of whorls (Enciso, 1983).

Enciso also measured the amount of genetic differentiation among the Tlaxcaltecan populations, utilizing Harpending and Jenkins' R_{st}, which is comparable to Wright's F_{st}. The R_{st} values for males and females were 0.028 and 0.021 compared with the minimal values based on dermatoglyphics, which were 0.013 and 0.010, respectively. Thus, these quantitative traits appear to be more conservative and show less differen-

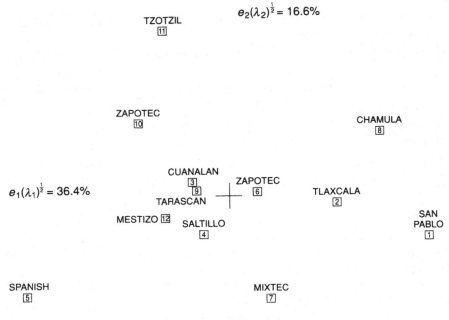

FIGURE 61 R-matrix plot of the first two principal scaled eigenvectors using digital pattern frequencies for males from 12 Mexican Indian populations (Enciso, 1983).

tiation than the simple Mendelian traits. In addition, the Tlaxcaltecan R_{st} values resemble those predicted by Cavalli-Sforza and Bodmer (1971) for six Mayan villages from Guatemala using migration matrices ($R_{st} = 0.013$).

Comparisons of Tlaxcaltecan populations with other Mexican groups are shown in Figs 61 and 62 from Enciso (1983). Figure 61 is an R-matrix plot of the first two scaled eigenvectors calculated from frequencies of digital pattern from males. With the addition of more populations, only 53% of the variance is explained by the first two eigenvectors. San Pablo is furthest from the Spanish population, having been separated along the first eigenvector. The second axis appears to reflect geography. The plot of the female subdivisions provides separation based on degree of Spanish admixture (see Fig. 62).

Enciso (1983) measured the correspondence of distances calculated from genetic and dermatoglyphic data. Digital-pattern frequencies yielded the highest correlations with the genetic data, with $r = 0.96$. Palmar variables and a mixture of 28 quantitative traits gave the lowest correlations of 0.68 and 0.62, respectively. The apparent strength of these correlations probably reflects the small number of distances in the matrices.

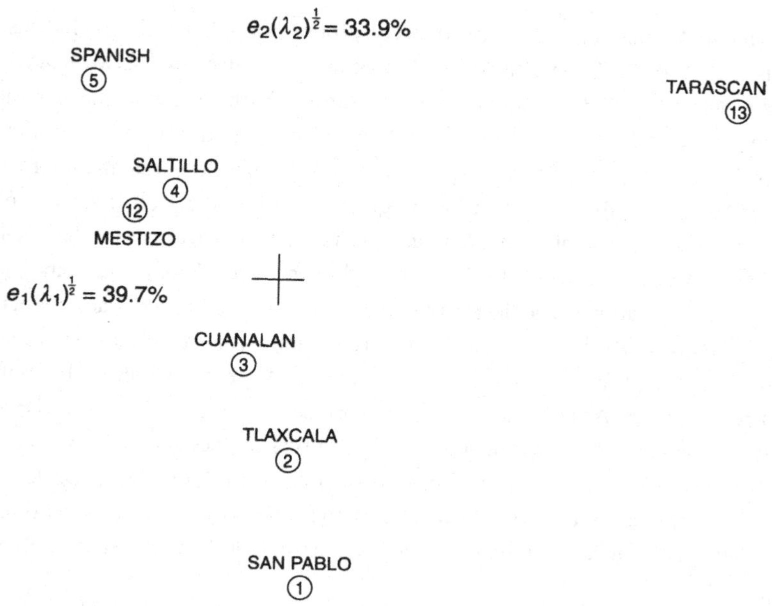

$$e_2(\lambda_2)^{\frac{1}{2}} = 33.9\%$$

SPANISH
⑤

TARASCAN
⑬

SALTILLO
⑫ ④
MESTIZO

$e_1(\lambda_1)^{\frac{1}{2}} = 39.7\%$ $+$

CUANALAN
③

TLAXCALA
②

SAN PABLO
①

FIGURE 62 R-matrix plot of the first two scaled eigenvectors using digital pattern frequencies for females from seven Mexican Indian populations (Enciso, 1983).

Calculation of the significance of these correlations using the Mantel tests yielded insignificant results (Duggirala and Crawford, 1994). In contrast, Neel *et al.* (1974) observed modest correlations between genetics and dermatoglyphics for a set of Yano-mama villages. They found intraclass correlations of 0.30 for genetics–dermatoglyphics but 0.08 for anthropometrics–dermatoglyphics.

South America The dermatoglyphics of South American Indian populations vary widely in frequencies of various traits both within and between groups. There are at least 25 major studies of the dermatoglyphics of South American Indian populations. Space does not permit me to list all of the different frequencies of traits such as finger patterns, total ridge counts, palmar patterns, and modal types of the C line. Instead, I shall refer to several extensive syntheses that summarize these data. This section focuses on studies that use dermal traits to provide insight into population structure and evolutionary process. Coope and Roberts (1971), Salzano and Callegari-Jacques (1988), and Garruto *et al.* (1979) have reviewed the dermatoglyphics literature for South America. Pena *et al.* (1972) described appreciable heterogeneity among the Ge-speaking Cayapo, Caingang, and Xavante populations. Similarly, Dennis *et al.*

(1978) documented significant differences between groups of the Mato Grosso Indians.

Rothhammer et al. (1973) examined the dermatoglyphic differences between seven Yanomama Indian villages. They noted a greater degree of among-population dermal variability in females than males, which may reflect female migration into the Yanomama villages. They further concluded that features distinguishing the Yanomama from other South American tribes include a low frequency of whorls, a high frequency of arches, and a low frequency of thenar/first interdigital palmar patterns (Rothhammer et al., 1973). Rothhammer et al. (1977) analyzed these same Yanomama data for differentiation of populations. The degree of concordance between genetics and dermatoglyphics varies with level of population differentiation. This association remains insignificant until the racial hierarchical level is attained. At the racial level, the rank correlation reaches 0.88. At the village level, the correlation is 0.16, and at the village cluster level it reaches 0.20. The tribal level of concordance goes up to 0.25. These results in part confirmed the findings of Neel et al. (1974) that, on the village level, the greatest congruence is found (0.34) when dermatoglyphic traits were of a quantitative nature and a generalized distance measure was used. Rothhammer et al. (1973) conclude

> 'The correspondence between marker gene and dermatoglyphic matrices increases with the order of differentiation. . . . it is difficult to postulate that differential selection has operated either on genetic marker gene systems or on dermatoglyphic traits' (p. 57).

6.5   DENTITION

Introduction

Teeth, because of their hardness, preservation and genetic etiology, have contributed much to the understanding of human evolution. The dentitions from members of early Asian and American Indian populations have provided useful information about the peopling of the New World and the subsequent differentiation of New World from Asian populations. Two basic approaches have been used to describe dental variation. These are: (1) the comparison of discrete dental morphological traits (odontology) in different populations (see Fig. 63 for scoring methods used in odontology); and (2) the use of dental measurements (odontometrics) to characterize populations. Both of these methods have been employed in studies of the peopling of the New World.

Utilizing dental measurements from six populations – Japanese, Ainu, Pima Indians, Australian Aborigines and Americans of European and African descent – Hanihara

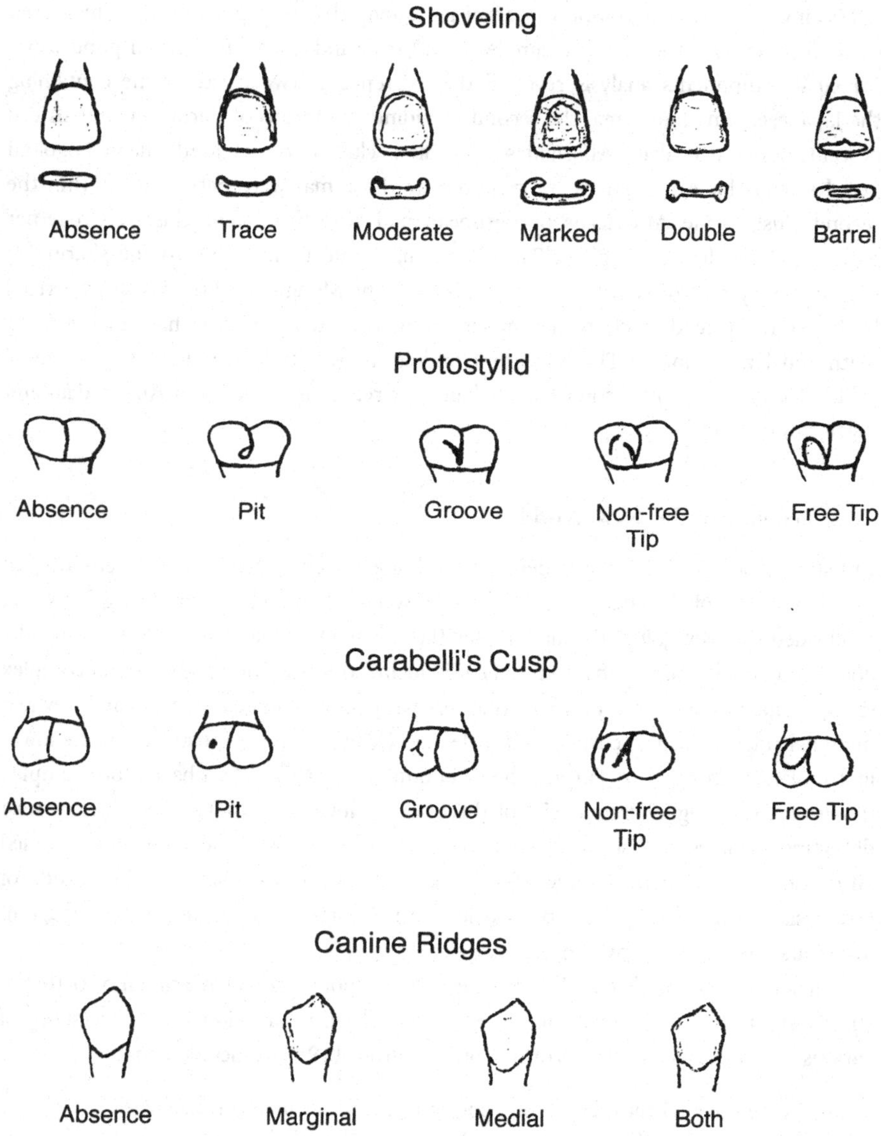

Shoveling

| Absence | Trace | Moderate | Marked | Double | Barrel |

Protostylid

| Absence | Pit | Groove | Non-free Tip | Free Tip |

Carabelli's Cusp

| Absence | Pit | Groove | Non-free Tip | Free Tip |

Canine Ridges

| Absence | Marginal | Medial | Both |

FIGURE 63 Scoring standards used for the discrete dental traits in the characterization of Tlaxcaltecan populations (Baume, 1976).

(1979) investigated the evolutionary affinities among the six populations. He measured mesiodistal crown diameters in samples of 20 male individuals from each population. Principal components analysis revealed the existence of two clusters, one containing the Japanese, Ainu and Pima, the second including Americans of European and African descent, and Australian Aborigines. The first cluster represented the Mongoloid peoples, and the Amerindians (represented by the Pima) were part of it. Within the second cluster, the Americans of European and African descent clustered together with the Australian Aborigines. This clustering is most likely due to the sharing of approximately 30% of genes through admixture. The Mongoloid cluster is characterized by larger front teeth and smaller molar teeth; the second cluster has smaller front teeth and larger molars. The important finding in this study is that, like the genetic traits, dental traits reflect the close phylogenetic relationship between Amerindian and Asian populations.

Peopling of the New World

Christy Turner (1987) utilized dental morphological characteristics in attempting to reconstruct the phylogenetic relationships between Asian and Amerindian groups. He subdivided the Mongoloid dental complex (first defined by Hanihara 1969) into sundadonts and sinodonts. Hanihara et al. (1975) included in the Mongoloid dental complex shovel-shaped upper central incisors (incisors with concave lingual surfaces and elevated marginal ridges), cusp 6, cusp 7, and deflecting wrinkles, and protostylids on the lower first molars. According to Turner, the sinodonts of Northeast Asia have more complex dentitions, with higher frequencies of three-rooted lower first molars, first molars with deflecting wrinkles, and shovel-shaped incisors (Turner, 1990). The sundadonts, named after the now submerged Sunda Shelf, occur in Southeast Asia and the islands of Indonesia and Melanesia. They possess less complex teeth with a higher percentage of lower first molars with two roots.

Turner (1987) stated that all of the New World populations have sinodont patterns, suggesting a Northeast Asian origin. However, there is variation in the incidence of various dental traits in New World groups. Turner (1987) concluded that

> 'Because dental microevolution appears to be rather uniform through-
> out the world, the dental differences between northeast Asians and
> American Indians can be used to calculate when these two populations
> separated. My estimate is that sinodonts first crossed over to Alaska
> not much before 15 000 years ago.'

Unfortunately, Turner failed to provide a means for estimating elapsed time by using distributions of dental traits. In another article (Turner, 1985), he attributed to

Smith a method for calculating a multivariate mean measure of divergence (MMD) of populations separated for a known period, which he divides by known time since separation to obtain a rate of dental microdifferentiation. Unfortunately no mathematical derivation of this method was provided so that the reader could more fully appreciate its elegance. This omission is particularly distressing since Smith's work is not listed in the bibliography. In order to set a dental clock in motion, evidence must be gathered that discrete traits evolve at constant rates. Quite to the contrary, dental evidence from Pleistocene remains suggests variable rates of evolution over longer periods (D. Frayer, personal communication). In addition, changes over an accurately determined age must be used to calibrate the temporal scale. I remain unconvinced that the dentition provides any evidence in support of a date of 15 000 years ago for the peopling of the New World.

The dentitions of all the indigenous populations of the New World have been lumped into the sinodont division of the Mongoloid complex. Turner (1987) argued for the existence of three New World clusters: (1) Aleut–Eskimo, characterized by high frequencies of three-rooted lower first molars and single-rooted upper first premolars and relatively low frequencies of incisor shoveling, and Carabelli's cusp; (2) most of the North American Indians and all the South American Indians were defined by Turner as a single 'homogeneous division, trait frequencies tend to be opposite of what is found in the Aleuts-Eskimo'; (3) Northwest Coast and Alaskan interior Indians, who are described as intermediate between the first two groups.

Turner offers three explanations for this variation: (1) a single migration followed by differentiation; (2) two migrations with the intermediate Northwest Coast – Alaska interior populations resulting from hybridization; (3) three separate migrations from Siberia. He rejected the first explanation by claiming that it requires that these groups in the New World be subjected to environmentally related selective pressures, and that the dental clusters do not correlate spatially with specific environments. The possible action of stochastic processes was ignored. Relethford (1991) has shown that quantitative traits are subject to genetic drift, much as are Mendelian traits, and has brought into question the belief that polygenic traits are more stable and somehow immune to the action of stochastic processes. Turner rejected the second alternative explanation because of the distribution of alleles of the ABO blood group system. He preferred the three separate migrations from Asia, not because of the dental evidence alone but because this alternative is supported by linguistic, archeological and genetic evidence. According to Zegura (1985), the genetic data by themselves do not indicate a three-migration origin. The archeological evidence clearly supports an early Indian and a late Eskimo migration, but there is insufficient evidence for a third migration. Greenberg's (1987) linguistic synthesis has also been under fire, with most scholars accepting the Eskimo–Aleut and Na-Dene families but strongly condemning his grouping of all other New World languages into a single Amerindian linguistic family. This so-called consensus on the three-wave peopling of the New World was crafted using a number of individual data sets, each leaning on the other and each unable to stand careful scrutiny on its own merit.

Geographic variation

Arctic and Subarctic populations

One of the more complete dental studies of a Subarctic group was done by Moorrees, who was part of an expedition to the Aleutian Islands in 1948. He published a volume devoted to Aleut dentition, which includes a description of discrete traits and odontometrics in a sample of 94 individuals (Moorrees, 1957). He scored the usual discrete traits such as shovel-shaped incisors and Carabelli's cusp. As among other Asians, the occurrence of shoveling of the maxillary incisors was high in the sample of Aleuts. The proportion of Aleuts with a pronounced shovel shape of the central incisors was higher (63%) than that found in an Eskimo population (38%) by Hrdlička (1920). Moorrees noted the faint presence of Carabelli's cusp in 8 out of 60 Aleuts (13.3%). This low frequency of Carabelli's cusp had been noted in Eskimo and Amerindian populations (Dahlberg, 1951; 1980). Moorrees also examined the Aleut dentition for an assortment of other discrete traits such as the number of cusps of the mandibular second premolar and maxillary molars, groove patterns on the mandibular molars and supernumerary teeth, and dental agenesis. In addition to calculating frequencies of discrete dental traits, Moorrees performed the standard measurements. He measured the mesiodistal length, buccolingual width, crown height and root length of Aleut teeth. Comparisons of tooth sizes of the Aleuts with other populations was of limited utility because multivariate statistics were not in use at that time. Moorrees did conclude that the Aleut odontometric data supported Butler's (1939) field theory in that the crown diameters are less variable in the mesial teeth when compared with the more distal teeth of each dental field. In spite of the statistical and computational limits of the day, Moorrees' study of Aleut dentition provided considerable information about population variability.

In an appendix to his article, Turner (1985) provided a summary of the frequencies of 27 discrete dental traits in 15 grouped samples. Included in these groupings are Eskimos, who exhibit some unique frequencies that distinguish them from other New World groups. For example, they have low frequencies of winged upper first incisors (UI1), and 3-rooted second upper molars (UM2). Cluster analysis revealed an early Eskimo branch coming off the New World tree, suggesting their dental uniqueness from other populations and their relative within-group similarity. Similarly, Matis and

Zwemer (1971), through stepwise discriminant analysis, demonstrated that Eskimos and Amerindians are dentally distinguishable between 90% and 100% of the time. However, the discrimination among Indian populations was successful only 60% of the time.

North America

In 1920, Hrdlička published his now classic study of shovelshaped incisors, showing that American Indians, Eskimos and Chinese all exhibit a high frequency of the shoveling trait (see Fig. 63). Africans and Europeans have lower frequencies of this trait. Sofaer et al. (1972) noted a high frequency of shoveling in the Southwest Amerindians, with some degree of palatal shoveling of the upper central incisor occuring in 97.3% of the Papago and 94.4% of the Pima. This form of shoveling is found in 85% of Asians and 17% of Europeans.

Other discrete dental traits have been used as indicators of affinities among populations. For example, the protostylid (a supernumerary cusp found on the interior portion of the lower molar) occurs in Amerindian and Asian populations (see Fig. 63). Dahlberg (1950) found population variation in this trait among the Pima Indians of Arizona. Similarly, Sofaer et al. (1972) found the frequencies of protostylids to vary from 0.1% in the Zuni to 7.8% in the Pima Indians.

Sofaer et al. (1972) utilized the dental morphology of Southwest Amerindian populations for purposes of deriving phylogenetic relationships. They scored dental casts of Papago, Pima and Zuni Indians for ten discrete traits. The traits scored included shoveling (both palatal and labial shoveling of the upper central incisor, and the presence of a barrel-shaped upper lateral incisor), Carabelli's cusp, number of cusps on the upper second molar, protostylid, and numbers of various grooves and cusps on the lower first and second molars (see Fig. 63). They found moderately good correspondence between known genetic differences and differences in dental morphology when traits whose scoring was least subjective were employed. They cautioned researchers who use discrete dental traits to carefully standardize scoring procedures lest the results reflect subjective biases more than meaningful phylogenetic relationships.

A follow-up study of discrete dental traits among a large sample of Pima Indians ($n = 1528$) compared the crown-trait frequencies with those of 13 other Southwest Indian samples. Scott et al. (1983) utilized the method of Harpending and Jenkins for reducing frequencies to maps of three dimensions. This analysis indicated that the Pima showed closest affinities to the Papago and Hopi (all three of these groups belong to the Uto-Aztecan linguistic family). The Pima differed most from the Athapaskans, Yuman and Zuni. This study demonstrated that the distribution of patterns of dental

Table 32. *The percentages of ten discrete dental traits in each of the four Tlaxcaltecan populations*

Trait	Affected state	Cuanalan	Saltillo	San Pablo	Tlaxcala
Shoveling	All degrees	87.9%	68.1%	82.5%	73.9%
Protostylid	All degrees	8.3	2.5	2.2	2.3
Carabelli's cusp (M^1)	All degrees	55.4	50.3	50.3	49.3
M^1 cusp no.	4 cusps	100.0	90.3	100.0	100.0
M^2 cusp no.	4 cusps	43.7	49.6	53.4	47.0
M^1 cusp pat.	Y-cusp	80.9	80.0	96.3	97.7
M^2 cusp pat.	Y-cusp ·	16.2	30.5	73.7	53.5
Mand. canine	Lateral and central extension inc. diff. ht.	27.9	30.0	46.9	44.8
Max. canine ridges	Presence of any ridges	67.1	69.0	77.6	77.1
Mand. canine ridges	Presence of any ridges	75.5	72.7	85.7	93.1

Source: After Baume and Crawford (1978).

morphology of southwestern populations corresponded closely to linguistic subdivisions.

Central America

The Tlaxcaltecan Indian populations provided an opportunity to compare dental traits in groups of known historical and well-defined genetic relationships. Four of these Amerindian populations were examined for variation in discrete traits and odontometric measures. Table 32 summarizes the frequencies of ten different discrete traits in four Tlaxcaltecan populations.

There was considerable variation in the frequencies of these discrete dental traits in the four Tlaxcaltecan populations. However, there were no significant differences in distributions of discrete traits between the sexes. It is interesting to note that although San Pablo del Monte is the 'most Indian' of the four groups it does not exhibit the highest incidence of shoveling. That distinction goes to Cuanalan. There is no form of shoveling in 22.5% of the residents of Saltillo. However, Cuanalan exhibited the highest incidence of Carabelli's cusp, considered a European marker, and protostylid. Comparisons of distributions of all ten traits in the four populations by means

Table 33. *Measures of divergence between pairs of*
Tlaxcaltecan populations

Populations	Sanghvi's X^2
Cuanalan–Saltillo	1.14
Cuanalan–San Pablo	4.91
Cuanalan–Tlaxcala	3.30
Saltillo–San Pablo	2.69
Saltillo–Tlaxcala	1.49
San Pablo–Tlaxcala	0.45

Source: Baume and Crawford (1978).

of the Kruskal–Wallis test indicated that the frequencies of all but two of the traits were significantly different (Baume and Crawford, 1978). Saltillo, on the basis of the lower frequency of dental shoveling, is most widely divergent from the other groups.

A comparison of genetic distances, measured on the basis of the frequencies of discrete dental traits, suggested that dentition provides similar patterns of affinities to genetic markers. San Pablo and Tlaxcala were most similar, with a X^2 measure of divergence of 0.45, while Cuanalan and San Pablo are most dissimilar with a distance of 4.91. Cuanalan and Tlaxcala show the next greatest difference, with a X^2 of 3.30. Table 33 summarizes the measures of divergence between the pairs of populations.

A study of dental-trait asymmetry (a statistical comparison of the presence or absence of specific discrete traits on the right and left sides of the mandible or maxilla) in the four Tlaxcaltecan populations showed that slight differences exist between the populations. Based on a comparison of average asymmetry, San Pablo, the least admixed of the four groups, had the lowest mean asymmetry (Baume and Crawford, 1981). On first thought, this finding suggested that the least admixed groups should be least asymmetric because there would be less disruption of the gene complexes that have evolved over thousands of years. Hybridization should break up some of these gene complexes and this would be evidenced by higher levels of asymmetry. However, this explanation was disproven when we compared the dental-trait asymmetry in Black Caribs and Creoles with those in the Tlaxcaltecan populations. The more highly admixed Caribs and Creoles exhibited much lower asymmetry than the Indian groups (Baume and Crawford, 1979). Apparently, for discrete dental traits, the degree of admixture is independent of the disruption of growth and development of these traits that leads to dental asymmetry.

Odontometric analyses of Tlaxcaltecan dental casts revealed considerable microdifferentiation among the four populations (Lin, 1976). However, this differentiation may be largely a function of the amount of admixture with Spaniards and, possibly, Africans. Dahlberg (1963) characterized Amerindian and African dentitions as being of large size. However, Hanihara (1979) asserted that the Mongoloid complex differed from the European–African one in that Amerindians and Asians have larger front teeth and smaller molars. The least admixed population, San Pablo, when compared with the other three Tlaxcaltecan populations, had intermediate measurements for most of the traits. Of the two transplanted populations, one had both the smallest and the other the largest dentitions (O'Rourke and Crawford, 1980).

Other studies of Central American Indians yielded frequencies of various discrete dental traits. Thus, Escobar et al. (1977) scored dental traits during intra-oral examinations of a sample of 540 Queckchi Maya from Guatemala. They utilized standard scoring techniques, but no interobserver scoring error was calculated. Dental casts from 114 Tarahumara Indians and 63 Mestizos of northern Mexico were collected and analyzed by Snyder et al. (1969). They compared the crown diameters for the Tarahumara and the Mestizos and found no clear-cut differences in size.

A recent study of dentition among four pre-Hispanic Mexican regional skeletal excavations demonstrated the existence of considerable variation in dental morphology (Haydenblit, 1996). The author refers to these four series (Tlatilco, 1300–800 BC; Cuicuilco, 800–100 BC; Monte Alban, 500 BC–AD 700; Cholula, AD 550–750) as 'populations'! Considering the time differentials and regional variation (e.g. Valleys of Puebla, Mexico versus Oaxaca), calling these dental collections populations somewhat stretches the concept of population beyond its broadest definition. However, this skeletal series does contain a solid sample size ($n = 209$) that were scored for 28 dental features. This study demonstrates the problems of studying dental variation on the one hand while on the other trying to pigeon-hole the results into vague morphological categories or patterns such as 'Sinodont' or 'Sundadont'. Haydenblit does point out that in these Mexican dental collections only 27% of the traits examined exhibit frequencies consistent with Sinodont variation.

South America

The frequencies and patterns of distribution of discrete dental traits have been studied in few South American Indian populations. Rothhammer et al. (1971) observed significant differences between males and females in the frequencies of shovel-shaped incisors in three mixed populations in Chile. Most likely, these differences were related to varying degrees of European admixture of the three Indian populations. Intervillage and intertribal variation in dental morphology was examined for seven Yanomama villages and one tribal sample of Makiritare (Brewer-Carias et al., 1976). Dental morphology was represented by eight standard, discrete traits. Some of the most marked variation between villages was in the frequency of shoveling, which ranged from 31% in one village to 97% in another. The frequency of hypocone reduction also varied significantly, from 18% to 68%, among villages. Some differences in the frequencies of discrete traits, such as winging of the incisors and shovel-shaped incisors, was noted when comparing Makiritare with Yanomama dentitions. The most interesting facet of this study was the comparison between village distance matrices for dental traits, genetic markers and geographic distance by Spearman rank-order correlation. The correlation between genetic and dental distance matrices was a fairly high 0.597. Distances based on dental traits fit better with geographic distances (0.492) than did genetic distances (0.310). This study showed that dental traits can provide reliable measures of population affinities. The measures of relationships of dental with genetic traits, reported by Brewer-Carias *et al.* for the Yanomama and Makiritare, are more reliable than those obtained by Sofaer *et al.* (1972) for Southwest Amerindian groups because of the larger number of populations compared.

A study of tooth size in prehistoric, coastal Peruvian groups concluded that teeth increased in size during a 9000-year period (Scott, 1979). This result is in contrast to the patterns of tooth reduction observed in most parts of the world. Examination of the data reveals that a sample size of 58 adults was used in reaching this conclusion. When this sample is further subdivided by four temporal periods, 2–3 teeth are the basis for such a trend. Using these 2–3 measurements Scott calculated not only average sizes but also standard deviations! To state that there is an increase in tooth size over 9000 years on that evidence stretches credulity to the limit.

6.6   SKIN COLOR

Introduction

Skin color is a highly visible human characteristic which has been traditionally used for racial classification. Early physical anthropologists tried to match standardized color tiles with shades of human skin. These measures of color arbitrarily divided the observed variation into discrete and artificial categories. Thus, the range of a continuous trait was subjectively subdivided, each subdivision being typified as the average color of some members of a race or by the percentages of different color matches within a population.

Skin color is a function of three pigments (Blum, 1961): (1) hemoglobin, which varies in color, depending upon whether it is in the veins (where it is reduced and burgundy wine in color) or the arteries (where it is oxygenated and red in color); (2)

melanin, a brown pigment found in granular form in the epidermis, where its concentration is the primary determinant of skin color; and (3) carotene, a yellow pigment located in the skin and the fat of an individual.

Various theories have been proposed to explain the wide range of variation in human skin color. These include Loomis' (1967) theory concerning pigmentation and the relationship between penetration of sunlight and vitamin-D synthesis. Loomis assumed that a balance exists between the amount of ambient sunlight and melanin concentration. Too much sunlight would cause hypervitaminosis D; insufficient sunlight would cause rickets and osteomalacia. Thus, the theory goes, natural selection operating through these conditions would result in the production of a skin-color gradient from the Equator (darkest skin) to the northerly regions (lightest skin). Although osteomalacia can reduce fertility by affecting the female pelvis, the flip side of this theory does not work. Hypervitaminosis occurs only when there is consumption of too much vitamin D, which can cause calcification of soft tissues. Simple exposure to sunlight fails to cause hypervitaminosis. Another popular explanation for the variation in skin color is Blum's (1961) theory that heavy melanin concentration protects individuals from skin cancer. It is true that individuals with darker skins have a lower risk of skin cancer; however, this form of cancer would be a poor selective agent because it generally develops after the age of reproduction. In addition, individuals with darker skins might be at a disadvantage in the tropics because of the heat stress associated with failure to reflect, as efficiently as light skin, the infrared portion of the spectrum.

In the early 1950s, with the development of reflectometers, a trait that previously could be treated only as discrete could be measured on a continuous scale (Lasker, 1954). The reflectometer permits the measure of the amount of light (of known wavelengths) reflected by an individual's skin. The lighter the skin, the greater the reflectance; the darker the skin, the less reflectance.

A number of investigators have tried to work our the genetic basis of skin color. For example, Curt Stern (1953) estimated that three to four additive loci were responsible for the skin-color variation in American Blacks. In a subsequent study of Black Caribs, Byard, Lees and Relethford (1984) inferred greater complexity in the genetics of skin color and estimated that from one to six loci were involved, depending on the number of phenotypic classes employed in the analysis.

Variation in New World populations

To date, little information has been published on skin-color reflectometry of American Indians. The only published EEL (British reflectometer) studies of Amerindians are those of Harrison and Salzano (1966), on the Caingang and Guarani Indians of Brazil, and Weiner *et al.* (1963), on the Aguarana Indians of Peru. These data are further limited in that the same filters were not used in both studies.

Conway and Baker (1972) published measurements, made with a Photovolt reflectometer, of four subgroups of Quechua Indians (males, females, urban and rural) of Peru. They found a range of reflectance among the Quechua similar to that observed in other Amerindians. For example, the mean upper-arm reflectance of adult males of Nunoa was 20.5% when a blue filter was used, 18.3% with a green filter, and 42.3% with a red filter. However, they noticed that Nunoa males were generally darker than females, possibly resulting from mate selection in favor of lighter female skin color. The urban residents were lighter than those individuals from rural areas (most likely a consequence of admixture with Spaniards).

Admixture and skin color

Lasker (1954) was one of the pioneers in the application of skin-color reflectance to questions of admixture. He found, in a Mexican village, that individuals with more Spanish ancestry were lighter than members of the more homogeneous Indian population. His work was followed by Relethford and Lees (1981), who utilized skin-color reflectance data (collected using a Photovolt reflectometer) to estimate admixture in several subpopulations of transplanted Tlaxcaltecan Indians of Saltillo. These investigators tested the correspondence of estimates of admixture based upon skin color with estimates based upon genetic markers. Both measures of admixture yielded similar results. Thus, estimates of admixture based upon polygenic traits such as skin color (which, in areas covered by clothing, remains relatively unchanged through adult life) appear to be reliable.

Similar estimates of admixture based upon skin color reflectometry were done for Black Carib populations in Belize. Byard *et al.* (1984a) compared mean reflectance curves of Europeans, Indians, Creoles, Caribs, Africans and Fali (an ethnic group from Africa). The Europeans reflected the highest percentage of light at all wavelengths (from almost 36% at 430 nm wavelength to almost 65% at 670 nm), and the Fali and other Africans reflected the least amount of light. The Fali reflected approximately 8% at 430 nm and 20% at 670 nm (see Fig. 64). The Creoles exhibited the highest coefficient of variation at all wavelengths, followed by the Black Caribs; the Europeans were the least variable. It is interesting to note that the greatest variation is found in the two most admixed populations, the Creole and the Black Caribs.

Estimates of admixture using skin-color reflectometry compare closely with those based on gene frequencies. The estimates based on these two data sets are compared in Table 34. The biracial model – one in which the founding population was made up of Africans and Indians only – indicates a high concordance of estimates of admixture

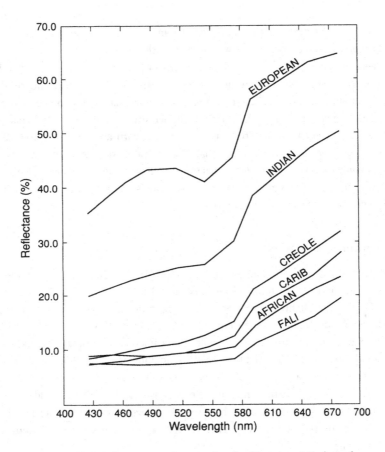

FIGURE 64 Mean reflectance curves comparing the skin color of Black Caribs and Creoles with African, European, and Indian populations (Byard *et al.*, 1984a).

Table 34. *A comparison of admixture estimates for Black Caribs of Belize based upon skin-color reflectometry and the frequencies of serological markers*

Method	% African	% Indian	% European
Skin color average biracial model	74	26	—
Genetic biracial	78	22	—
Skin color average triracial	73	19	8
Genetic triracial	76	20	4

Source: After Byard *et al.* (1984a).

based on skin-color reflectometry and genetic marker frequencies. Using skin-color, the triracial model slightly overestimates the European component at the expense of the African contribution to the Black Caribs. The Indian contribution to the Black Carib gene pool was of the same magnitude regardless of whether skin-color reflectometry or genetic markers were used (19–20%); the European component is estimated to be 8% and 4%, respectively.

6.7   CONGRUENCE BETWEEN MORPHOLOGY, GENETICS AND GEOGRAPHY

Table 35 summarizes the observed relationships between morphological traits, such as dermatoglyphics, odontometrics and anthropometrics, and non-biological factors, such as the geography, language and ethnohistory of Amerindian populations. With few exceptions, anthropometrics and geography provide relatively high correlations (see Table 35). The exceptions are the Caingang Indians ($r_s = -0.257$) and the Tlaxcaltecans ($r_s = 0.257$ for males and $r_s = 0.200$ for females). These correlations were based upon only four communities (Salzano et al., 1980a; Crawford et al., 1979). Considering the ecosensitivity of the anthropometric measurements, these high correlations were expected. The correlations between anthropometric and genetic distances are moderate to low, probably reflecting the effects of the environment on the concordances between the matrices. There is considerable variation in the magnitude of the correlations between dermatoglyphic and geographic distances (see Table 35). This variation is most probably the result of the 'noise' produced by the variety of dermal traits employed and the numbers of populations used in the comparisons. Overall, the correlations between dermatoglyphics and genetics are relatively high. Comparisons of the correlations between distances based upon various morphological traits provide a series of low correlations.

6.8   CONCLUSIONS

This chapter documents the high level of morphological variation observed in the Eskimos and Amerindians of the New World. There are at least three sources of this variation. First, there is the variation brought into the New World from Siberia by the founding populations. This variation is evident when contemporary Eskimos are compared with Amerindian groups. In some respects, the Eskimos have closer affinities to Siberian Asian populations than to Amerindian groups. Second, variation has developed through the fission and divergence of the founding populations into the wide array of contemporary populations with their diverse genetic and morphological charac-

Table 35. *Relationship between morphology, genetics and non-biological traits*

Values within parentheses are the number of groups studied; r, product-moment correlation coefficient; t, Kendall coefficient of rank correlation (tau); r_s, Spearman's coefficient of rank correlation; r_c, Correlation as measured by MATFIT.

Test of comparison	Population	Correlation	Reference
Anthropometrics and geography			
	Yanomama (19)	$r_s = 0.80$	Spielman (1973a,b)
	Caingang (4)	$r_s = -0.26$	Salzano et al. (1980a)
	Yanomama (7)	$r_s = 0.29$	Neel et al. (1974)
Anthropometrics and genetics			
	S.A. Indians	$t = 0.32$	Da Rocha et al. (1974)
	Yanomama (7)	$t = 0.30$	Neel et al. (1974)
	Caingang (4)	$r_s = 0.83$	Salzano et al. (1980a)
	Yanomama (19)	$r_s = 0.19$	Spielman (1973)
	Tlaxcaltecans (4)[a]		
	Male	$r_s = 0.26$	Crawford et al. (1979)
	Female	$r_s = 0.20$	Duggirala and Crawford (1994)
Dermatoglyphics and geography			
	Yanomama (7)	$t = 0.30$	Neel et al. (1974)
	Black Caribs (5)		Lin et al. (1984)
	Male	$r_c = 0.38$	
	Female	$r_c = 0.64$	
	S.A. Indians (8)	$t = 0.06$	Rothhammer et al. (1979)
	N.A. Indians (9)		Crawford and Duggirala (1992)
	Male	$r = 0.22$	
	Female	$r = 0.44$	
	Combined (10)	$r = 0.32$	
	Tlaxcaltecans (4)		Enciso (1983)
	Male		
	Dig. Pat.	$r_c = 0.65$	
	Dig. RCs	$r_c = 0.92$	
	Palmar	$r_c = 0.67$	
	28 traits	$r_c = 0.90$	

Table 35 *(cont.)*

Test of comparison	Population	Correlation	Reference
	Female		
	Dig. Pat.	$r_c = 0.97$	
	Dig. RCs	$r_c = 0.62$	
	Palmar	$r_c = 0.51$	
	28 traits	$r_c = 0.43$	
Dermatoglyphics and genetics			
	Eskimo (4)		Crawford and Duggirala
	Male	$r = 0.41$	(1992)
	Female	$r = 0.85$	
	Combined	$r = 0.55$	
	Yanomama (7)	$t = 0.34$	Neel *et al.* (1974)
	S.A. Indians (8)	$t = 0.25$	Rothhammer *et al.* (1979)
	N.A. Indians		Crawford and Duggirala
	Male (9)	$r = 0.47$	(1992)
	Female (9)	$r = 0.49$	
	Combined (10)	$r = 0.44$	
	Tlaxcaltecans (4)		Enciso (1983)
	Male		
	Dig. Pat.	$r_c = 0.86$	
	Dig. RCs	$r_c = 0.96$	
	Palmar	$r_c = 0.70$	
	28 traits	$r_c = 0.83$	
	Female		
	Dig. Pat.	$r_c = 0.92$	
	Dig. RCs	$r_c = 0.97$	
	Palmar	$r_c = 0.72$	
	28 traits	$r_c = 0.83$	
Dermatoglyphics and linguistics			
	S.A. Indians (8)	$t = 0.07$	Rothhammer *et al.* 1979
	N.A. Indians		Crawford and Duggirala
	Male (9)	$r = 0.22$	(1992)
	Female (9)	$r = 0.19$	
	Combined (10)	$r = 0.30$	

Table 35 (cont.)

Test of comparison	Population	Correlation	Reference
Odontometrics and geography			
	Yanomama (7)	$r_s = 0.49$	Brewer-Carias et al. (1976)
	Amerindians (8)		Harris and Nweeia (1980)
	Size	$r_s = 0.09$	
	Shape	$r_s = -0.11$	
Odontometrics and genetics			
	Yanomama (7)	$r_s = 0.60$	Brewer-Carias et al. (1976)
	S.W. Indians (3)	$t = 0.33$	Sofaer et al. (1972)
	Tlaxcaltecans[a]		
	Male (4)	$r_s = -0.43$	Crawford et al. (1979)
	Female (4)	$r_s = 0.35$	O'Rourke and Crawford (1980)
			Duggirala and Crawford (1994)
Pigmentation and genetics			
	Tlaxcaltecans		Relethford and Lees (1981)
	8 Populations	$r_s = 0.52$	
	7 Populations	$r_s = 0.96$	
Anthropometrics and dermatoglyphics			
	Yanomama (7)	$t = 0.08$	Neel et al. (1974)
	Tlaxcaltecans (4)[a]		Duggirala and Crawford (1994)
	Male	$r_s = 0.37$	
	Female	$r_s = 0.60$	
Anthropometrics and odontometrics			
	Tlaxcaltecans (4)[a]		Duggirala and Crawford (1994)
	Male	$r_s = -0.66$	
	Female	$r_s = -0.75$	

Table 35 *(cont.)*

Test of comparison	Population	Correlation	Reference
Dermatoglyphics and odontometrics			
	Tlaxcaltecans (4)[a]		Duggirala and Crawford
	Male	$r_s = -0.54$	(1994)
	Female	$r_s = -0.52$	

Dig. Pat.: Digital patterns = % of loops, whorls and arches per finger.
Dig. Rcs.: Digital ridge counts = ridge counts between triradii on fingers.
Palmar: Palmar ridge counts = number of ridges between triradii on palms.
[a] Correlations calculated by Duggirala and Crawford (1994).
Source: After Duggirala and Crawford (1994).

teristics. These are the characteristics that bear the 'made in America' label. Some of these traits may have been increased in frequency by selection, whereas others may have increased in frequency as a consequence of the founder effect. Newman's (1953) observations concerning the geographic (north–south) gradient of quantitative traits, such as stature, in Amerindians provided us with an excellent example of gene–environment interactions that occurred after the peopling of the New World. Third, variation also resulted from gene flow into Indian populations from European settlers and African slaves brought to the New World. The considerable amount of gene flow has confounded many of the pre-Columbian patterns of population affinities and differences among Native American populations.

Although morphological traits have been 'tainted' by their earlier associations with racial typologies, these polygenic traits have proved to be useful in studies of population structure (Blangero, 1990; Crawford, 1976; Williams-Blangero et al., 1990). For example, skin-color variation in hybridized populations has been successfully utilized to measure gene flow in Tarascan and Tlaxcaltecan Indian and Belizean Black Carib populations. Dental morphology has provided insight into the peopling of the Americas. Dermatoglyphic variation in Eskimo populations confirms the affinities predicted on the basis of genetics, ethnohistory and linguistics.

Much of the variation observed in complex, multifactorial traits is likely a consequence of gene–environment interactions. The statistical approaches used to dissect quantitative traits yield approximate results. Often these multifactorial traits are treated as 'black boxes', which eventually will be pried open by molecular genetics. The interface between molecular genetics and quantitative trait loci (QTLs) in humans is becoming a reality and will forever alter the use of morphological traits in the reconstruction of human phylogeny. The QTLs can be dissected by linkage analysis, allele-sharing methods, and association studies (Lander and Schork, 1994). Once a chromosomal region is located, the gene(s) responsible for the phenotype and the gene product can be identified by positional cloning. Through these steps the mechanism of the gene action that interacts with environmental factors can eventually be deduced. We are on the threshold of a new era with the capability of understanding the nature and mechanism of genetic–environmental interactions and their responsibility for the production of complex human phenotypes, such as dermal ridges, body morphology, dentition, and pigmentation.

7 The survivors

7.1 INTRODUCTION

In this final chapter the biomedical effects of acculturation and the accompanying changes in lifestyle of the Amerindians who survived the depopulation are considered. The interaction of culture change with the underlying genetics of native people provides a unique opportunity to assess and better understand the mechanisms involved in the etiology of complex diseases. In addition, research into unique Native American responses to various degenerative diseases may shed some light on the mechanisms of interactions between genetic and various environmental factors. The evolutionary consequences of increased genetic admixture between Amerindian populations (resulting from tribes with separate ethnohistories and geographic isolation being forced to coexist in newly created reservations) are also considered in this chapter. The massive infusion of genes from Africans and Europeans into Native American gene pool has undoubtedly altered the risks of diseases in admixed families.

The occupation of the Americas by peoples of Siberian ancestry for possibly 30 000–40 000 years was dramatically disrupted within the past 25 generations (500 years) by a violent collision between Old and New World cultures (see Crawford, 1992a,b). The consequences of this confrontation and the conquest that followed have been, and still are, devastating to the numerical balance that had existed between the environment and the Amerindian populations (see chapter 2). From a total Indian population perhaps approaching 44 million persons at contact, conquest and its numerous sequelae hammered the population down to fewer than 10 million at the nadir. Only recently has the surviving Amerindian population recovered and in some regions of the Americas, such as the Valley of Tlaxcala, begun to increase and perhaps even surpass its pre-Columbian levels (see chapter 2, Fig. 13). Superior arms, slavery to plantation and mine work, the persistence of Old World infectious disease imports, territorial aggression and the resulting deterioration of food economies reduced the North American Indian population to its lowest level at the turn of this century. In addition, forced acculturation (from the imposition of a foreign religion to the cultivation of land by former hunters and gatherers) and loss of traditional patterns of subsistence and culture had highly deleterious effects that threatened the very existence of the American Indian. Those Amerindian populations that survived the initial contact passed through a tight selective 'bottleneck' that must have altered their genetic constitution. The survivors were forced to lodge in the lowest socioeconomic stratum, and this deterioration in the quality of life was reflected in their elevated morbidity and mortality statistics.

7.2   THE EFFECTS OF RESERVATIONS

After conquest, the Amerindian survivors were segregated on reservations in the United States and in reserves in Canada. In the Americas, only the United States and Canada have reservations and a reserve system of allotting land to the natives. Brazil has its Xingu National Park, which harbors a variety of tropical forest tribes. Originally, the reservation and reserve systems were permitted neither autonomy nor self-rule, but outside rule, and decision-making by bureaucracies such as the BIA (Bureau of Indian Affairs) were imposed on the native peoples.

Marshall Newman (personal communication) noted that the United States and Canada have reservation–reserve systems, whereas those countries south of the border do not. He attributed this dichotomy to the definition of the term 'Indian'. The North Americans use essentially a biological definition (based primarily upon physical appearance), whereas Latin Americans utilize a cultural one (inability to speak Spanish or Portugese, failure to wear Western clothes, a lack of formal education). A cultural definition provides more vertical mobility than a biological one. Given some financial success, or some education, an Indian can enter another social category and become a Mestizo or Ladino. In those Latin American countries that had originally contained a large Indian population, almost everyone in the contemporary society has some Indian ancestry. Thus, if most nationals are partly Indian there is often no thought of either restitution or reservations.

Forcing the Amerindians to settle on reservations, thus taking them off their natural range, has had profound biological and social consequences. Often they were moved out of the ecosystems to which they had culturally and biologically become adapted, onto barren, unproductive tracts of land, and they were poorly supported by the whites' dole system. When physically active people are removed from their traditional lifestyles to a more sedentary existence, aboriginal food-procuring pursuits end, and the cheapest high-carbohydrate and saturated-fat foods are substituted. The stage is set for dramatic physical deterioration. The surviving American natives continued to be besieged by Old World diseases, the effects of which were further exacerbated by poverty and the absence of preventive medicine. Only in the past few decades has the Indian Health Service provided adequate reservation facilities and in some regions the Native Americans have organized their own health corporations (e.g. the Norton Sound Health Corporation) thus markedly reducing their own morbidity and mortality.

7.3   HERITAGE OF CONQUEST: DISEASE

Healthwise, the heritage of conquest includes an increased incidence of a series of degenerative conditions such as non-insulin-dependent diabetes mellitus (NIDDM), gall-bladder diseases, hypertension and cerebrovascular disease (stroke), coronary heart disease (CHD), and alcoholism. The late-onset diabetes (NIDDM), gallstones and gall-bladder cancer have been packaged together as the New World syndrome (Weiss *et al.*, 1984). In a volume on ethnicity and disease, Polednak (1989) listed the disease patterns that appear to be distinctive in contemporary Amerindians and Eskimos. He grouped the diseases into those with high and those with low prevalence (see Table 36).

This list of diseases in Amerindians and Eskimos contains a melange of persistent infectious conditions (such as tuberculosis, otitis media, plague, bacterial meningitis, hepatitis), together with cancers (melanoma, colon, kidney, cervical, stomach, naso-pharyngeal, breast, leukemia, lymphoma), and complex, polygenic diseases (hypertension, NIDDM, rheumatoid arthritis, IHD, congenital hip dysplasias, and alcoholism). This list reveals the distinctiveness of disease patterns in New World populations, particularly when compared to those from the Old World. Yet some diseases present at high frequencies in Native Americans can be traced to Asia where they are also prevalent. These diseases are: nasopharyngeal cancer in Eskimos (high incidence in several regions of China), alcoholism (highly prevalent in Siberian indigenous populations), and otitis media (high incidence in northerly Siberian groups). The prevalence of several diseases, such as tuberculosis, hepatitis, and alcoholism, is exacerbated by the poverty and living conditions experienced by the majority of contemporary Amerindians. In the case of tuberculosis, there appears to be some susceptibility in Amerindian populations despite its well-documented pre-Columbian presence in the New World (see chapter 2). It is worth noting that the New World populations are heterogeneous in their disease patterns, since regional, sexual, and ethnic differences exist. The regional variation in disease patterns can be explained by a combination of factors, such as living conditions, socioeconomic variability, dietary patterns, and genetic factors. For example, the recent hantavirus outbreak in the Southwestern Native American communities can be traced to the presence of animal hosts, and the living

Table 36. *A comparison of disease prevalences in contemporary Amerindians and Eskimos*

High frequency	Low frequency
Amerindians	
Plague (S.W. United States)	Duodenal ulcer
Otitis media	Melanoma of skin
Tuberculosis	Lung cancer
Congenital hip disorders	Hodgkin's disease (males)
Sudden Infant Death Respiratory	Colon cancer (S. & C. Amer.)
Syndrome (SIDRS)	Hypertension (isolated S. American
Alcoholism	groups)
Diabetes (NIDDM)	
Rheumatoid arthritis	
Chorioepithelioma	
Kidney cancer (Canada)	
Gall bladder/bile duct cancer	
Cervical cancer (Colombia, Chile, Central	
America)	
Stomach cancer	
Primary open-angle glaucoma	
Eskimos	
Iron-deficiency anemia	Ischemic heart disease (IHD)
Bacterial meningitis	Prostate cancer
Hepatitis B	Breast cancer
SIDS (Alaska)	Leukemia (males)
Salivary gland tumors	Lymphoma (males)
Esophageal cancer (Alaska)	Osteoarthritis
Nasopharyngeal cancer	

Source: After Polednak (1989).

conditions that made the transmission possible. Certain malignancies and cancers, involving the cervix, esophagus, and kidneys, have particularly high prevalence in specific geographic regions of the Americas. For example, Table 36 indicates that the disease patterns observed in Inuit (Eskimo) populations are distinct from the diseases listed for the Amerindians.

New World syndrome

In some Amerindian societies the consumption of foods rich in carbohydrates and lard, plus prolonged inactivity due to the loss of traditional food procurement practices, has led to severe obesity, often beginning in early childhood. That is not to say that all Indians are fat, but during the past three generations there has been a tendency towards obesity, particularly in a number of Southwestern tribes (Price *et al.*, 1993). Hall *et al.* (1992) documented an increase in obesity among the Navajo from 18.9% of the population suffering excessive fat during the 1955–1961 period to 43.7% obesity observed during a health survey in 1988. In this health survey of the Navajo the prevalence of non-insulin-dependent diabetes mellitus (NIDDM) jumped from 1.2% to 12.4%. Obesity reduces the desire as well as the ability to exercise and this sedentism further compounds the health problems. Clinically, obesity is often associated with late-onset diabetes, gallstones and cancer of the bladder, and with cardiovascular disease. The appearance of this constellation of diseases has prompted Weiss *et al.* (1984) to propose the existence of a New World syndrome in those tribes of Amerindians who have experienced major dietary changes from the traditional one of high protein and low carbohydrate to one of high refined carbohydrates and high saturated fats. This New World syndrome consists of: late-onset diabetes (NIDDM), gall bladder diseases, and cardiovascular disease.

Diabetes mellitus

Kirk *et al.* (1985:119) describe diabetes mellitus as a heterogeneous disorder in which all of the affected phenotypes 'share an inability to metabolize glucose normally'. The clinical severity of diabetes is a function of the degree of this inability. Initially, at least two forms of this illness were recognized: (1) insulin-dependent diabetes mellitus (IDDM); and (2) non-insulin-dependent diabetes mellitus (NIDDM). Later, the World Health Organization (1980) and then the National Diabetes Data Group (1986) adopted a more rigorous set of criteria and approved a new terminology for diabetes mellitus. They classified diabetes into four categories: (1) insulin-dependent diabetes mellitus (IDDM), caused by the failure of the beta cells of the pancreas to produce insulin, also called type-1 diabetes; (2) non-insulin-dependent diabetes mellitus (NIDDM), also termed type-2 diabetes, characterized by excess concentrations of insulin in the blood; (3) gestational diabetes (GDM), a category of this disease (probably heterogeneous in etiology) that occurs during pregnancy and causes a high risk of fetal and neonatal clinical complications; and (4) diabetes associated with or caused by other conditions, such as pancreatic diseases, drugs and chemicals. Diabetes is most accurately diagnosed by overnight fasting, followed by an oral ingestion of 75 g of glucose, thus 'loading' the glucose metabolism of the suspected diabetic. Multiple measures of plasma glucose (mM) at several intervals after loading usually reveals diabetes through the persistence of highly elevated plasma concentrations of glucose. A plasma concentration of glucose in excess of 11.1 mM two hours after the oral ingestion of the glucose load is the primary WHO criterion for diabetes mellitus. Classic symptoms of clinical diabetes include excessive thirst, urination, and hunger with weight loss (Szathmary, 1994).

The association between upper-body fatness and risk of NIDDM has been known for almost 40 years (Vague, 1956). A recent study of Mexican American children with a diabetic proband indicates the prevalence of excessive obesity (defined by a body-mass index, BMI, greater than or equal to the ninety-fifth percentile of a reference population) two or three times that of white children of comparable age and sex (Chakraborty et al., 1993). These higher rates of obesity of children are predictive of a future excessive upper body mass and higher risk of NIDDM in adult Mexican Americans. In fact, diabetes mellitus is the fifth leading cause of death for 65 year old and older American Indians and Alaskan Natives (Indian Health Service, 1991). The death rate for all forms of diabetes combined is 288.6 per 100 000 population, three times the rate observed in all US ethnicities combined (95.1 per 100 000). These data suggest that Amerindians have a special predisposition to diabetes mellitus.

Shepard and Rode (1996) reviewed the prevalence of diabetes mellitus in circumpolar native populations and found a trend similar to that described for the more southerly native populations. Accompanying the nutritional changes observed in some subarctic populations is an increase in obesity and NIDDM. The Yupik-speaking Eskimos have the lowest prevalence of diabetes in the state of Alaska, whereas the coastal Amerindian groups exhibited the highest (Schraer et al., 1988). The overall age-adjusted prevalence of diabetes in Alaskan natives is 1.36%, compared with 2.47% for the US as a whole. Similarly, the Canadian Inuit exhibit a much lower prevalence of diabetes mellitus than the Cree and Ojibwa Indians. In all cases diabetes is age-dependent and is highly correlated with body mass and obesity.

Thrifty gene hypothesis

Neel (1962) first proposed a 'thrifty gene' hypothesis to explain the persistence of high prevalence of late-onset diabetes, despite its debilitating effects in southwestern Amerindians. Neel argued that under feast or famine conditions associated with hunting and gathering societies the diabetic genotype must have confered a selective advantage in order for it to survive. During seasonal feast conditions, individuals possessing the thrifty gene secrete insulin rapidly in response to elevated blood glucose levels. This 'quick insulin trigger' facilitates the storage of glucose in the form of triglycerides in fat cells. Consequently, during periods of famine these individuals would metabolize their fat reserves in order to generate energy and consequently have a higher likelihood of survival than their slimmer counterparts. With agriculture and an ample food supply, persons with the quick insulin trigger would secrete excess insulin, which would eventually exhaust the beta cells of the pancreas, resulting in late onset diabetes.

In 1982 Neel updated his 'thrifty gene' hypothesis by suggesting that the overstimulation of the beta cells of the pancreas resulted in a decrease of the number of receptors on target cells, thus inhibiting glucose transport into peripheral tissues. This mechanism would eventually result in obesity, hyperinsulinemia, and hyperglycemia, similar to the form of NIDDM observed among the Pima Indians.

A comparison between Pima Indians of Arizona (with a more 'affluent' or western lifestyle) and Pima Indians of Mexico (living a more traditional lifestyle) indicates that NIDDM and obesity are highly infrequent among the Mexican Pima (Ravussin et al., 1994). Despite the shared genetic predisposition, the differences in lifestyle (i.e. the environmental factors) protect against the development of obesity and NIDDM. Here is an example of a genetic predisposition for a complex disease that, because of environmental conditions, is not expressed phenotypically.

Szathmary (1990) noted that the thrifty gene model was developed under the assumption that carbohydrate intake exceeded daily energy requirements. Judging from contemporary nutritional patterns, the Siberian migrants into the New World must have had a diet that was predominantly meat and fat, with few calories coming from carbohydrates. Under such conditions, selection would favor individuals with enhanced gluconeogenesis (production of glucose from amino-acid precursors) and free fatty acid release (Young, 1993). Szathmary (1987) points out that the majority of Amerindian populations studied to date fail to exhibit the fasting hyperinsulinemia that is predicted by the thrifty gene hypothesis as seen among the Pima (Szathmary, 1986b). She proposes the exploration of mechanisms other than a thrifty gene to explain the survival of hunters in nutritionally demanding environments.

Gall-bladder bisorders

Weiss (1985) has further developed the gall-bladder cancer component of the New World syndrome as an example of a 'late-stage phenotypic amplification product' of exposure to recent environmental change. This concept of phenotypic amplification can be defined as the exaggeration or amplification of normal genetic variants into chronic diseases under new environmental conditions. Weiss goes on to posit that some elements of the diet interact with the genetic substrate present in certain New

World peoples. He bolsters these arguments with the fact that gall-bladder cancer is 2–6 times as frequent in Amerindians as in other United States populations.

Do the Amerindians have a genetic predisposition for this New World syndrome, or is this complex of traits simply an organismic, physiological response to severe obesity and high lipid concentrations in the blood? When gall-bladder disease rates are compared between Indian, Hispanic, and Anglo populations, the observed incidence among the Hispanics is intermediate (Weiss, 1985). One interpretation of these data is that the Hispanic gene pool contains a proportion of genes (possibly up to 40%) that are at risk for gall-bladder disease, therefore an intermediate incidence of the disease is observed. However, this evidence can also be explained entirely by cultural or physiological factors. The incidence of this disease may represent that proportion of individuals or families within a population who follow particular dietary patterns that may result in obesity. Hence, the higher incidence of gall-bladder disease, high cholesterol concentrations and diabetes in Amerindians and Hispanics may reflect lifestyle rather than a unique genetic predisposition. If a genetic predisposition does exist, it cuts across populations and is shared by most members of our species.

Siberian indigenous populations

Because the Amerindians originated from Asia and shared common ancestry, could the genetic predisposition to gall-bladder disease and diabetes mellitus in Amerindians be traced in contemporary Siberians? Do Siberians experiencing similar acculturation exhibit the New World syndrome? Unfortunately, reliable statistical information on the incidence of gall-bladder stones and cancer, and of diabetes, in Siberian indigenous groups does not exist. Nikitin (1985) attempted to summarize the health problems of indigenous populations compared with Russian migrants to Siberia and noted a higher incidence of heart disease and hypertension among the tundra-dwelling Chukchi, who subsist on European diets (high sugar and saturated fats), when compared with coastal Chukchi and Eskimos, who follow traditional dietary patterns. He also observed a lower incidence of diabetes in the Evenks, Yakuts and Chukchi when compared with those living in Russia. A higher incidence of impaired glucose tolerance was reported in the indigenous population of Chukotka (5%), in contrast to 16% in Russian migrants (Astakhova et al., 1991; Stepanova and Shubnikov, 1991). Although Nikitin (1985) reviews the prevalence of cancers observed among Siberians, there is no mention of gall-bladder cancer as a major problem. Discussions with the physicians who accompanied our research team to the Evenki brigades convinced me that (based upon their clinical experiences) gall-bladder disease and diabetes are infrequent phenomena among native Central Siberian populations. The few morbidity records that I have been able to examine further supported the apparent absence of the New World disease syndrome in Siberia. This absence of the syndrome is most likely the result of the comparatively high-protein diet and low intake of carbohydrates when contrasted with the Pima Indians of the American Southwest.

Coronary heart disease (CHD)

Mortality from heart disease

The leading causes of death in American Indians and Alaska Natives (1986–1988) continue to be diseases of the heart. This high death rate persists in Native American populations despite the dramatic decreases observed in CHD mortality in the United States as a whole in recent years. The death rate provided by the Indian Health Service is 238.7 per 100 000 population. This rate is slightly lower than the death rate for 'all races' (ethnicities) in the United States, who experience 270.9 deaths from heart disease per 100 000 population. However, cardiovascular diseases are not the leading cause of death in all US ethnicities; this dubious distinction goes to malignant neoplasms, which kill 301.7 persons per 100 000. Gillum (1988) attributes more than one in four deaths of American Indians to cardiovascular disease. Regional or tribal variation in cardiovascular disease has been recognized, with the highest rates among Northern Plains Indians (Coulehan and Welty, 1990). Such high mortality rates from cardiovascular disease of Amerindians are not surprising when the synergistic effects of risk factors such as obesity, hypercholesterolemia, inadequate medical care and insufficient physical activity are considered. Gillum (1988) reports that Native Americans who smoke also have high levels of abdominal obesity and high plasma fibrinogen, but low to moderate mean serum cholesterol concentrations. This combination of high fibrinogen, obesity and smoking appears to increase the risk of cardiovascular disease. There is a strong association between cigarette smoking and atherosclerosis (hardening of the inner lining of the arteries, resulting from an accumulation of cholesterol) because smoking affects platelet function, increases the risk of thrombosis, and increases oxidation of lipoproteins (US Department of Health and Human Services, 1994).

An extensive body of literature links the observed variation in lipid concentrations in the plasma to risk of death from coronary heart disease (CHD). High concentrations of total cholesterol (TC), low-density lipoprotein cholesterol (LDL-C) and triglycerides (TG) apparently increase the risk of CHD. In contrast, a negative correlation has been noted between CHD and high-density lipoprotein cholesterol (HDL-C). LDLs deliver cholesterol and its metabolites to cells by binding to specific receptors on the cell surface. High concentrations of LDL are implicated in the development of atherosclerosis (Badimon *et al.*, 1992). HDLs, produced in the liver and small intestines, play a scavenger role in transport of excess free cholesterol from peripheral tissues, including arterial walls, back to the liver, where it is degraded (Katch and McArdle, 1993). There are a number of covariates associated with concentrations of lipids: age, sex, lifestyle, diet, socioeconomic factors, smoking, alcohol consumption, activity patterns, and psychosocial attributes (Duggirala, 1995). For example, the onset of heart disease in women relative to men is delayed by approximately 10 years (US Department of Health and Human Services, 1994). There is a beneficial effect of estrogen on lipid concentrations in premenopausal women (Miller, 1994). Diets high in saturated fat raise the serum concentrations of LDL (Ginsberg et al., 1995).

There is considerable variation in lipid profiles among contemporary Amerindian populations (Harris-Hooker and Sanford, 1994). Some native American groups have higher HDL-C concentrations, coupled with lower LDL-C in comparison with the general US population. Mancilha-Carvalho and Crews (1990) described the lipid concentrations in unacculturated Yanomama Indians, with traditional slash-and-burn agriculturalist diets. They practice no animal husbandry and consume lean meat caught by hunting. The Yanomama display low total cholesterol and triglycerides. However, Mexican Americans display TC and LDL-C concentrations similar to those of European Americans. Bang et al. (1980) attributed the low LDL-C concentrations among Inuit groups to the dietary intake of fish oils, containing omega-3 polyunsaturated fats (Bulliyya et al., 1990).

Apolipoproteins are components of plasma lipoprotein particles that play a significant role in the secretion, processing and catabolism of lipoproteins (Kamboh et al., 1996). Several apolipoprotein mutations in US and European populations appear to influence plasma concentrations of cholesterol and lipoproteins, leading to lipid disorders (Farese et al., 1992). Kamboh and his associates (1990, 1991, 1996) have investigated the relationship between three apolipoprotein genes (APOE, APOH and APOA4) and their relationship with plasma quantitative traits associated with CHD in Amerindian and Siberian populations. APOE is a ligand for two cell-surface receptors: low-density lipoprotein receptor (LDL) and LDL-related protein receptor (LRP). It mediates the uptake by cells of apoE-containing lipoprotein particles. APOH is believed to be involved in triglyceride metabolism, but its precise function has not been determined (Davignon et al., 1988). APOA4 apparently plays an important role in the metabolism of HDL and triglyceride-rich lipoprotein particles (Goldberg et al., 1990). Crews et al. (1993) found among the Yanomama two of the three common alleles observed in European populations at the APOE locus. APOE*3 was present at a frequency of 0.844 and APOE*4 had a frequency of 0.156. These frequencies for the Yanomama were close to those observed among the Evenki: APOE*3 = 0.843 and APOE*4 = 0.153 (Kamboh et al., 1996). One Evenki heterozygote exhibited a 2–4 genotype (i.e. had one APOE*2 gene and one APOE*4 gene), suggestive of admixture with a Russian settler. In contrast to the absence of variation at the APOH locus of the Yanomama, the Evenki are highly polymorphic and exhibit frequencies of APOH*1 = 0.013; APOH*2 = 0.788; and APOH*3 = 0.199. In comparison to Europeans, the Evenki have a significantly lower frequency of the APOH*1 allele and a higher frequency of the APOH*3 allele. The Yanomama exhibit unique gene and genotypic frequencies at the APO-A4 locus, such as the absence of the APO-A4*2 allele. In addition, unlike in European populations where the APOE*4 allele is associated with higher cholesterol levels, among the Yanomama and the Evenki this association does not exist (Crews et al., 1993; Kamboh et al., 1996). Similarly, in the Evenki there is no association between any of the APOE genotypes and lipid concentrations. However, at the APO-A4 locus, the HincII polymorphism at codon 127 showed a significant impact on triglyceride variation in the Evenki sample (Kamboh et al., 1996).

Evenki lipid concentrations levels and lifestyle

Recent US–Canadian–Russian series of expeditions to the Evenki reindeer herders of Central Siberia have afforded an opportunity to investigate the relationship between the nutritional patterns of indigenous groups living under diverse ecological conditions and their subsequent serum lipid concentrations (Leonard et al., 1992). The Evenki are a Tungusic-speaking group distributed widely over the boreal forest (taiga) of Central to Eastern Siberia. They were once reindeer hunters who, over time, adopted a breeding and herding subsistence (Vasilevich, 1946; Okladnikov, 1964). During Stalinist times, they were forcibly reorganized into cooperative settlements and herding groups called brigades. We compared the Evenki living in brigades with those residing in villages for a number of traits: age, stature, weight, BMI (body-mass index), sum of skinfolds, total cholesterol, high-density and low-density lipoproteins and triglyceride concentrations. Although the male village-dwellers were slightly younger, they exhibited more body fat, greater weight and significantly higher concentrations of total cholesterol and low-density lipoproteins (Leonard et al., 1994a). The female comparison, in part suffering from smaller samples, failed to show statistically significant higher concentrations of the lipids in village residents. A multiple regression analysis of predictors of total cholesterol and LDL cholesterol among males revealed that weight, age, and residence were significant predictors for total cholesterol; all of the same variables except weight predicted the concentrations of LDL cholesterol. Differences in activity patterns explain some of the variation in lipid concentrations between Evenki living in the brigades and those living in villages. In the brigades, men do most of the herding while women are responsible for maintaining the camp. Thus, male activity patterns differ markedly with location of residence, whereas the women's activities are comparable in both brigades and villages. This gender difference in activity patterns provides an explanation for why residence location is a significant predictor of cholesterol levels in Evenki males but not in Evenki females (Leonard et al., 1994a). Here is an excellent example of the reduction in activity resulting from acculturation affecting a phenotypic parameter such as lipid concentrations (Katzmarzyk et al., 1994a, b).

The Evenki herders of Siberia display remarkably low cholesterol and triglyceride concentrations for a population that derives much of its diet from animal foods. Their total cholesterol and triglyceride concentrations are among the lowest observed in human populations, with the exception of the pastoralist Masai of Kenya who have cholesterol concentrations of 3.49 ± 0.88 mM versus the Evenki 3.62 ± 0.67 (Leonard et al., 1994a; Biss et al., 1971). Dietary factors may also contribute to the differences in serum lipid concentrations between brigade- and village-living individuals. Evenki residing in cooperative villages consume higher quantities of carbohydrates than do those living in the brigades.

Lipid concentrations in circumpolar populations

Shephard and Rode (1996) recently summarized the available information about lipid concentrations in circumpolar populations of Canada, the United States, and Russia. They stated that up to World War II the indigenous populations such as the Inuits exhibited low averages for serum cholesterol and low rates of hyperlipidemia. However, with acculturation and the accompanying nutritional changes a progressive increase in cholesterol and triglycerides was noted (Draper, 1976, 1980). This increase in lipid concentrations has been interpreted by Rode et al. (1995) as resulting from a reduction in physical activity and a decreased consumption of omega-3 fatty acids associated with eating less raw fish. Shephard and Rode (1996) concluded that high levels of physical activity, consumption of animal protein with a low fat content, and a high intake of omega-3 fatty acids protected the circumpolar inhabitants from heart disease.

Obesity, leptin and lifestyle

Recent identification of the obese (ob) gene mutation in inbred mouse strains provides a connection between genetics, diet, and the regulation of body fat (Friedman and Leibel, 1992). The gene product of the ob gene is leptin, a protein secreted by the cells of the fatty tissue that signals to the hypothalamus the size of fat stores and the necessity to control energy intake (Zhang et al., 1994; Hamilton et al., 1995). In both obese humans and mice, plasma leptin concentrations were highly correlated with body-mass index (BMI), measured as mass divided by the square of height/body length, respectively (Maffei et al., 1995). Unlike the inbred strains of mice, to date only one mutation has been detected in the leptin gene in humans. However, a strong positive correlation between serum leptin concentrations and the percentage of body fat in obese and lean human subjects was recently described (Considine et al., 1996). These authors concluded that the most obese persons were insensitive to endogenous leptin.

My recent collaborative research on leptin levels has shown considerable populational and sexual dimorphic differences between Mexican Americans and the Evenki (Comuzzie et al., 1997). Despite matching these two ethnic samples by age, sex and BMI, interpopulational differences in leptin concentrations were observed. Serum leptin concentrations were twice as high in Mexican American males as in Evenki males (2.57 ng ml^{-1} and 1.25 ng ml^{-1}, respectively) and eight times higher in Mexican American women when compare with Evenki women (10.16 ng ml^{-1} and 1.21 ng ml^{-1}, respectively). The statistical comparisons between ethnic groups, within the sexes were significant ($p = 0.002$ and $p = 0.001$). These extreme differences in leptin concentrations between the Evenki and Mexican Americans are most likely the result of diversity in dietary and activity patterns. The Evenki consume a relatively low-fat, non-Western diet (high in protein but low in fat) and maintain high levels of physical activity. Although no sexual dimorphism in leptin concentrations in Evenki was observed, the sexes differ considerably in BMI and physical activities. The Evenki males are physically more active and display lower body fat (Leonard et al., 1994a). Thus, populational concentrations of leptin may not be entirely driven by body fat concentrations but may also be affected by genetic factors.

Blood pressure and hypertension

Blood pressure is a measure of left ventricular contraction of the heart (systolic pressure) and the resistance of the arteries to the flow of blood (diastolic pressure). Systolic blood pressure (measured by a sphygmomanometer) above 160 mm of mercury (Hg) and/or diastolic pressure above 95 mmHg, by definition, signifies hypertension. Hypertension is classified as primary (or essential) or secondary depending on whether the underlying causes are known. For example, individuals with some forms of kidney disease that elevate blood pressure are diagnosed as secondary hypertensives. Essential hypertension affects approximately 20% of adults in the United States (Kim et al., 1995).

Variation in blood pressure is the product of both genetic and environmental factors (Miall, 1967; Feinleib et al., 1979). The genetic portion of the total variation (heritability) for blood pressure varies between 13 and 30% in the Black Carib population of St. Vincent Island (Hutchinson et al., 1984). Apparently, the genetic factors involved in essential hypertension operate through the renin–angiotensin system (RAS). This system regulates the sodium and fluid balance in the organism through several mechanisms: it (1) stimulates reabsorption of sodium; (2) stimulates aldosterone production by the adrenal glands; and (3) influences the hemodynamics and vascular tone (Ito *et al.*, 1995). Changes in the activity of the RAS system (i.e. variation in genes encoding renin, angiotensinogen, angiotensin-converting enzyme, and angiotensin receptors) may be implicated in the etiology of hypertension in human populations (Ito *et al.*, 1995).

Statistical connections between elevated blood pressure and stress have been made for a number of human populations (McGarvey and Baker, 1979; Ward *et al.*, 1979). Elevation of blood pressure has been attributed to an assortment of social and psychological variables (termed stressors), such as conflict between newly imposed and traditional societal norms, or the imposition of new roles and statuses caused by colonialization or subjugation (Scotch, 1963; Sever *et al.*, 1980; Ward *et al.*, 1979). Considering the social and psychological disruption due to the conquest of the New World and its sequelae, it may be useful to examine the variation in blood pressure among the Amerindian survivors.

Amerindian populations of the Americas are highly variable in their blood-pressure distributions. For example, Mayberry and Lindemann (1963) and DeStefano *et al.* (1979) observed similar blood-pressure patterns in Native North Americans and European Americans. Strotz and Shorr (1973) found that hypertension was elevated among the Papago adults, with a rate of 20% among 4440 persons tested. The Navajos of Arizona and New Mexico exhibit some elevation of systolic blood pressure as a function of age. The diastolic pressure is modified in adulthood and either remains constant or increases slightly (DeStefano *et al.*, 1979). In addition to age variation among the Navajo, some distinctive differences may be observed when comparing female and male blood-pressure patterns (see Fig. 65).

Crews and Mancilha-Carvalho (1994) examined the correlates of blood pressure in a relatively unacculturated Brazilian Indian tribe, the Yanomama of Northwestern Brazil. This population, with a diet low in salt and saturated fats, exhibits mean blood-pressure values significantly below those noted for non-isolated populations in both industrial and developing countries. Although the blood pressure (systolic mean 100.9 mmHg, s.d. 10.3; diastolic mean 67.8 mmHg, s.d. 8.5, is exceptionally low, there is a moderate to high association of body habitus with blood-pressure. Thus, age and sex are predictors of blood pressure variability among the Yanomami, despite their low mean systolic and diastolic blood pressures.

In indigenous circumpolar populations both the systolic and diastolic blood pressures were found to be uniformly low (Shephard and Rode, 1996). The Igloolik Inuit of Canada display low blood pressure except for the oldest age cohort (70–79 years of age), which has an elevated pressure (mean of 138 ± 13). Shephard and Rode compared the blood pressure of the Inuit with that exhibited by the Nganasan of the Tamyr Peninsula. Given the population fragmentation, severe alcoholism, and

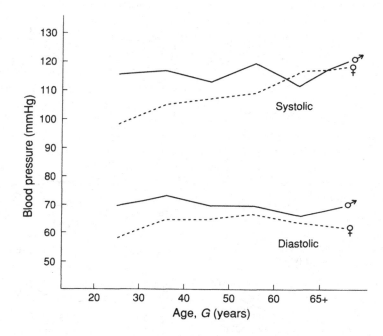

FIGURE 65 A plot of the relationship between systolic and diastolic blood pressure and age among Navajo males and females (DeStefano *et al.* (1979).).

acculturation, it is not surprising that the average blood pressures among the Ngansan were substantially higher than among the Igloolik Inuit. Despite the small Nganasan sample sizes, a substantial proportion appear to be hypertensive.

Hypertension is a major risk factor for coronary heart disease (CHD), cerebrovascular disease (stroke), and kidney failure (Sharlin *et al.*, 1994). In 1983, age-adjusted cardiovascular disease mortality rates for northern Plains Amerindians were 1.5 times higher than for the overall United States population. Similarly, the incidence of hypertension (13.9%) among a sample of the northern Plains Amerindians was also found to be higher than in the overall US population (Deprez *et al.*, 1985). Kraus (1954) reported hypertension morbidity rates among the Native peoples of the southwestern regions of the United States of 7 per 1000 for Papagos, 4 per 1000 for Pimas, and 2 per 1000 for Apaches. These Amerindian hypertension rates are many times lower than what is seen in European Americans, with estimates of 25 per 1000.

The blood-pressure distributions of the Black Caribs of St. Vincent Island, a triracial hybrid population, are similar to those exhibited by Afro-Americans from Muscogee County, Georgia (Hutchinson, 1983; Hutchinson and Crawford, 1981). Both the systolic and diastolic pressures increase as a function of age (see Fig. 66). This pattern

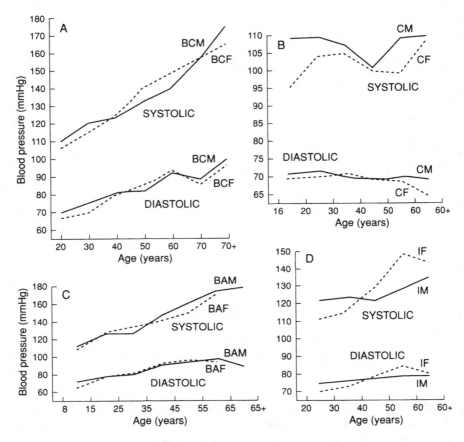

FIGURE 66 Blood-pressure distribution by age and sex compared for (A) the Black Caribs of St. Vincent; (B) the Carajas of Brazil; (C) Black Americans from Muscogee County, Georgia; and (D) the Ilora of Western Nigeria (Hutchinson and Crawford (1981).).

contrasts with the Carajas of Brazil, who exhibit no increase in diastolic blood pressure with age and only a small increase in systolic arterial pressure. The Ilora of Western Nigeria exhibit no increase in diastolic blood pressure as a function of age, but display a significant rise in systolic pressure in females. The relationship between hypertension and degree of individual African admixture was tested but found not to be statistically significant. Hutchinson (1986) tested the possible relationship between elevated blood pressure and an assortment of social variables in the Black Caribs of St. Vincent Island and found some statistically significant relationships. She noted that single individuals have higher blood pressures than married subjects, and that males who never attend church have higher blood pressure levels than their more religious counterparts. Among males, as the level of education (measured in numbers of years) increases, blood pressure is equally elevated. Hutchinson (1986) attributes these relationships between blood pressure and social variables to stress, associated with deviations from acceptable social patterns among the Black Caribs of St. Vincent Island (Hutchinson and Byard, 1987).

Alcoholism

Alcoholism has been described as one of the most common human diseases, with an incidence ranging between 3 and 5% among males and between 0.1 and 1% among females (Goodwin and Guze, 1989). It is a serious health and societal problem that plagues almost all human societies and particularly the Amerindians and their Siberian relatives. Alcoholism may be defined as the inability to consume and 'handle' alcohol. Hill (1978) points out that alcohol abuse is associated with almost 75% of all Indian arrests in the United States, and with an estimated 94% of suicides on reservations. Chronic liver disease and cirrhosis is the ninth leading cause of death among American Indians and Alaskan Natives (65 years of age and older) with a rate of 54.6 per 100 000 (Indian Health Service, 1991). This rate is significantly higher than the mortality from liver disease observed in all combined ethnic groups of the United States (33 per 100 000). The age-adjusted alcoholism death rates for American Indians and Alaskan Natives were 33.9 deaths per 100 000 population or 5.4 times the US 'all races' rate of 6.3 (Indian Health Service, 1991). The age-specific alcoholism death rates for Indian males were higher than the rates for females in all age cohorts except 15–24 years of age.

Although similar problems associated with alcoholism are acknowledged by Russian authorities about indigenous Siberian peoples, reliable statistical data on prevalence and mortality are unavailable. Interviews with the Evenki concerning causes of death revealed that almost every family had been touched by alcohol-related deaths. Numerous family members were remembered who froze in the taiga, drowned in a river (although only one small stream transects their territory), died from cirrhosis of the liver or met some other alcohol-related death. Homicide, particularly between family members and involving alcohol consumption, is extremely common. Returning to a settlement in Siberia after an absence of one year was a particularly sad experience for me because of the numerous cases of homicide, suicide and accidents that claimed the lives of friends and informants.

Avksentyuk et al. (1995) have reported some questionnaire-generated data on the drinking patterns of Chukchi and Eskimos of Chukotka, Siberia. No lifetime or current abstainers were self-reported in a sample of 139 men, and only 6% of the women (sample size 162) did not drink. Among the men 48% were regular lifetime drinkers and almost 9% of the women acknowledged being regular lifetime imbibers. Given the heterogenous cultures of the Eskimos and Chukchi it is difficult to assess the reliability of these summary figures. None the less, the average amount of alcohol consumed at a typical drinking occasion is 178 g, a hefty intake considering the relatively small body mass of the indigenous Siberian groups. More than 50% of the men reported alcohol-related symptoms such blackouts, loss of control, and morning hangovers that require a morning drink. These data support the subjective observations that alcohol abuse is common in Siberian populations.

To what extent is alcohol abuse genetic, and to what extent is this susceptibility the result of cultural fragmentation and the relative inexperience of native people in the consumption of distilled alcohol? As poignantly explained by native American writer Lame Deer (1992):

> 'I sure wish they'd teach us how to drink. When you buy a camera or
> a tape recorder, it always comes with a little booklet which tells you
> how to use it, but when they brought us the white man's whiskey,
> they forgot the instruction book. This has caused us no end of
> trouble.'

Certainly, cultural and psychological factors are important contributors to the style of drinking; for example, whether binge or steady drinking is the norm in a society is primarily a cultural phenomenon. Social pressures may be exerted on an individual to drink with a group through a complex series of motivations. For example, a 'time-out' theory has been proposed to explain intoxication; it provides an excuse to avoid social sanctions that would normally be applied to unacceptable behavior. Violence is often blamed not on the moral character of the perpetrator but on his or her drunk character (Hill, 1978). The consumption of alcohol unites adolescent male peers and also serves social functions associated with solidarity in adult society. The sociocultural factors that influence the drinking behavior of Amerindians, Siberians or members of any society are numerous and interrelate with poverty, unemployment, repressed aggression and boredom, and may even have religious connotations.

Given this complex behavior with multiple etiology, why should we believe that genetic factors contribute to the risk of alcoholism? Cannot all the behavioral variation associated with the abuse of alcohol be explained by societal problems or individual psychological difficulties and by the fact that alcohol serves as an escape from poverty and pain? What is the evidence for genetic involvement in alcoholism? Obviously, genetics alone does not cause a complex behavior such as alcoholism. Not everyone with a family history of alcohol abuse becomes an alcoholic (Devor, 1993). However, relatives of alcoholics have been shown to exhibit unique responses to certain stimuli.

A study by Wolff (1973) on the genetics of alcohol response, measured through changes in skin reflectance, observed a high incidence of intense facial flushing (more than 80%) in American Indians and Asians in response to drinking alcohol with levels adjusted by body mass. Only 5% of the Caucasians tested experienced a similar flushing response (Wolff, 1973). This reddening of the face was accompanied by an increase in blood pressure and the temperature of the cheek. Here was population variation in the physiological response to alcohol consumption. This finding was coupled with the research by Fenna et al. (1972) that had concluded that Indians and Eskimos require 40% longer to metabolize alcohol than does a white control group. This finding was disputed by Bennion and Li (1976), who found no differences in alcohol metabolism between American Indians and Americans of European ancestry when the dosages were adjusted for body mass.

Because the flushing responses in Asians and Amerindians could not be simply explained by differential alcohol metabolic rates between these populations, a number of studies focused on the enzymes responsible for the oxidative metabolism of ethanol. There are two enzymes found in multiple forms (alcohol dehydrogenase and aldehyde dehydrogenase) that are responsible for ethanol oxidation to acetate (Goldman, 1993). The alcohol is absorbed through the gastrointestinal tract and distributed in total body water (Bosron et al., 1988a). The rate of absorption and the distribution of the alcohol is dependent upon the diet, smoking, age, and chronic alcohol consumption (Reed, 1978). Most of the ethanol metabolism occurs in the liver through its oxidation to acetaldehyde by alcohol dehydrogenase, ADH (with the aid of other catalysts); the acetaldehyde is oxidized to acetate by aldehyde dehydrogenase, ALDH (Li, 1983). In humans, genetically determined isoenzymes of ADH have been identified with different kinetic properties capable of effecting the metabolism of ethanol (Reed et al., 1976; Bosron and Li., 1986). Electrophoretically, several alcohol dehydrogenase alleles can be identified (see Table 37). Although the genes associated with alcohol metabolism code for specific enzymes, there is considerable variation in their frequencies among human populations. The ADH*2 allele (which codes for the beta 2 subunit) varies in frequency among Asian populations (0.02 among the Chukchi to 0.73 among a Korean sample) and renders the enzyme more active (Thomasson et al., 1997). This form of ADH tends to produce higher acetaldehyde concentrations and may protect against heavy drinking. According to Thomasson et al. (1991) the ADH2*2 allele is lower in frequency in alcoholic Chinese than in non-alcoholic Chinese. This allele is low in frequency in northern indigenous Siberian populations and high in China, Japan and Korea, but is absent in the Native peoples of the Americas.

Some genetic variability has been described in human populations at the aldehyde dehydrogenase locus ALDH. The ALDH2*2 allele renders the mitochondrial ALDH2 form essentially inactive and is apparently responsible for the alcohol flush reaction.

Table 37. *Frequencies (%) of ADH alleles in human populations*

Populations	ADH2 locus			ADH3 locus	
	ADH2*1	ADH2*2	ADH2*3	ADH3*1	ADH3*3
White American	>95	<5	<5	50	50
White European	90	10	<5	60	40
Chinese, Japanese	35	65	<5	95	5
Black American	85	<5	15	85	15
Evenki[a]	90	7	3	—	—
Nivkh[a]	91	9	0	—	—
Buryats[a]	77	23	—	83	17
Altai[a]	99	<1	—	19	81
Chukchi[a]	100	0	—	—	—
Sioux[b]	100	0	—	—	—
Navajo[c]	100	0	—	—	—
Northwest Coast[d]	98	2	—	—	—
Southeast Alaska[d]	100	0	—	—	—

[a] Thomasson *et al.* (1997). [b] Rex *et al.* (1985). [c] Bosron *et al.* (1988b). [d] Chen *et al.* (1992). *Source*: After Bosron *et al.* (1988b) and Thomasson *et al.* (1997).

The elevated blood acetaldehyde concentrations not only causes the flushing response but also modifies the amount of ethanol that can be consumed. This allele is either absent or at low frequencies in Siberian and Native American populations, but is relatively frequent among Chinese and Japanese (Thomasson *et al.*, 1997). Apparently the ALDH2*2 allele provides a protective effect against the development of alcoholism in the Chinese and Atayal groups of Taiwan (Thomasson *et al.*, 1997). Novoradovsky *et al.* (1995) have recently reported a number of mutations in Native American populations, including a new allele ALDH2*3, caused by a silent mutation, which is found at a high frequency (0.04) among the Pima.

Because the physiological processes of alcohol metabolism and its molecular control are well documented, the behavioral phenotype of alcoholism or alcohol abuse provides an excellent candidate for understanding the interrelationships between society, behavior, environment, and genetics. Studies of adopted children of alcoholics also suggest that the susceptibility of individuals to alcholism is, in part, genetically controlled (Goodwin, 1987; Cloninger, 1987). Twin studies indicate a high concordance for alcoholism and the complications arising from excessive drinking such as cirrhosis of the liver. The observed variation in the metabolism of alcohol, together with cultural and psychological covariates and concentrations of serotonin (a neurotransmitter) all interact and contribute to the risk of alcoholism. Despite the major breakthroughs in the biology of alcohol-abuse phenotypes, the environmental components should not be overlooked. Culture defines the form of the drinking patterns, the availability of alcohol, and the type of alcohol abuse. In addition to the two most common forms described in western societies, Type I or Type II, Bau and Salzano (1995) have described a third form in Brazil that may be intermediate. Thus, the environmental components should be quantified as covariates of the phenotype instead of being viewed as a 'black box' defined statistically as unity minus the genetic component.

7.4   HYBRIDIZATION

One of the most dramatic results of the collision between the New and Old Worlds was the massive gene flow (genetic hybridization) that took place between the European settlers, the Amerindian natives and the African slaves. In addition, as survivors from various Amerindian tribes were forced to coexist on reservations, tribal genetic distinctions (often the result of thousands of years of reproductive isolation) began to blur. This intertribal genetic hybridization was documented by Hulse for the Hupa Indian reservation. Intergroup genetic variation has diminished and those tribes experiencing the longest period of acculturation share the most European and African genes and thus resemble each other genetically. The genetic results of this massive hybridization included the increase of genetic heterozygosity (as exemplified by the Black Caribs). Given an ever-changing environment, those populations containing the most genetic variation should have an evolutionary advantage. New combinations of genes are 'tested' by the environment, possibly resulting (through natural selection) in better-adapted populations. On the other hand, the enormous genetic bottleneck that the Amerindian populations experienced apparently reduced some of the genetic variation present at Contact. Undoubtedly, some genetic combinations that resulted from thousands of years of genetic adaptation to unique New World environments were lost as the selective forces were modified by the European conquerors and settlers.

A number of genetic epidemiological studies have been conducted on various admixed populations made up of parental groups with genetic predispositions to specific diseases. For example, a survey has been conducted on Mexican-American populations for the risk of non-insulin-dependent diabetes mellitus (NIDDM) (Chakraborty et al., 1986). A significant relation between the degree of Amerindian admixture and the prevalence of NIDDM was observed in the Mexican-American population of San Antonio, Texas. This finding further supports numerous studies that have argued for a genetic basis of the high incidence of NIDDM in Amerindians

(Neel, 1962; Weiss et al., 1984). However, a recent comparison of the three methods used by researchers for assessing Amerindian admixture in Mexican Americans (genealogical approach, skin-color reflectometry, and genetic markers) showed no association of NIDDM disease with degree of admixture in males. These three measures of admixture were also poorly correlated with each other. Mitchell et al. (1993) conclude that 'the three measures considered may assess different dimensions of admixture . . . .' Considering the high degree of African admixture reported for northern Mexico, it is surprising that Mitchell et al. used only a biracial model of admixture for the Mexican Americans of San Antonio (Crawford, 1976; Crawford et al., 1979). A more appropriate model should have been triracial and included African, Spanish and Amerindian parental populations.

Similarly, Hanis et al. (1986) examined the possible relationship between estimates of individual admixture and risk from chronic diseases such as diabetes and gall-bladder disease. The individual admixture estimates were based upon 16 blood-group and protein loci, and a biracial hybrid model was applied to a population of Mexican-Americans from Starr County, Texas. The individual estimates were unrelated to the probability of being diabetic and marginally related to gall-bladder disease. Hanis et al. concluded that the independent assortment of loci precludes the use of this method, unless the loci used for estimating admixture are either linked to the disease or involved in its etiology (such as DNA candidate genes). An alternative explanation is that the biracial admixture model that was imposed on a triracially hybridized population caused too much statistical noise for the detection of a significant pattern. Hutchinson and Crawford (1981) also failed to demonstrate a statistically significant relation between risk of hypertension and individual estimate of African admixture (grouped into quartiles).

7.5   EPILOG

The 'native' inhabitants of the Americas have been swept along by two turbulent demographic and evolutionary events which have sculpted their gene pools and forever altered their societies. The first of these events was the peopling of the New World, in which Asians expanded into the Americas. Most likely this expansion was fueled by population pressures that forced Siberian groups across Beringia into two unpopulated continents containing a vast assortment of ecological niches as varied as the tundra, tropical forests, and deserts of the Americas. These populations were reproductively and culturally isolated from the remainder of the world for more than 30 000 years. Some gene flow must have continued from the Old World at various times and in varying magnitudes. Following the earliest Asian settlers (Paleo-Indians) came the

Na-Dene-speakers, and finally the Eskimos–Aleuts. In addition, small trickles of Asian Eskimos continued to cross the Bering Strait and intermingle with related groups on St. Lawrence Island, the Diomedes and the Alaskan mainland. Thus, some genes continued to flow from the Old World and may have prevented the further genetic differentiation of the New World populations. Evolutionarily, this peopling of the Americas can be viewed as a natural 'experiment' with genes made in Asia being 'tested' by the geographic diversity and selective forces of the New World.

The second unique historical event was the collision of the two worlds that had evolved separately for thousands of years. This contact resulted in the initial reduction and genetic bottleneck of the indigenous populations. The populations of the Americas were drastically reduced by disease, warfare, and slavery until the extinction of some groups and the attainment of a population nadir towards the end of the nineteenth century. This population reduction has forever altered the genetics of the surviving groups, thus complicating any attempts at reconstructing the pre-Columbian genetic structure of most New World groups. Massive population movements from Europe and Africa followed the Conquest of the New World and created admixed populations of Afro-Americans and Hispanic Americans. From the evolutionary viewpoint, these new combinations of genes 'tested' by environments that differ from their place of origin resulted in some success stories and some failures. The Garifuna (Black Caribs) amalgam of genes and cultures allowed them to be highly successful in their colonization of coastal Central America. However, many other groups quietly and tragically disappeared without much fanfare.

Despite this tragic history of oppression and death, the surviving North Amerindians have rallied and numerically attained pre-Contact population levels. Apparently, the improvement of health facilities and medical care has stemmed the downward spiral of population decline and lowered the morbidity and mortality rates of native Americans. Currently, with the decreases in mortality and increases in fertility, the Amerindians are the fastest-growing ethnic group in the United States. With recent developments in native businesses, successful litigation against the United States government concerning past treaties and rights, and overall improvements in their economic conditions, the future for the natives of the Americas looks much brighter. It is hoped that world opinion will help the Amerindian enclaves in Latin America survive the massive intrusions and destruction of their rain forest. As a species, we will be poorer if we continue to lose the ever-dwindling human biological and cultural diversity: our evolutionary heritage.

References

Adovasio, J.M. & Carlisle, R.C. (1986). Pennsylvania pioneers. *Natural History*, 12, 20–7.

Alexseyeva, T.I., Alexseyev, V.P., Spitsyn, V.A., Krukovskaya, O.B., Boeva, S.B. & Irissova, O.V. (1978). Enzymes and other blood proteins and differentiation of the populations of Northeastern Asia (some results of geneticoanthropological investigations). *Voprosi Antropologii*, 58, 3–19 (in Russian).

Alexseyev, V.P. (1979). Anthropometry of Siberian peoples. In *The First Americans: Origins, Affinities, and Adaptation*, ed. W.S. Laughlin and A.B. Harper, pp. 57–80. New York: Gustav Fischer.

Allen, F.H., Diamond, L.K. & Niedziela, B. (1951). A new blood-group antigen. *Nature*, 167, 482.

Allison, M.J., Daniel, M. & Alejandro, P. (1974a). A radiographic approach to childhood illness in pre-Columbian inhabitants of southern Peru. *American Journal of Physical Anthropology*, 40, 409–16.

Allison, M.J., Gullermo, F. & Santoro, C. (1982). The pre-Columbian dog from Arica, Chile. *American Journal of Physical Anthropology*, 59, 299–304.

Allison, M., Hossaini, A., Munizaga, J. & Fung R. (1978). ABO blood groups in Chilean and Peruvian mummies. *American Review of Respiratory Diseases*, 44, 55–62.

Allison, M., Mendoza, D. & Pezzia, A. (1973). Documentation of a case of tuberculosis in pre-Columbian America. *American Review of Respiratory Diseases*, 107, 985–91.

Allison, M.J., Pezzia, A. & Gerszten, E. (1974b). Infectious diseases in pre-Columbian inhabitants of Peru. *American Journal of Physical Anthropology*, 41, 468.

Allison, M.J., Pezzia, A., Gerszten, E. & Mendoza, D. (1974c). A case of Carrion's disease associated with human sacrifice from the Huari culture of Southern Peru. *American Journal of Physical Anthropology*, 41, 295–300.

Allison, M.J., Pezzia, A., Hasegawa, I. & Gerszten, E. (1974d). A case of hookworm infestation in a precolumbian. *American Journal of Physical Anthropology*, 41, 103–6.

Alper, C. (1976). Inherited structural polymorphism in human C2: Evidence for genetic linkage between C2 and BF. *Journal of Experimental Medicine*, 144, 1111–15.

Alper, C.A., Boerusch, T. & Watson, L.J. (1972). Genetic polymorphism in human glycine-rich beta glycoprotein. *Journal of Experimental Medicine*, 135, 68–80.

Altland, K., Bucher, R., Kim, T.W., Busch, H., Brockelmann, C. & Goedde, H.W. (1969). Population genetics studies on pseudocholinesterase polymorphism in Germany, Czechoslovakia, Finland and among Lapps. *Humangenetik*, 8, 158–61.

Alvarado, A.L. (1970). Cultural determinants of population stability in the Havasupai Indians. *American Journal of Physical Anthropology*, 33, 9–14.

Ammerman, A.J. & Cavalli-Sforza, L.L. (1984). *The Neolithic Transition and the Genetics of Population in Europe*. Princeton: Princeton University Press.

Anderson, S., Bankier, A.T., Barrell, B.G., De Bruijn, M.H.L., Coulson, A.R., Drouin, J., Eperon, I.C., Nierlich, D.P., Roe, B.A., Sanger, F., Schreier, P.H., Smith, A.J.H., Staden, R. & Young, I.G. (1981). Sequence and organization of the human mitochondrial genome. *Nature*, 290, 457–74.

Aoki, K. (1993). Modelling the dispersal of the first Americans through an inhospitable ice-free corridor. *Anthropological Sciences*, 101, 79–90.

Araujo, A.J.G., Ferreira, L.F., Confalonieri,

U.E.C. & Chame, M. (1988). Hookworms and the peopling of America. *Cad. Saude Publica*, 40, 226–33.

Arends, T. & Gallango, M.L. (1972). Aloalbuminemia: Su distribucion en Poblaciones Venezolanas. *Acta Cientifica Venezolana*, 22, 191–5.

Arends, T., Weitkamp, L., Gallango, R., Neel, J.V. & Schultz, J. (1970). Gene frequencies and microdifferentiation among the Makiritare Indians. II. Seven serum protein systems. *American Journal of Human Genetics*, 22, 526–32.

Arriaza, B. (1993). Seronegative spondyloarthropathies and diffuse idiopathic skeletal hyperostosis in acient Northern Chile. *American Journal of Physical Anthropology*, 91, 263–78.

Arriaza, B., Allison, M. & Gerszten, E. (1988). Maternal mortality in pre-Columbian Indians of Arica, Chile. *American Journal of Physical Anthropology*, 77, 35–71.

Ashburn, P.M. (1974). *The Ranks of Death. A Medical History of the Conquest of America.* Toronto: Coward-McCann.

Astakhova, T., Rjabikov, A., Astakhov, V., Bondareva, Z., Lutova, F. & Bulgakov, Y. (1991). Risk factors and non-communicable disease in native residents and newcoming population of Chukotka. In *Circumpolar Health 90*, ed. B.D. Postl *et al.*, pp. 408–9. Winnipeg: Canadian Society for Circumpolar Health.

Auer, J. (1950). Fingerprints in Eskimos of the Northwest territories. *American Journal of Physical Anthropology*, 8, 485–8.

Auger, F., Jamison, P.L., Balslev-Jorgensen, J., Lewin, T., DePena, J.F. & Skrobak-Kaczynski, J. (1980). Anthropometry of circumpolar populations. In *The Human Biology of Circumpolar Populations*, ed. F.A. Milan, pp. 213–55. Cambridge: Cambridge University Press.

Avksentyuk, A.V., Kurilovich, S.A., Duffy, L.K., Segal, B., Voevoda, M.I. & Nikitin, Y.P. (1995). Alcohol consumption and flushing response in natives of Chukotka, Siberia. *Journal of Studies on Alcohol*, 56, 194–201.

Bada, J., Schroeder, R.A. & Carter, G.F. (1974). New evidence for the antiquity of man in North America deduced from aspartic acid racemization. *Science*, 184, 791–3.

Badimon, J.J., Fuster, V. & Badimon, L.L. (1992). Role of high density lipoproteins in the regression of atherosclerosis. *Circulation*, 86, 86–94.

Bailliet, G., Rothhammer, F., Carnese, F.R., Bravi, C.M. & Bianchi, N.O. (1994). Founder mitochondrial haplotypes in Amerindian populations. *American Journal of Human Genetics*, 55, 27–33.

Baker, B.J. & Armelagos, G. (1988) The origin and antiquity of syphilis. *Current Anthropology*, 29, 707–37.

Baker, P.T., Hanna, J.M. & Baker, T.S. (1986). *The Changing Samoans: Behavior and Health in Transition*. New York: Oxford University Press.

Baker, P.T. & Little, M. (1976). *Man in the Andes. A Multidisciplinary Study of High-Altitude Quechua.* Stroudsburg: Dowden, Hutchison and Ross.

Barbujani, G., Oden, N.L. & Sokal, R.R. (1989). Detecting regions of abrupt change in maps of biological variables. *Systematic Zoology*, 38, 376–89.

Bark, J.E., Harris, M.J. & Firth, M. (1976). Typing of the common phosphoglucomutase variants using isoelectric focusing. A new interpretation of the phosphoglucomutase system. *Journal of the Forensic Science Society*, 16, 115.

Barinaga, M. (1996). An intriguing new lead on Huntington's disease. *Science*, 271, 1233–4.

Barrantes, R., Smouse, P.E., Neel, J.V., Mohrenweiser, H.W. & Gershowitz, H. (1982). Migration and genetic infrastructure of the Central American Guaymi and their affinities with other tribal groups. *American Journal of Physical Anthropology*, 58, 201–14.

Bau, C.H.D. & Salzano, F.M. (1995). Alcoholism in Brazil: The role of personality and susceptibility to stress. *Addiction*, 90, 693–8.

Baume, R.M. (1976). *Discrete Dental Trait Analysis of Five Genetically Related Mexican Communities.* Lawrence, Kansas (unpublished MA thesis).

Baume, R.M. & Crawford, M.H. (1978). Discrete dental traits in four Tlaxcaltecan Mexican populations. *American Journal of Physical Anthropology*, 49, 351–60.

Baume, R.M. & Crawford, M.H. (1979). Discrete dental trait asymmetry in Mexico and Belize. *Journal of Dental Research*, 58, 1811.

Baume, R.M. & Crawford, M.H. (1981). Discrete dental trait asymmetry in Mexican and Belizean groups. *American Journal of Physical Anthropology*, 52, 315–21.

Beals, K.L. & Kelso, A.J. (1975). Genetic variation and cultural evolution. *American Anthropologist*, 77, 566–79.

Bender, K. & Frank, R. (1974). Esterase D-polymorpismus: Darstellung in der Hochspannungs-electrophorese und Mitteilung von Allelhaufigkeiten. *Humangenetik*, **23**, 315–18.

Bennion, L.J. & Li, T.-K. (1976). Alcohol metabolism in American Indians and Whites. Lack of racial differences in metabolic rate and liver alcohol dehydrogenase. *New England Journal of Medicine*, **294**, 9–13.

Bernal, J.E., Papiha, S.S., Keyeux, G., Lanchbury, J.S. & Mauff, G. (1985). Compliment polymorphism in Colombia. *Annals of Human Biology*, **12**, 261–5.

Bernstein, F. (1931). Die geographische Verteilung der Blutgruppen und ihre anthropologische Bedeutung. In *Comitato Italiano per Studio dei Problemi della Populazione*, ed. C. Gini, pp. 227–43. Rome: Instituto Poligrafico de Stato.

Bernstein, F. (1924). Ergebnisse einer biostatistischen Zusammenfassenden Betrachtung uber die erblichen Blutstrukturen des Menschen. *Klinische Wochenschrift*, **3**, 1495–7.

Bevilaqua, L.R.M., Mattevi, V.S., Ewald, G.M., Salzano, F.M., Coimbra, C.E.A., Santos, R.V. & Hutz, M.H. (1995). Beta-globin gene cluster haplotype distribution in five Brazilian Indian tribes. *American Journal of Physical Anthropology*, **98**, 395–401.

Bianchi, N.O. & Rothhammer, F. (1995). Reply to Torroni and Wallace. *American Journal of Human Genetics*, **56**, 1236–8.

Bias, W.B. & Migeon, B.R. (1967). Haptoglobin: A locus on the D 1 chromosome? *American Journal of Human Genetics*, **19**, 393.

Birdsell, J. (1941). A preliminary report of the trihybrid origin of the Australian aborigines. *American Journal of Physical Anthropology*, **28**, 6.

Birdsell, J. (1951). The problem of the early peopling of the Americas as viewed from Asia. In *Physical Anthropology of the American Indian*, ed. W. Laughlin, pp. 1–68. New York: Viking.

Bischoff, J.L. & Rosenbauer, R.J. (1981). Uranium series dating of human skeletal remains from the Del Mar and Sunnyvale Sites, California. *Science*, **213**, 1003–6.

Biss, K., Ho, K.-J., Mikkelson, B., Lewis, L.A. & Taylor, C.B. (1971). Some unique biological characteristics of the Masai of East Africa. *New England Journal of Medicine*, **284**, 694–9.

Black, F.L., Berman, L.L. & Gabbay, Y. (1980a).

HLA antigens in South American Indians. *Tissue Antigens*, **16**, 368–76.

Black, F.L. & Salzano, F.M. (1981). Evidence for heterosis in the HLA system. *American Journal of Human Genetics*, **33**, 894–9.

Black, F.L., Salzano, F.M., Layrisse, Z., Franco, M.H.L.P., Harris, N.S. & Weimer, T. (1980b). Restriction and persistence of polymorphisms of HLA and other blood genetic traits in the Parakana Indians of Brazil. *American Journal of Physical Anthropology*, **52**, 119–32.

Black, F.L., Santos, S.E.B., Salzano, F.M., Callegari-Jacques, S.M., Weimer, T.A., Franco, M.H.L.P., Hutz, M.H., Rieger, T., Kubo, R.R., Mestriner, M.A. & Pandey, J.P. (1988). Genetic variation within the Tupi linguistic group: New data on three Amazonian tribes. *Annals of Human Biology*, **15**, 337–51.

Blanco, R. & Chakraborty, R. (1975). Genetic distance analysis of twenty-two South American Indian populations. *Human Heredity*, **25**, 177–93.

Blangero, J. (1990). Population structure analysis using polygenic traits: Estimation of migration matrices. *Human Biology*, **62**, 27–48.

Blum, H.F. (1961). Does the melanin pigment of human skin have adaptive value? *Quarterly Review of Biology*, **36**, 50–63.

Blumenbach, J.F. (1775). *De Generis Humani varietate nativa*. Gottingen.

Boas, F. (1895). Anthropometrical observations on the Mission Indians of Southern California. *Proceedings of the American Association for the Advancement of Science*, **44**, 261–69.

Boas, F. (1911). Changes in bodily form in the descendants of immigrants. In *Abstract of the Report on Changes in Bodily Form of Descendants of Immigrants*. Washington, D.C.: U.S. Government Printing Office.

Boas, F. (1912). Changes in bodily form of descendants of immigrants. *American Anthropologist*, **14**, 530–62.

Bodmer, W.F. (1975). Evolution of HL-A and other major histocompatibility systems. *Genetics*, **79**, 293–304.

Bodmer, J. & Bodmer, W.F. (1973). Population genetics of the HLA system. A summary of data from the Fifth International Histocompatibility Testing Workshop. *Israel Journal of Medical Science*, **9**, 1257–68.

Bogdan, G. & Weaver, D.S. (1988). Possible treponematosis in human skeletons from a pre-Columbian ossuary of Coastal North Carolina. *American Journal of Physical Anthropology*, 75(2), 187–8.

Borah, W. (1962). Population decline and the social and institutional changes of New Spain in the middle decades of the sixteenth century. *Internationalen Amerikanisten-kongresses*, 34, 172–8.

Borah, W. & Cook, S.F. (1963). *The Aboriginal Population of Central Mexico on the Eve of the Spanish Conquest*. Berkeley: Ibero-Americana.

Borhegi, S.F. & Scrimshaw, H.S. (1957). Evidence for pre-Columbian goiter in Guatemala. *American Antiquity*, 23, 144–56.

Bosron, W.F. and Li, T.-K. (1986). Genetic polymorphism of human liver alcohol and aldehyde dehdrogenases, and their relationship to alcohol metabolism and alcoholism. *Hepatology*, 6, 502–10.

Bosron, W.F., Lumeng, L. & Li, T.-K. (1988a). Genetic polymorphism of enzymes of alcohol metabolism and susceptibility to alcohol liver disease. *Molecular Aspects of Medicine*, 10, 147–58.

Bosron, W.F., Rex, D.K., Harden, C.A., Li, T.-K. & Akerson, R.D. (1988b). Alcohol and aldehyde dehydrogenase isoenzymes in Sioux North American Indians. *Alcoholism: Clinical and Experimental Research*, 12, 454–5.

Bourke, J. B. (1967). A review of the paleopathology of the arthritic diseases. In *Diseases in Antiquity*, ed. D. Brothwell & A.T. Sandison, pp. 352–70. Springfield, Ill.: Charles C. Thomas.

Bowcock, A.M., Bucci, C., Hebert, J.M., Kidd, J.R., Kidd, K.K., Friedlaender, J.S. & Cavalli-Sforza, L.L. (1987). Study of 47 DNA markers in five populations from four continents. *Gene Geography*, 1, 47–64.

Boyd, W. (1952). *Genetics and the Races of Man*. Boston: Little, Brown and Co.

Boyd, W. & Boyd, L. (1937). Blood grouping tests on 300 mummies. *Journal of Immunology*, 32, 307–19.

Bray, W. (1988). The palaeoindian debate. *Nature*, 332, 107.

Brennan, E.R. (1983). Factors underlying decreasing fertility among the Garifuna of Honduras. *American Journal of Physical Anthropology*, 60, 177.

Brewer-Carias, C.A., Le Blanc, S. & Neel, J.V. (1976). Genetic structure of a tribal population, the Yanomama Indians. XIII. Dental microdifferentiation. *American Journal of Physical Anthropology*, 44, 5–14.

Brown, W.M., Prager, E.M., Wang, A. & Wilson, A.C. (1979). Rapid evolution of animal mitochondrial DNA. *Proceedings of the National Academy of Sciences USA*, 76, 1967–71.

Brues, A.M. (1963). Stochastic tests of selection in the ABO blood groups. *American Journal of Physical Anthropology*, 21, 287–99.

Bryan, A.L., Casamiquela, R.M., Cruxent, J.M., Gruhn, R. & Ochsenius, C. (1978). An El Jobo mastodon kill at Taima-taima, Venezuela. *Science*, 200, 1275–7.

Buffon, C.D.G. (1749). *Historie Naturelle*. Paris.

Buikstra, J. (1976). *Hopewell in the Lower Illinois Valley*. Evanston: Northwestern Archeological Program Scientific Papers.

Bullen, A.K. (1972). Paleoepidemiology and distribution of prehistoric treponemiasis (syphilis) in Florida. *Florida Anthropologist*, 25, 133–74.

Bulliyya, G., Reddy, K.K., Reddy, G.P.R., Reddy, P.C., Reddanna, P. & Kumari, K.S. (1990). Lipid profiles among fish-consuming coastal and non-fish-consuming inland populations. In *Diet and Disease*, ed. G.A. Harrison & J.C. Waterlow, pp. 307–25. Cambridge: Cambridge University Press.

Burgess, S.M. (1974). *The St. Lawrence Islanders of Northwest Cape: Patterns of Resource Utilization*. (Ph.D. Dissertation.) Ann Arbor, Michigan: University Microfilms.

Butler, P.M. (1939). Studies of the mammalian dentition. Differentiation of the post-canine dentition. *Proceedings of the Zoological Society of London*, 109, 1–36.

Byard, P.J. (1981). *Population History, Demography and Genetics of the St. Lawrence Island Eskimos*. (Unpublished Ph.D. Dissertation.) Lawrence: University of Kansas.

Byard, P.J. & Crawford, M.H. (1991). Founder effect and genetic diversity on St. Lawrence Island, Alaska. *Homo*, 41, 219–27.

Byard, P.J. & Lees, F.C. (1981). Estimating the number of loci determining skin colour in a hybrid population. *Annals of Human Biology*, 8, 49–58.

Byard, P.J., Lees, F.C. & Relethford, J.H. (1984a). Skin color of the Garifuna of Belize. In *Current Developments in Anthropological Genetics*. Vol. 3. *Black Caribs: A Case Study in Biocultural Adaptation*, ed. M.H. Crawford, pp. 149–68. Plenum Press.

Byard, P.J., Schanfield, M.S. & Crawford, M.H.

(1984b). Admixture and heterozygosity in West Alaskan populations. *Journal of Biosocial Science*, 15, 207–16.

Callegari-Jacques, S.M., Salzano, F.M., Constans, J. & Maurieres, P. (1993). GM haplotype distribution in Amerindians: Relationship with geography and language. *American Journal of Physical Anthropology*, 90, 427–44.

Callegari-Jacques, S.M., Salzano, F.M., Weimer, T.A., Hutz, M.H., Black, F.L., Santos, S.E.B., Guerreiro, J.F., Mestriner, M.A. & Padney, J.P. (1994). Further blood genetic studies on Amazonian diversity – data from four Indian groups. *Annals of Human Biology*, 21, 465–81.

Campuzano, V., Montermini, L., Molto, M., Pianese, L., Cossee, M., Cavalcanti, F., Monros, E., Rodius, F., Duclos, F., Monticelli, A., Zara, F., Canizares, J., Koutnikova, H., Bidichandani, S., Gellera, C., Brice, A., Trouillas, P., De Michele, G., Filla, A., De Frutos, R., Palau, F., Patel, P., Di Donato, S., Mandel, J.-L., Cocozza, S., Koenig, M. & Pandolfo, M. (1996). Friedreich's ataxia: Autosomal recessive disease caused by intronic GAA triplet repeat expansion. *Science*, 271, 1423–5.

Cann, R.L. (1982). *The Evolution of Human Mitochondrial DNA*. (Ph.D. Dissertation.) Berkeley: University of California.

Cann, R.L. (1985). Mitochondrial DNA variation and the spread of modern populations. In *Out of Asia*, ed. E. Szathhmary & R.L. Kirk, pp. 113–22. Canberra: Australian National University.

Carpentero, D.J. & Viana, E.J. (1980). Hipotesis sobre el desarrollo de trypanosomiasis americana. In *Carlos Chagas (1897–1934) y la Trypanosomiasis Americana*, pp. 73–92. Quito: Casa de la Cultura Ecuatoriana.

Case, J.T. & Wallace, D.C. (1981). Maternal inheritance of mitochondrial DNA polymorphisms in cultured human fibroblasts. *Somatic Cell Genetics*, 7, 103–8.

Cavalli-Sforza, L.L. & Bodmer, W.F. (1971). *The Genetics of Human Populations*. San Francisco: W.H. Freeman.

Cavalli-Sforza, L.L. & Edwards, A.W.F. (1967). Phylogenetic analysis: Models and estimation procedures. *American Journal of Human Genetics*, 19, 223–57.

Cavalli-Sforza, L.L., Menozzi, P. & Piazza, A. (1993). Demic expansions and human evolution. *Science*, 259, 639–46.

Cavalli-Sforza, L.L., Menozzi, P. & Piazza, A.

(1994). *The History and Geography of Human Genes*. Princeton, N.J.: Princeton University Press.

Cavalli-Sforza, L.L., Piazza, A., Menozzi, P. & Mountain, J. (1988). The reconstruction of human evolution: Bringing together genetic, archaeological, and linguistic data. *Proceedings of the National Academy of Sciences, USA*, 85, 6002–6.

Centerwall, W.R. (1968). A recent experience with measles in a virgin-soil population. *Biomedical Challenges Presented by the American Indian*, 165, pp. 77–81. Pan American Health Organization.

Chagas, C. (1946). Sobre un Trypanosoma de tatu (Tatusia novemcinctus). *Brazil Medicina*, 26, 305–6.

Chagnon, N.A. (1968). *Yanomamo: The Fierce People*. New York: Holt, Rinehart and Winston.

Chakraborty, R. (1975). Estimation of race admixture - a new method. *American Journal of Physical Anthropology*, 42, 507–12.

Chakraborty, R. (1986). Gene admixture in human populations: Models and predictions. *Yearbook of Physical Anthropology*, 29, 1–43.

Chakraborty, R., Blanco, R., Rothhammer, F. & Llop, E. (1976). Genetic variability in Chilean Indian populations and its association with geography, language, and culture. *Social Biology*, 23, 73–81.

Chakraborty, R., Ferrell, R.E., Stern, M.P., Haffner, S.M., Hazuda, H.P. & Rosenthal, M. (1986). Relationship of prevalence of non-insulin dependent diabetes mellitus to Amerindian admixture in the Mexican Americans of San Antonio, Texas. *Genetic Epidemiology*, 3, 435–54.

Chakraborty, B., Mueller, W.H., Joos, S.K., Hanis, C.L., Barton, S.A. & Schull, W.J. (1993). Obesity and upper body fat distribution in Mexican American children from families with a diabetic proband. *American Journal of Human Biology*, 5, 575–86.

Chapman, M. (1980). Infanticide and fertility among Eskimos: A computer simulation. *American Journal of Physical Anthropology*, 53, 317–27.

Chen, L.Z., Easteal, S., Board, P.G. & Kirk, R.L. (1990). Evolution of beta-globin haplotypes in human populations. *Molecular Biology and Evolution*, 7, 423–37.

Chen, S-H., Zhang, M. & Scott, C.R. (1992).

Gene frequencies of alcohol dehydrogenase$_2$ and aldehyde dehydrogenase$_2$ in Northwest Coast Amerindians. *Human Genetics*, **89**, 351–2.

Chiasson, L.P. (1960). Finger-print pattern frequencies in the Micmac Indians. *Canadian Journal of Genetics and Cytology*, **2**, 184–8.

Chiasson, L.P. (1963). Gene frequencies of the Micmac Indians. *Journal of Heredity*, **54**, 229–36.

Chlachula, J. (1996). Environments du Pleistocene final et occupacion Paleo-Americaine du Sudouest de l'Alberta, Canada. *L'Anthropologie*, **100**, 83–131.

Chrisman, D., MacNeish, R.S., Mavalwala, J. & Savage, H. (1996). Late Pleistocene human friction skin prints from Pendejo Cave, New Mexico. *American Antiquity*, **61**, 357–76.

Cloninger, C.R. (1987). Neurogenetic adaptive mechanism in alcoholism. *Science*, **236**, 410–16.

Coca, A.F. & Deibert, O. (1923). A study of the occurence of the blood groups among the American Indians. *Journal of Immunology*, **8**, 487–91.

Cockburn, T.A. (1961). The origin of treponematoses. *Bulletin of the World Health Organization*, **24**, 221–8.

Cockburn, T.A. (1963). *The Evolution and Eradication of Infectious Disease*. Baltimore: Johns Hopkins Press.

Cockburn, T.A. (1967). Paleoepidemiology. In *Infectious Diseases: Their Evolution and Eradication*, ed. T.A. Cockburn, pp. 50–65. Springfield: Charles C. Thomas.

Cockburn, T.A. (1971). Infectious diseases in ancient populations. *Current Anthropology*, **12**, 45–62.

Comas, J. (1971). Anthropometric studies in Latin American Indian populations. In *The Ongoing Evolution of Latin American Populations*, ed. F.M. Salzano, pp. 333–404. Springfield: Charles C. Thomas.

Comuzzie, A.G. (1993). *Genomic, Genetic, and Morphological Variation in a Sample of Modern Evenki and Their Relationship with Other Indigenous Siberian Populations*. (Ph.D. Dissertation.) Lawrence: University of Kansas.

Comuzzie, A.G., Crawford, M.H., Mitchell, B.D., Mahaney, M.C., Blangero, J., Leonard, W.R., Stern, M.P. & MacCluer, J. (1997). A comparison of serum leptin levels between Mexican-American and Siberian indigenous populations. (unpublished manuscript).

Comuzzie, A.G., Duggirala, R., Leonard, W.R. & Crawford, M.H. (1995). Population relationships among historical and modern indigenous Siberians based on anthropometric characters. *Human Biology*, **67**, 459–80.

Confalonieri, U.E.C., Ferreira, L.F. & Araujo, A.J.G. (1991). Intestinal helminths in lowland South American Indians: Some evolutionary interpretations. *Human Biology*, **63**, 863–74.

Considine, R.V., Sinha, M.K., Heinman, M.L., Kriaciunas, A., Stephens, T.W., Nyce, M.R., Ohannesian, J.P., Marco, C.C., McKee, L.J., Bauer, T.L. & Caro, J.F. (1996). Serum immunoreactive-leptin concentrations in normalweight and obese humans. *New England Journal of Medicine*, **334**, 292–5.

Constans, J., Hazout, S., Garruto, R.M., Gajdusek, D.C. & Spees, E.K. (1985). Population distribution of the human vitamin D binding protein: Anthropological considerations. *American Journal of Physical Anthropology*, **68**, 107–22.

Constans, J., Salzano, F.M. & Black, F.L. (1988). Gc subtypes: New data and distribution comparisons with HLA in Amerindians. *International Journal of Anthropology*, **3**, 9–17.

Conway, D.L. & Baker, P.T. (1972). Skin reflectance of Quechua Indians: The effects of genetic admixture, sex and age. *American Journal of Physical Anthropology*, **36**, 267–82.

Cook, S.F. (1973). The significance of disease in the extinction of New England Indians. *Human Biology*, **45**, 485–508.

Cook, S.F. & Borah, W. (1971). *Essays in Population History: Mexico and the Caribbean*. Berkeley: University of California Press.

Cook, S.F. & Simpson, L.B. (1948). *The Population of Central Mexico*. Ibero-Americana, No. 31. Berkeley: University of California Press.

Coope, E. & Roberts, D.F. (1971). Dermatoglyphic studies of populations in Latin America. In *The Ongoing Evolution of Latin American Populations*, ed. F.M. Salzano, pp. 405–54. Springfield: Charles C. Thomas.

Corcoran, P.A., Allen, F.H., Allison, A.C. & Blumberg, B.S. (1959). Blood groups of Alaskan Eskimos and Indians. *American Journal of Physical Anthropology*, **17**, 187–93.

Cordova, M.S., Lisker, R. & Loria, A. (1967). Studies on several genetic hematological traits of the Mexican population. II. Distribution of blood group antigens in twelve Indian tribes.

American Journal of Physical Anthropology, 26, 55–66.

Corvello, C.M., Franco, M., Salzano, F.M., Black, F.L. & Santos, S.E. (1989). GC polymorphism investigated by isoelectric focusing: A study in South American Indians. *Revista Brasileira Genetica*, 12, 133–43.

Coulehan, J.L. & Welty, T.K. (1990). Cardiovascular disease. In *Indian Health Conditions, An Indian Health Service Report*, pp. 71–98. Rockville, MD.: Department of Human Health Services.

Cowgill, U.M. (1966). The season of birth in man: The northern New World. *Kroeber Anthropological Society Papers1, No. 35*. Berkeley.

Crawford, M.H. (1967). *Re-examination of the Taxonomy and Phylogeny of the Hominoidea Based Upon Experimental Data*. (Unpublished Ph.D. Dissertation.) Seattle: University of Washington.

Crawford, M.H. (1973). The use of genetic markers of the blood in the study of the evolution of human populations. In *Methods and Theories of Anthropological Genetics*, ed. M.H. Crawford & P.L. Workman, pp. 19–38. Albuquerque: University of New Mexico Press.

Crawford, M.H. (1976). *The Tlaxcaltecans: Prehistory, Demography, Morphology, and Genetics*, Publications in Anthropology 7. Lawrence: University of Kansas Press Anthropology Series.

Crawford, M.H. (1977). Human population study. *Science*, 196, 153–5.

Crawford, M.H. (1978). Population dynamics in Tlaxcala, Mexico: The effects of gene flow, selection, and geography on the distribution of gene frequencies. In *Genetic Studies of Human Populations*, ed. C. Otten, Ben Hamed & R. Meier, pp. 215–25. The Hague: Mouton Press.

Crawford, M.H. (1983). The anthropological genetics of the Black caribs (Garifuna) of Central America and the Caribbean. *Yearbook of Physical Anthropology*, 25, 155–86.

Crawford, M.H. (1984). *Current Developments in Anthropological Genetics*. Vol. 3. *Black Caribs: A Case Study of Biocultural Adaptation*. New York: Plenum Press.

Crawford, M.H. (1992). When worlds collide. *Human Biology*, 64, 271–80.

Crawford, M.H. (1992). *Antropologia Biologica de los Indios Americanos*. Madrid: Editorial Mapfre.

Crawford, M.H. & Bach Enciso, V. (1982). Population structure of the circumpolar people of Siberia, Alaska, Canada, and Greenland. In *Current Developments in Anthropological Genetics*. Vol. 2. *Ecology and Population Structure*, ed. M.H. Crawford & J.H. Mielke, pp. 51–91. New York: Plenum Press, New York.

Crawford, M.H. & Comuzzie, A.G. (1989). Genetic and morphological variation in the Black Carib populations of St. Vincent and Livingstone, Guatemala. *Collegium Anthrolpologicum*, 13, 51–61.

Crawford, M.H., Comuzzie, A.G., Leonard, W.R. & Sukernik, R.I. (1994). Molecular genetics, protein variation, and population structure of the Evenki. In *Isozymes: Roles in Evolution, Genetics, Gene Mapping, and Physiology*, ed. C. Markert, J.G. Scandalios, H.A. Lim & O.L. Serov, pp. 227–41. New Jersey: World Scientific Publication.

Crawford, M.H. & Devor, E.J. (1990). Population structure and admixture in transplanted Tlaxcaltecan populations. *American Journal of Physical Anthropology*, 52, 485–90.

Crawford, M.H. & Duggirala, R. (1992). Digital dermatoglyphic patterns of Eskimo and Amerindian populations: Relationship between dermatoglyphic, genetic, linguistic and geographic distances. *Human Biology*, 64, 683–704.

Crawford, M.H., Dykes, D.D. & Polesky, H.F. (1989). Genetic structure of Mennonite populations of Kansas and Nebraska. *Human Biology*, 61, 493–514.

Crawford, M.H., Dykes, D.D., Skradski, K. & Polesky, H.F. (1979). Gene flow and genetic microdifferentiation of a transplanted Tlaxcaltecan Indian population: Saltillo. *American Journal of Physical Anthropology*, 50, 401–12.

Crawford, M.H., Dykes, D.D., Skradsky, K. & Polesky, H.F. (1984). Blood group, serum protein, and red cell enzyme polymorphisms, and admixture among the Black Caribs and Creoles of Central America and the Caribbean. In *Current Developments in Anthropological Genetics*. Vol. 3. *Black Caribs. A Case Study in Biocultural Adaptation*, ed. M.H. Crawford, pp. 303–33. New York: Plenum Press.

Crawford, M.H. & Gmelch, G. (1975). Demography, ethnohistory, and genetics of the Irish Tinkers. *Social Biology*, 21, 321–31.

Crawford, M.H., Gonzalez, M.S., Schanfield, M.S., Dykes, D.D., Skradski, K. & Polesky, H.F. (1981a). The Black Caribs (Garifuna) of

Livingston, Guatemala: Genetic markers and admixture estimates. *Human Biology*, 53, 87–104.

Crawford, M.H., Leonard, W. & Sukernik, R.I. (1992). Biological diversity and ecology in the Evenks of Siberia. *Man and the Biosphere Northern Science Network Newsletter*, 1, 13–14.

Crawford, M.H., Leyshon, W.C., Brown, K., Lees, F. & Johnson, R.S. (1974). Human biology of Tlaxcala: I. Blood group, serum, red cell frequencies, and genetic distances of the Indian populations of Mexico. *American Journal of Physical Anthropology*, 42, 1–18.

Crawford, M.H., Lin, P.M. & Thippeswamy, G. (1976a). Quantitative analysis of dermatoglyphics of four Tlaxcaltecan populations. In *The Tlaxcaltecans. Prehistory, Demography, Morphology, and Genetics*, Publications in Anthropology 7, ed. M.H. Crawford, pp. 120–44. Lawrence: University of Kansas Press.

Crawford, M.H., Mielke, J.H., Devor, E.J., Dykes, D.D. & Polesky, H.F. (1981b). Population structure of Alaskan and Siberian indigenous communities. *American Journal of Physical Anthropology*, 55, 167–86.

Crawford, M.H., Williams, J. & Duggirala, R. (1997a). Population structure of indigenous Siberian populations and the peopling of the New World. In *Horizons of Anthropology*, ed. A. Derevayanko, V. Tishkov & T. Alexseeva. Moscow: Nauka. (In press.)

Crawford, M.H., Williams, J. & Duggirala, R. (1997b). Genetic structure of Siberian indigenous populations. *American Journal of Physical Anthropology*. (In press.)

Crawford, M.H., Workman, P.L., McLean, C. & Lees, F.C. (1976b). Admixture estimates and selection in Tlaxcala. In *The Tlaxcaltecans: Prehistory, Demography, Morphology and Genetics*, Publications in Anthropology 7, ed. M.H. Crawford, pp. 161–8. Lawrence: University of Kansas Press.

Crews, D.E. & Mancilha-Carvalho, J.J. (1994). Correlates of blood pressure in Yanomami Indians of northwestern Brazil. *Ethnicity and Disease*, 3, 362–71.

Crews, D.E., Kamboh, M.I., Mancilha-Carvalho, J.J. & Kottke, B. (1993). Population genetics of apolipoprotein A-4, E, and H polymorphisms in Yanomami Indians of northwestern Brazil: Associations with lipids, lipoproteins, and

carbohydrate metabolism. *Human Biology*, 65, 211–24.

Crosby, A.W. (1974). *The Columbian Exchange. Biological and Cultural Consequences of 1492*. Westport: Greenwood Press.

Crow, J.F. (1958). Some possibilities for measuring selection intensities in man. *Human Biology*, 30, 1–13.

Crow, J.F. (1966). The quality of people: Human evolutionary changes. *Bioscience*, 16, 863–7.

Crow, J.F. (1989). Update to some possibilities for measuring selection intensities in man. *Human Biology*, 61, 776–80.

Cummins, H. (1941). Dermatoglyphics in North American Indians and Spanish-Americans. *Human Biology*, 13, 177–88.

Cummins, H. & Fabricius-Hansen, V. (1946). Dermatoglyphics in Eskimos of West Greenland. *American Journal of Physical Anthropology*, 4, 395–402.

Cummins, H. & Goldstein, M.S. (1932). Dermatoglyphics in Commanche Indians. *American Journal of Physical Anthropology*, 17, 229–35.

Cummins, H. & Midlo, C. (1943). *Fingerprints, Palms and Soles*. Philadelphia: Blakiston.

Cummins, H. & Steggerda, M. (1936). Finger prints in Maya Indians. *Measures of Men*, 7, 107–24.

Custodio, R. & Huntsman, R.G. (1984). Abnormal hemoglobins among the Black Caribs. In *Current Developments in Anthropological Genetics*. Vol. 3. *Black Caribs: A Case Study in Biocultural Adaptation*, ed. M.H. Crawford, pp. 335–44. New York: Plenum Press.

Cutbush, M., Mollison, P.L. & Parkin, D.M. (1950). A new human blood group. *Nature*, 165, 188.

Cybulski, J.S. (1988). Brachydactyly, a possible inherited anomaly at prehistoric Prince Rupert Harbour. *American Journal of Physical Anthropology*, 76, 363–75.

Dahlberg, A.A. (1951). The dentition of the American Indian. In *The Physical Anthropology of the American Indian*, ed. W.S. Laughlin, pp. 138–76. New York: Viking Fund Inc.

Dahlberg, A.A. (1963). Analysis of the American Indian dentition. In *Dental Anthropology*, ed. D.R. Brothwell, pp. 149–77. New York: Pergamon Press.

Dahlberg, A.A. (1980). Craniofacial studies. In *The Human Biology of Circumpolar Populations*, ed. F.A. Milan, pp. 169–92. Cambridge: Cambridge University Press.

Danforth, M.E., Collins Cook, D. & Knick, S.G. (1994). The human remains from Carter Ranch Pueblo, Arizona. *American Antiquity*, 59, 88-101.

Da Rocha, F.J., Spielman, R.S. & Neel, J.V. (1974). A comparison of gene frequency and anthropometric distance matrices in seven villages of four Indian tribes. *Human Biology*, 46, 295-310.

Davidson, W. (1984). The Garifuna in Central America: Ethnohistorical and geographical foundations. In *Current Developments in Anthropological Genetics. Vol. 3. Black Caribs: A Case Study in Biocultural Adaptation*, ed. M.H. Crawford, pp. 13-35. New York: Plenum Press.

Davignon, J., Gregg, R.E. & Sing, D.F. (1988). Apolipoprotein E polymorphism and atherosclerosis. *Arteriosclerosis*, 8, 1-21.

Debets, G.F. (1947). *The Selkups*. Vol. 2. Moscow: Trudy. (In Russian.)

Debets, G.F. (1951). *Anthropological Investigations in the Kamchatka Region*. Moscow: Trudy. (In Russian.)

Decastello, A.A. & Sturli, A. (1902). Ueber die Isoagglutinine im Serum gesunder und kranker Menschen. *Munchener Medizinische Wochenschrift*, 49, 1090-5.

Degos, L., Colombani, J., Chaventre, A., Bengston, B. & Jacquard, A. (1974). Selective pressure on HLA polymorphism. *Nature*, 249, 62-3.

Denevan, W.M. (1976). *The Native Population of the Americas in 1492*. Madison: University of Wisconsin Press.

Denevan, W.M. (1992). The pristine myth: The landscape of the Americas in 1492: The Americas Before and After 1492: Current Geographic Research. *Annals of American Association of Geography*, 82, 369-85.

Denevan, W.M. (1992). Native American populations in 1492: Recent research and a revised hemispheric estimate. In *The Native Population of the Americas*. 2nd Edition, ed. W.M. Denevan, pp. xvii-xxix. Madison: University of Wisconsin Press.

Dennis, R.L.H., Sunderland, E., Rosa, P.J. & Lightman, S. (1978). The digital and palmar dermatoglyphics of the Brazilian Mato Grosso Indians. *Human Biology*, 50, 325-42.

Deprez, R.D., Miller, E. & Hart, S.K. (1985). Hypertension prevalence among Penobscot Indi-ans of Indiand Island, Maine. *American Journal of Public Health*, 75, 653-4.

Destefano, G., Coulehan, F.J. & Wiant, M.K. (1979). Blood pressure survey on the Navajo Indian Reservation. *American Journal of Epidemiology*, 109, 335-45.

Devor, E.J. (1993). Why there is no gene for alcoholism. *Behavioral Genetics*, 23, 145-51.

Devor, E.J., Crawford, M.H. & Bach Enciso, V. (1984). Genetic population structure of the Black Caribs and Creoles. In *Current Developments in Anthropological Genetics. Vol. 3. Black Caribs: A Case Study in Biocultural Adaptation*, ed. M.H. Crawford, pp. 365-80. New York: Plenum Press.

Devor, E.J., McGue, M., Crawford, M.H. & Lin, P.M. (1986). Transmissible and non-transmissible components of anthropometric variation in the Alexanderwohl Mennonites. II. Resolution by path analysis. *American Journal of Physical Anthropology*, 69, 83-92.

Dewey, W.J. & Mann, J.D. (1967). Xg blood group frequencies in some further populations. *Journal of Medical Genetics*, 4, 12-5.

Dillehay, T.D. (1987). By the Banks of the Chinchihuapi. *Natural History*, 4, 8-12.

Dillehay, T.D. & Collins, M.B. (1988). Early cultural evidence from Monte Verde in Chile. *Nature*, 322, 150-2.

Dobyns, H.F. (1966). Estimating aboriginal American populations: An appraisal of techniques with a new hemispheric estimate. *Current Anthropology*, 7, 395-416.

Dobyns, H.F. (1993). Disease transfer at Contact. *Annual Review of Anthropology*, 22, 273-91.

Dobyns, H.F. & Swagerty, W.R. (1983). *The Numbers Become Thinned: Native American Population Dynamics in Eastern North America*. Knoxville: University of Tennessee Press.

Doeblin, T.D., Ingall, A.B., Pinkerton, P.K., Dronamraju, K.R. & Bannerma, R.M. (1968). Genetic studies of the Seneca Indians: Haptoglobins, transferrins, G-6-PD deficiency, hemoglobinopathy, color blindness, morphological traits and dermatoglyphics. *Acta Genetica Basel*, 18, 251-60.

d'Orbigny, A. (1839). *L'homme Americain de l'Amerique Meridionale*. Paris.

Dow, M.M., Cheverud, J.M. & Friedlaender, J.D. (1987). Partial correlations of distance matrices in studies of population structure. *American Journal of Physical Anthropology*, 72, 343-52.

Downey, J.E. (1927). Types of dextrality among

North American Indians. *Journal of Experimental Psychology*, 10, 478–88.

Draper, H.H. (1976). A review of nutritional research. In *Circumpolar Health*, ed. R.J. Shephard & S. Itoh, pp. 120–9. Toronto: University of Toronto Press.

Draper, H.H. (1980). Nutrition. In *Human Biology of Circumpolar Populations*, ed. F.A. Milan, pp. 257–84. London: Cambridge University Press.

Driver, H.E. (1968). On the population nadir of Indians in the United States. *Current Anthropology*, 9, 30.

Dubos, R. (1965). *Man Adapting*. New Haven: Yale University Press.

Ducros, J. (1978). Finger ridge counts of the Ammassalimiut Eskimo of Greenland and other Eskimo population groups: The founder effect and interbreeding. *Man*, 13, 651–6.

Ducros, J. & Ducros, A. (1972). Nombre de cretes et dessins des doigts D'Ammassalimiut (Scoreby-Sund, Groenland Oriental) et des populations Eskimo et asiatiques. *L'Anthropologie*, 76, 711–26.

Duggirala, R. (1995). *Cultural and Genetic Determinants of Lipids and Lipoproteins in the Mennonite Community*. Lawrence: University of Kansas. (Ph.D. Dissertation.)

Duggirala, R. & Crawford, M.H. (1994). Human population structure: Use of quantitative characters. In *Human Genetics*, ed. P. Kumar Seth & S. Seth, pp. 33–63. New Delhi: Omega Scientific Publishers.

Dumond, D.E. (1976). An outline of the demographic history of Tlaxcala. In *The Tlaxcaltecans. Prehistory, Demography, Morphology and Genetics*, Publications in Anthropology 7, ed. M.H. Crawford, pp. 13–23. Lawrence: University of Kansas Press.

Dumond, D.E. (1987). A reexamination of Eskimo-Aleut prehistory. *American Anthropologist*, 89, 32–56.

Dykes, D.D., Crawford, M.H. & Polesky, H.F. (1983). Population distribution in North and Central America of PGM1 and Gc subtypes as determined by isoelectric focusing (IEF). *American Journal of Physical Anthropology*, 62, 137–45.

Dykes, D.D. & Polesky, H.F. (1984). Review of isoelectric focusing for Gc, PGM, Tf, and Pi subtypes: Population distributions. CRC 1. *Critical Reviews in Clinical Laboratory Sciences*, 20, 115–51.

Dykes, D.D. & Polesky, H.F. & Crawford, M.H. (1981). Properdin factor B (Bf) distribution in North and Central American populations. *Electrophoresis*, 2, 320–3.

Early, J.D. & Peters, J.F. (1990). *The Population Dynamics of the Mucajai Yanomama*. New York: Academic Press.

Eaton, J.W. & Mayer, A.J. (1953). The social biology of very high fertility among the Hutterites. *Human Biology*, 25, 206–64.

Eickstedt, E.F. von (1934). *Rassenkunde und Rassengeschichte der Menschheit*. Stuttgart.

Elston, R. (1971). The estimation of admixture in racial hybrids. *Annals of Human Genetics*, 35, 9–17.

Elting, J.J. & Starna, W.A. (1984). A possible case of pre-Columbian treponematosis from New York state. *American Journal of Physical Anthropology*, 65, 267–73.

El-Najjar, M.Y. (1979). Human treponematosis and tuberculosis: Evidence from the New World. *American Journal of Physical Anthropology*, 51, 599–618.

El-Najjar, M.Y., Lozoff, B. & Ryan, D.J. (1975). The paleoepidemiology of porotic hyperostosis in the American Southwest: Radiological and ecological considerations. *American Journal of Physical Anthropology*, 44, 477–88.

Enciso, V.B. (1983). *Estimates of Genetic Distances and Population Structure from Blood and Dermatoglyphic Data*. Lawrence: University of Kansas. (Unpublished M.A. Thesis.)

Escobar, V., Conneally, P.M. & Lopez, C. (1977). The dentition of the Queckchi Indians. Anthropological aspects. *American Journal of Physical Anthropology*, 47, 443–52.

Eveleth, P.B. & Tanner, J.M. (1980). *Worldwide Variation in Human Growth*. Cambridge: Cambridge University Press.

Fagan, B.M. (1987). *The Great Journey: The Peopling of Ancient America*. New York: Thames and Hudson.

Farese, R.V., Linton, M.F. & Young, S.G. (1992). Apolipoprotein B gene mutations affecting cholesterol levels. *Journal of Clinical Medicine*, 231, 643–52.

Faulhaber, J. (1970). Anthropometry of living Indians. In *Handbook of Middle American Indians*. Vol. 9. *Physical Anthropology*, ed. T.D. Stewart, pp. 82–104. Austin: University of Texas Press.

Feinleib, M., Garrison, J., Stallones, L., Kannel, W.B., Castelli, W.P. & McNamara, P.M.

(1979). A comparison of blood pressure, total cholesterol, and cigarette smoking in parents in 1950 and their children in 1970. *American Journal of Epidemiology*, 110, 291-303.

Felstenstein, J. (1975). A pain in the torus: Some difficulties with models of isolation by distance. *American Naturalist*, 109, 359-68.

Fenna, D., Mix, L., Schaefer, O. & Gilbert, J. (1972). Ethanol metabolism in various racial groups. *Canadian Medical Association Journal*, 105, 472-5.

Ferreira, L.F., Araujo, A.J.G. & Confalonieri, U.E.C. (1980). The finding of eggs and larvae of parasitic helminths in archaeological material from Unai, Minas Gerais, Brazil. *Transactions of the Royal Society of Tropical Medicine and Hygiene*, 74, 798-800.

Ferreira, L.F., Araujo, A.J.G. & Confalonieri, U.E.C. (1983). The finding of helminth eggs in a Brazilian mummy. *Transactions of the Royal Society of Tropical Medicine and Hygiene*, 77, 65-7.

Ferreira, L.F., Araujo, A.J.G., Confalonieri, U.E.C. & Nunez, L. (1984). The finding of eggs of *Diphyllobothrium pacificum* in human coprolites (4100-1950 B.C.) from northern Chile. *Memoires Institute Oswaldo Cruz*, 79, 175-80.

Ferrell, R.E., Chakraborty, R., Gershowitz, H., Laughlin, W.S. & Schull, W.J. (1981). The St. Lawrence Island Eskimos: Genetic variation and genetic distance. *American Journal of Physical Anthropology*, 55, 351-8.

Field, L.L., Gofton, J.P. & Kinsella, T.D. (1988). Immunoglobulin (GM and KM) allotypes and relation to population history in native peoples of British Columbia: Haida and Bella Coola. *American Journal of Physical Anthropology*, 76, 155-64.

Fildes, R.A. & Parr, C.W. (1963). Human red cell phosphogluconate dehydrogenase. *Nature*, 200, 890.

Firschein, I.L. (1961). Population dynamics of the sickle-cell trait in Black Caribs of British Honduras, Central America. *American Journal of Human Genetics*, 13, 233-54.

Firschein, I.L. (1984). Demographic patterns of the Garifuna (Black Caribs) of Belize. In *Current Developments in Anthropological Genetics*. Vol. 3. *Black Caribs: A Case Study in Biocultural Adaptation*, ed. M.H. Crawford, pp. 67-94. New York: Plenum Press.

Fisher, R.A., Race, R.R. & Taylor, G.L. (1944).

Mutation and the rhesus reaction. *Nature*, 153, 106.

Fitch, W. & Neel, J.V. (1969). Phylogenetic relationships of some Indian tribes of Central and South America. *American Journal of Human Genetics*, 21, 384-94.

Fladmark, K.R. (1979). Routes: Alternative migration corridors for early man in North America. *American Antiquity*, 44, 55-69.

Fladmark, K.R. (1986). Getting One's Berings. *Natural History*, 11, 8-19.

Flickinger, S.A.M. (1975). *A Comparative Dermatoglyphic Study of the Apache, Choctaw, and Navajo Indians*. University of Southern Mississippi. (Ph.D. Dissertation.)

Flint, J., Boyce, A.J., Martinson, J.J. & Clegg, J.B. (1989). Population bottlenecks in Polynesia revealed by minisatellites. *Human Genetics*, 83, 256-63.

Fonseca, O. (1969). Parasitismo e migracoes humanas pre-historicas. In *Estudos de Prehistoria Geral e Brasileira*, pp. 1-346. Sao Paulo.

Foote, D.C. (1965). *Exploration and Resource Utilization in Northwestern Arctic Alaska Before 1855*. Montreal: McGill University. (Ph.D. Dissertation.)

Fox Lalueza, C. (1996). Mitochondrial haplogroups in four tribes from Tierra del Fuego-Patagonia: Inferences about the peopling of America. *Human Biology*, 68, 855-71.

Freeman, D. (1983). *Margaret Mead and Samoa: The Making and Unmaking of an Anthropological Myth*. Cambridge: Harvard University Press.

Friedlaender, J.S. (1975). *Patterns of Human Variation. The Demography, Genetics, and Phenetics of Bougainville Islanders*. Cambridge: Harvard University Press.

Friedman, J.M. & Leibel, R.D. (1992). Tackling a weighty problem. *Cell*, 69, 217-20.

Fry, G.F. & Moore, J.G. (1969). Enterobius vermicularis: 10,000 years of human infection. *Science*, 166, 1620.

Gabaldon, A. (1965). Leading causes of death in Latin America. *Components of Population Change in Latin America, Milbank Memorial Quarterly*, 43, 242-57.

Gage, T.B. (1988). Mathematical hazard models of mortality: An alternative to model life tables. *American Journal of Physical Anthropology*, 76, 429-41.

Gage, T.B., Dyke, B. & Riviere, P.G. (1984). Estimating mortality from two censuses: An application to the Trio of Surinam. *Human Biology,* 56, 489–502.

Gage, T.B., McCullough, J.M., Weitz, C.A., Dutt, J.S. & Abelson, A. (1989). Demographic studies and human population biology. In *Human Population Biology,* ed. M.A. Little & J.D. Haas, pp. 45–65. Oxford: Oxford University Press.

Gallango, M.L. & Suinaga, R. (1978). Uridine monophosphate kinase polymorphism in two Venezuelan population. *American Journal of Human Genetics,* 30, 215–18.

Gallo, G., Buhagiar, M., Cuschieri, A. & Viviani, F. (1997). Huntington's chorea (HD) in Malta: Epidemiology and origins. *Human Biology.* (Submitted.)

Garn, S.M. (1958). *Methods for Research in Human Growth.* Ann Arbor: University of Michigan Press.

Garruto, R.M., Plato, C.C., Hoff, C.J., Newman, M.T., Gajdusek, D.C. & Baker, P.T. (1979). Characterization and distribution of dermatoglyphic features in Eskimo and North, Central and South American Indian populations. In *Dermatoglyphics – Fifty Years Later,* ed. W. Wertelecki & C.C. Plato, pp. 277–334, Birth Defects: Original Article Series, Vol. XV, No. 6, Washington, D.C.: National Foundation.

Geerdink, R., Nijenhuis, L.E., Van Loghem, E. & Sjoe, E.L.F. (1974). Blood groups and immunoglobulin groups in Trio and Wajana Indians from Surinam. *American Journal of Human Genetics,* 26, 581–7.

Gershowitz, H. (1959). The Diego factor among Asiatic Indians, Apaches, and West African Negroes: Blood types of Asiatic Indians and Apaches. *American Journal of Physical Anthropology,* 17, 195–200.

Gershowitz, H. & Neel, J.V. (1978). The immunoglobulin allotypes (Gm and Km) of twelve Indian tribes of Central and South America. *American Journal of Physical Anthropology,* 49, 289–302.

Giblett, E.R. (1958). Js, a 'new' blood group antigen found in Negroes. *Nature,* 181, 1221–2.

Giblett, E.R. (1969). *Genetic Markers in Human Blood.* Oxford: Blackwell.

Giblett, E.R., Anderson, J.E., Cheng, S.-H., Teng, Y.-S. & Cohen, F. (1974). Uridine monophosphate kinase: A new genetic polymorphism with possible clinical implications. *American Journal of Human Genetics,* 26, 627–35.

Gillum, R.F. (1988). Ischemic heart disease mortality in American Indians, United States, 1969–71 and 1979–81. *American Heart Journal,* 115, 1141–4.

Ginsberg, H.N., Karamally, W., Siddiqui, M., Hollerman, S., Tall, A.R., Blaner, W.S. & Ramakrishnan, R. (1995). Increases in dietary cholesterol are associated with modest increases in both LDL and HDL cholesterol in healthy young women. *Arteriosclerosis Thrombosis Biology,* 15, 169–78.

Goedde, H.W., Benkmann, H.-G., Agarwal, D.P. & Kroeger, A. (1977). Genetic studies in Ecuador: Acetylator phenotypes, red cell enzyme and serum protein polymorphisms of Shuara Indians. *American Journal of Physical Anthropology,* 47, 419–26.

Goedde, H.W., Rothhammer, F., Benkmann, H.-G. & Bogdanski, P. (1984). Ecogenetic studies in Atacameno Indians. *Human Genetics,* 67, 343–46.

Goldberg, I.J., Scheraldi, C.A., Yacoub, K.L., Saxena, U. & Bisgaier, C.L. (1990). Lipoprotein apoC-II activation of lipoprotein lipase: Modulation by apolipoprotein A-IV. *Journal of Biological Chemistry,* 265, 4266–72.

Goldman, D. (1993). Genetic transmission. In *Recent Developments in Alcoholism.* Vol. 2. *Ten Years of Progress,* ed. M. Galanter, pp. 231–48. New York: Plenum Press.

Goldstein, D.B., Linares, A.R., Cavalli-Sforza, L.L. & Feldman, M.W. (1995). An evaluation of genetic distances for use with microsatellite loci. *Genetics,* 139, 463–71.

Goldstein, M.S. (1943). *Demographic and Bodily Changes in Descendants of Mexican Immigrants.* Austin: Institute of Latin American Studies, University of Texas.

Gonzalez, N.L. (1984). Garifuna (Black Carib) social organization. In *Current Developments in Anthropological Genetics.* Vol. 3. *Black Caribs: A Case Study in Biocultural Adaptation,* ed. M.H. Crawford, pp. 51–66. New York: Plenum Press.

Gonzalez, N.L. (1988). *Sojourners of the Caribbean: Ethnogenesis and Ethnohistory of the Garifuna.* Urbana: University of Illinois Press.

Goodwin, D.W. (1987). Adoption studies of alcoholism. *Progress in Clinical Biological Research,* 241, 61–70.

Goodwin, D.W. & Guze, S.B. (1989). *Psychiatric Diagnosis.* 4th edn. Oxford: Oxford University Press.

Grayson, D., (1987). quoted by R. Lewin, Science, **238**, 1230.

Greenacre, M.J. & Degos, L. (1977). Correspondence analysis of HLA gene frequency data from 124 population samples. *American Journal of Human Genetics*, **29**, 60–75.

Greenberg, J.H. (1987). *Language in the Americas*. Palo Alto: Stanford University Press.

Greenberg, J.H., Turrner, C.G. & Zegura, S.L. (1985). Convergence of evidence for the peopling of the Americas. *Collegium Antropologicum*, **9**, 33–42.

Greenberg, J.H., Turner, C.G. & Zegura, S.L. (1986). The settlement of the Americas: A comparison of the linguistic, dental, and genetic evidence. *Current Anthropology*, **27**, 477–97.

Greulich, W.W. & Pyle, S.I. (1959). *Radiographic Atlas of Skeletal Development of the Hand and Wrist*. Stanford: Stanford University Press.

Guerreiro, J.F., Figueiredo, M.S., Santos, S.E.B & Zago, M.A. (1992). Beta-globin gene cluster haplotypes in Yanomama Indians from the Amazon region of Brazil. *Human Genetics*, **89**, 629–31.

Guerreiro, J.F., Figueiredo, M.S. & Zago, M.A. (1994). Beta-globin gene cluster haplotypes of Amerindian populations from the Brazilian Amazon region. *Human Heredity*, **44**, 142–9.

Guidon, N. & Delibrias, G. (1985). Carbon-14 dates point to man in the Americas 32,000 years ago. *Nature*, **321**, 769–71.

Gullick, C.J.M.R. (1979). Ethnic interaction and Carib language. *Journal of Belizean Affairs*, **9**, 3–20.

Gullick, C.J.M.R. (1984). The changing Vincentian Carib population. In *Current Developments in Anthropological Genetics*. Vol. 3. *Black Caribs: A Case Study in Biocultural Adaptation*, ed. M.H. Crawford, pp. 37–50. New York: Plenum Press.

Gurtler, 1971, cited from Roychoudhury, A.K. and Nei, M. (1988).

Gutsche, B.B., Scott, E.M. & Wright, R.C. (1967). Hereditary deficiency of pseudocholinesterase in Eskimos. *Nature*, **215**, 322–3.

Gyllensten, U., Wharton, D., Josefsson, A. & Wilson, A.C. (1991). Paternal inheritance of mitochondrial DNA in mice. *Nature*, **352**, 255–7.

Haas, E.J.C., Salzano, F.M., Araujo, H.A., Grossman, F., Barbetti, A., Weimer, T.A., Franco, M.H., Verruno, L., Nasif, O., Morales, V.H. &

Arienti, R. (1985). HLA antigens and other genetic markers in the Mapuche Indians of Argentina. *Human Heredity*, **35**, 306–13.

Haas, J.D. (1980). Maternal adaptation and fetal growth at high altitude in Bolivia. In *Social and Biological Predictors of Nutritional Status, Physical Growth, and Neurological Development*, pp. 257–90. New York: Academic Press.

Halberstein, R.A. & Crawford, M.H. (1973). Human biology in Tlaxcala, Mexico: Demography. *American Journal of Physical Anthropology*, **36**, 199–212.

Halberstein, R.A. & Crawford, M.H. (1975). Demographic structure of a transplanted Tlaxcalan population in the Valley of Mexico. *Human Biology*, **47**, 201–32.

Halberstein, R.A., Nutini, H.G. & Crawford, M.H. (1973). Historical-demographic analysis of Indian populations in Tlaxcala, Mexico. *Social Biology*, **20**, 40–50.

Hall, T.R., Hickey, M.E. & Young, T.B. (1992). Evidence for recent increases in obesity and non-insulin dependent diabetes mellitus in a Navajo community. *American Journal of Human Biology*, **4**, 547–53.

Hamilton, B.S., Paglia, D., Kwan, D. & Deitel, M. (1995). Increased obese mRNA expression in omental fat cells from massively obese humans. *Nature Medicine*, **1**, 953–6.

Hanihara, K. (1969). Mongoloid dental complex in the permanent dentition. In *Proceedings VIIIth International Congress of Anthropological and Ethnological Sciences*, ed. S. Watanabe, Tokyo, pp. 298–300.

Hanihara, K. (1979). Dental traits in Ainu, Australian Aborigines, and New World populations. In *The First Americans: Origins, Affinities, and Adaptations*, ed. W.S. Laughlin & A.B. Harper, pp. 125–34. New York: Gustav Fischer.

Hanihara, K., Masuda, T., Tanaka, T. & Tamada, M. (1975). Comparative studies of dentition. In *Anthropological and Genetic Studies on the Japanese, Part III: Anthropological and Genetic Studies of the Ainu. JIBP Synthesis*, Vol. 2, ed. S. Watanabe, pp. 256–62. Tokyo: University of Tokyo Press.

Hanis, C.L., Chakraborty, R., Ferrell, R.E. & Schull, W.J. (1986). Individual admixture estimates: Disease associations and individual risk of diabetes and gallbladder disease among Mexican-Americans in Starr County, Texas. *Amer-*

ican Journal of Physical Anthropology, 70, 433–41.

Hanna, J.M. (1973). Ethnic groups, human variation, and alcohol use. In Cross-cultural Approaches to the Study of Alcohol: An Interdisciplinary Perspective, ed. M. Everett, J. Waddell & D. Heath, Chicago: Aldine, pp. 235–42.

Hansen, J.A., Lanier, A.P., Nisperos, B., Mickelson, E. & Dahlberg, S. (1986). The HLA system in Inupiat and Central Yupik Alaskan Eskimos. Human Immunology, 16, 315–28.

Harding, R.M. (1992). VNTRs in review. Evolutionary Anthropology, 1, 62–71.

Harihara, S., Saitou, N., Hirai, M., Gojobori, T., Park, K.S., Misawa, S., Ellepola, S.B., Ishida, T. & Omoto, K. (1988). Mitochondrial DNA polymorphism among Asian populations. American Journal of Human Genetics, 43, 134–43.

Harpending, H. & Jenkins, T. (1973). Genetic distance among southern African populations. In Methods and Theories of Anthropological Genetics, ed. M.H. Crawford & P.L. Workman, pp. 177–99. Albuquerque: University of New Mexico Press.

Harris, E.F. & Nweeia, M.T. (1980). Tooth size of Ticuna Indians, Columbia, with phenetic comparisons to other Amerindians. American Journal of Physical Anthropology, 53, 81–91.

Harris, H., Hopkinson, D.A., Luffman, J.E. & Rapley, S. (1968). Electrophoretic variation in erythrocyte enzymes. In Genetically Determined Abnormalities of Red Cell Metabolism. New York: Grune and Stratton.

Harris, H., Hopkinson, D.A. & Robson, E.B. (1962). Two-dimensional electrophoresis of pseudocholinesterase components in normal human serum. Nature, 196, 1296.

Harris, H., Hopkinson, D.A., Robson, E.B. & Whittaker, M. (1963). Genetical studies on a new variant of pseudocholinesterase detected by electrophoresis. Annals of Human Genetics, 26, 359–73.

Harris, M. (1964). Patterns of Race in the Americas. New York: Walker and Co.

Harris-Hooker, S. & Sanford, G.L. (1994). Lipids, lipoproteins and coronary heart disease in minority populations. Atherosclerosis, 108, S83–S104.

Harrison, G.A. & Salzano, F.M. (1966). The skin color of the Caingang and Guarani Indians of Brazil. Human Biology, 38, 104–11.

Haydenbilt, R. (1996). Dental variation among four prehispanic Mexican populations. Amer-

ican Journal of Physical Anthropology, 100, 225–46.

Haynes, C.V. (1969). The earliest Americans. Science, 166, 709–15.

Haynes, V. (1973). The Calico site: Artifacts or geofacts. Science, 181, 305–9.

Hedrick, P.W. (1971). A new approach to measuring genetic similarity. Evolution, 25, 276–80.

Hill, T.W. (1978). Drunken comportment of urban Indians: "Time-out" behavior? Journal of Anthropological Research, 34, 442–67.

Hirschfeld, J. (1959). Immunoelectrophoretic demonstration of qualitative differences in normal human sera and their relation to the haptoglobins. Acta Pathologica et Microbiologica Scandanavica, 47, 160.

Hirschfeld, J., Jonsson, B. & Rasmuson, M. (1960). Inheritance of a new group-specific system demonstrated in normal human sera by means of immunoelectrophoretic technique. Nature, 158, 123.

Holmes, W.H. (1925). The antiquity phantom in American archaeology. Science, 62, 256–8.

Holt, S.B. (1968). The Genetics of Dermal Ridges. Springfield: Charles C. Thomas.

Hooton, E.A. (1930). The Indians of Pecos Pueblo. A study of their skeletal remains. In Papers of the Philips Academy Southwestern Expedition, 4. New Haven: Yale University Press.

Hooton, E.A. & Dupertuis, C.W. (1955). The Physical Anthropology of Ireland, Volume 30. Cambridge: Peabody Museum Papers.

Hopkins, D.M. (1979). Landscape and climate of Beringia during late Pleistocene and Holcene time. In The First Americans: Origins, Affinities, and Adaptations, ed W.S. Laughlin & A.B. Harper, pp. 15–42. New York: Gustav Fischer.

Hopkins, D.M. (1967). The Bering Land Bridge. Stanford: Stanford University Press.

Hopkinson, D.A. & Harris, H. (1965). Evidence for a second 'structural' locus determining human phosphoglucomutase. Nature, 208, 410.

Hopkinson, D.A., Mestriner, M.A., Cortner, J. & Harris, H. (1973). Esterase D: A new human polymorphism. Annals of Human Genetics, 37, 119–37.

Hopkinson, D.A., Spencer, N. & Harris, H. (1963). Red cell acid phosphatase variants: A new human polymorphism. Nature, 199, 969–71.

Howell, N. (1976). Toward a uniformitarian theory of paleodemography. In *The Demographic Evolution of Human Populations*, ed. R.H. Ward & K.M. Weiss, pp. 25–40. New York: Academic Press.

Howells, W.W. (1989). *Skull Shapes and the Map. Craniometric Analyses in the Dispersion of Modern Man.* Cambridge: Peabody Museum, Harvard University.

Hrdlička, A. (1912). The problem of the unity or plurality and probable place of origin of American Aborigines. *American Anthropology*, 14, 9012.

Hrdlička, A. (1917). Preliminary report on finds of supposedly ancient human remains at Vero, Florida. *Journal of Geology*, 25, 43–51.

Hrdlička, A. (1920). Shovel shaped teeth. *American Journal of Physical Anthropology*, 3, 429–65.

Hrdlička, A. (1937). Early man in America: What have the bones to say? *Early Man*. Philadelphia.

Hrdlička, A. (1942). The problem of man's antiquity in America. *Proceedings of the Eighth American Scientific Congress II*, pp. 53–55.

Hudson, E.H. (1965). Treponematosis and man's social evolution. *American Anthropologist*, 67, 886–901.

Hulse, F.S. (1957). Linguistic barriers to gene flow. *American Journal of Physical Anthropology*, 15, 235–46.

Hulse, F.S. (1960). Ripples on a gene pool: The shifting frequencies of blood-type alleles among the Indians of the Hupa Reservation. *American Journal of Physical Anthropology*, 18, 141–52.

Hutchinson, J.F. (1983). *A Biocultural Analysis of Blood Pressure Variation Among the Black Caribs and Creoles of St. Vincent, West Indies.* Lawrence: University of Kansas. (Unpublished Ph.D. Dissertation.)

Hutchinson, J.F. (1986). Association between stress and blood pressure variation in a Caribbean population. *American Journal of Physical Anthropology*, 71, 69–80.

Hutchinson, J.F. & Byard, P.J. (1987). Family resemblance for anthropometric and blood pressure measurements in Black Caribs and Creoles from St. Vincent Island. *American Journal of Physical Anthropology*, 73, 33–40.

Hutchinson, J.F. & Crawford, M.H. (1981). Genetic determinants of blood pressure level among the Black Caribs of St. Vincent. *Human Biology*, 53, 453–66.

Hutchinson, J.F., Lin, P.M. & Crawford, M.H. (1984). Factors influencing blood pressure level among the Black Caribs of St. Vincent. *Current Developments in Anthropological Genetics. Vol. 3. Black Caribs: A Case Study in Biocultural Adaptation*, ed. M.H. Crawford, pp. 215–39. New York: Plenum Press.

Ikin, E.W., Mourant, A.E., Pettenkofer, H.J. & Blumenthal, G. (1951). Discovery of the expected haemagglutinin, anti-Fyb. *Nature*, 168, 1077.

Imbelloni, J. (1958). Nouveaux rapports à la classification de l'homme americain. *Congresso Internationall Americanistas* (Mexico), 107–36.

Indian Health Service. (1989). *Trends in Indian Health.* Washington D.C.: Department of Health and Human Services.

Indian Health Service. (1991). *Trends in Indian Health. 1991.* Washington, D.C.: Department of Health and Human Services.

Irving, W.M. (1987). New dates from old bones. *Natural History*, 2, 8–14.

Irving, W.N. & Harrington, C.R. (1973). Upper Pleistocene radiocarbon-dated artifacts from the Northern Yukon. *Science*, 179, 335–40.

Ishida, H. (1993). Population affinities of the Peruvian with Siberians and North Americans: A nonmetric cranial approach. *Anthropological Science*, 101, 47–64.

Ito, M., Oliverio, M.I., Mannon, P.J., Best, C.F., Maeda, N., Smithies, O. & Coffman, T.M. (1995). Regulation of blood pressure by the type 1A angiotensin II receptor gene. *Proceedings of the National Academy of Sciences, USA*, 92, 3521–5.

Jantz, R.L. (1995). Special Issue on the Population Biology of Late Nineteenth-Century Native North Americans and Siberians: Analyses of Boas's Data. *Human Biology*, 67, 337–516.

Jantz, R.L., Hunt, D.R., Falsetti, A.B. & Key, P.J. (1992). Variation among North Amerindians: Analysis of Boas' anthropometric data. *Human Biology*, 64, 435–61.

Javid, J. (1967). Haptoglobin 2-1 Bellevue, a haptoglobin beta chain mutant. *Proceedings of the National Academy of Sciences, USA*, 57, 920.

Jayle, M.F. & Moretti, J. (1962). Haptoglobin: Biochemical, genetic and physiopathological aspects. In *Progress in Hematology, Vol. 3*, ed. C.V. Moore & E.B. Brown, p. 342. New York: Grune & Stratton.

Jenkins, T. (1972). *Genetic Polymorphisms of*

Man in Southern Africa. University of London. (M.D. Thesis.)

Jett, S.C. (1978). Precolumbian transoceanic contacts. In *Ancient North Americans*, ed. J.D. Jennings, pp. 557–613. San Francisco: W.H. Freeman and Co.

Johansson, S.R. (1982). The demographic history of the native peoples of North America: A selective bibliography. *Yearbook of Physical Anthropology*, 25, 133–52.

Johnson, M.J., Wallace, D.C., Ferris, S.D., Rattazzi, M.C. & Cavalli-Sforza, L.L. (1983). Radiation of human mitochonrdia DNA types analyzed by restriction endonuclease cleavage patterns. *Journal of Molecular Evolution*, 19, 255–71.

Johnston, F.E., Alarcon, O., Benedict, F., Dary, M., Galbraith, M. & Gindhart, P. (1973). Albumin Mexico (Al^Mc) in the Guatemalan Highlands. *American Journal of Physical Anthropology*, 38, 27–30.

Johnston, F.E., Blumberg, B.S., Agarwal, S., Melartin, L. & Burch, T.A. (1969). Alloalbuminemia in Southwestern U.S. Indians: Polymorphism of albumin Naskapi and albumin Mexico. *Human Biology*, 41, 263–70.

Johnston, F.E. & Kensinger, K.M. (1971). Fertility and mortality differences and their implications for microevolutionary change among the Cashinahua. *Human Biology*, 43, 356–64.

Johnston, F.E. & Schell, L.M. (1979). Anthropometric variation of native American children and adults. In *The First Americans: Origins, Affinities, and Adaptations*, ed. W.S. Laughlin & A.B. Harper, pp. 275–91. New York: Gustav Fischer.

Jones, G.D. (1978). The ethnohistory of the Guale coast through 1684. In *The Anthropology of St. Catherine Island 1. Natural and Cultural History (Anthropological Papers of the American Museum of Natural History, 55)*, ed. D.H. Thomas, G.D. Jones, R.S. Durham & åarsen, pp. 178–210. New York: American Museum of Natural History.

Jorde, L.B. (1980). The genetic structure of subdivided human populations. In *Current Developments in Anthropological Genetics. Vol. 1. Theory and Methods*, ed. J.H. Mielke & M.H. Crawford, pp. 135–208. New York: Plenum Press.

Jorde, L.B. (1985). Human genetic distance studies: Present status and future prospects. *Annual Review of Anthropology*, 14, 343–73.

Kalmus, H., De Garay, A.L., Rodarte, U. & Cobo, L. (1964). The frequency of PTC tasting, hard ear wax, colour blindness, and other genetical characters in urban and rural Mexican populations. *Human Biology*, 36, 134–45.

Kalyanaramen, V.S., Sarngadharan, M.G. & Robert-Guroff, M. (1982). A new subtype of human T-cell leukemia virus (HTLV-II) associated with a T-cell variant of hairy cell leukemia. *Science*, 218, 571–3.

Kamboh, M.I., Bhatia, K.K. & Ferrell, R.E. (1990). Genetic studies of human apolipoproteins. XII. Population genetics of apolipoproteins in Papua New Guinea. *American Journal of Human Biology*, 2, 17–23.

Kamboh, M.I., Weiss, K.M. & Ferrell, R.E. (1991). Genetic studies of human apolipoproteins. XVI. APO E polymorphism and cholesterol levels in the Mayans of the Yucatan Peninsula, Mexico. *Clinical Genetics*, 39, 22–32.

Kamboh, M.I., Crawford, M.H., Aston, C.E. & Leonard, W.R. (1996). Population distributions of APOE, APOH, and APOA4 polymorphisms and their relationships with quantitative plasma lipid levels among the Evenki herders of Siberia. *Human Biology*, 68, 231–43.

Kaplan, B.A. (1954). Environment and human plasticity. *American Anthropologist*, 56, 780–800.

Karaphet, T.M., Sukernik, R.I., Osipova, L.P. & Simchenko, Y.B. (1981). Blood groups, serum proteins and red cell enzymes in the Ngansans (Tavghi) – Reindeer hunters from Taimir Peninsula. *American Journal of Physical Anthropology*, 56, 139–45.

Karafet, T., Zegura, S.L., Vuturo-Brady, J., Posukh, O. Osipova, L., Wiebe, V., Romero, F., Long, J., Harihara, S., Jin, F., Dashnyam, B., Gerelsaikhan, T., Omoto, K. & Hammer, M.F. (1997). Y chromosome markers and trans-Bering Strait dispersal. *American Journal of Physical Anthropology*, 102, 301–14.

Karp, G.W. & Sutton, H.E. (1967). Some new phenotypes of human red cell acid phosphatase. *American Journal of Human Genetics*, 19, 54–62.

Kasprisin, D.O., Crow, M., McClintock, C. & Lawson, J. (1987). Blood types of the native Americans of Oklahoma. *American Journal of Physical Anthropology*, 73, 1–7.

Katzmarzyk, P.T., Leonard, W.R., Crawford, M.H. & Sukernik, R.I. (1994a). Resting metabolic rate and daily energy expenditure among two indigenous Siberian populations. *American Journal of Human Biology*, 6, 719–30.

Katzmarzyk, P.T., Leonard, W.R., Crawford, M.H. & Sukernik, R.I. (1994b). Predicted maximal oxygen consumption of indigenous Siberians. *American Journal of Human Biology*, 6, 783–90.

Katch, F.I. & McArdle, W.D. (1993). *Introduction to Nutrition, Exercise, and Health*. Philadelphia: Lea and Febiger.

Kenen, R. & Hammerslough, C.R. (1987). Reservation and non-reservation American Indian mortality in 1970 and 1978. *Social Biology*, 34, 26–36.

Kidd, J.R., Black, F.L., Weiss, K.M., Balazs, I. & Kidd, K.K. (1991). Studies of three Amerindian populations using nuclear DNA polymorphisms. *Human Biology*, 63, 775–94.

Kim, H.-S., Krege, J.H., Kluckman, K.D., Hagaman, J.R., Hodgin, J.B., Best, C.F., Jennette, J.C., Coffman, T.M., Maeda, N. & Smithies, O. (1995). Genetic control of blood pressure and the angiotensinogen locus. *Proceedings of the National Academy of Sciences, USA*, 92, 2735–9.

Kirk, R.L., Serjeantson, S.W., King, H. & Zimmet, P. (1985). The genetic epidemiology of diabetes mellitus. In *Diseases of Complex Etiology in Small Populations: Ethnic Differences and Research Approaches*, ed. R. Chakraborty & E. Szathmary, pp. 119–46. New York: Liss Press.

Knussmann, R. (1969). Interkorrelationen im Hautleistensystem des Menschen und ihre faktorenanalytische Auswertung. *Humangenetik*, 4, 221–43.

Koeberle, F. (1968). Chagas' disease and Chagas' syndromes. The pathology of American trypanosomiases. *Advances in Parasitology*, 6, 63–116.

Konigsberg, L.W. (1988). Migration models of prehistoric postmarital residence. *American Journal of Physical Anthropology*, 77, 471–82.

Konigsberg, L.W. & Ousley, S.D. (1995). Multivariate quantitative genetics of anthropometrics from the Boas data. *Human Biology*, 67, 435–61.

Korey, K.A. (1978). A critical appraisal of methods for measuring admixture. *Human Biology*, 50, 343–60.

Kraus, B.S. (1954). *Indian Health in Arizona*. Tucson: University of Arizona Press.

Krieger, A.D. (1964). Early man in the New World. In *Prehistoric Man in the New World*, eds. J.D. Jennings and E. Norbeck, pp. 23–81. Houston: University of Houston Press.

Krieger, H., Morton, N.E., Mi, E., Azevedo, E., Freire-Maia, A. & Yasuda, N. (1965). Racial admixture in northeastern Brazil. *Annals of Human Genetics*, 29, 113–25.

Kroeber, A.L. (1939). Cultural and natural areas of native North America. *University of California Publications in American Archeology and Ethnology*, 38. Berkeley: University of California Press.

Krogman, W.M. (1972). *Child Growth*. Ann Arbor: University of Michigan Press.

Kunz, M.L. & Reanier, R.E. (1994). Paleoindians in Beringia: Evidence from Arctic Alaska. *Science*, 263, 660–2.

Lalouel, J.M. (1977). The conceptual framework of Malecot's model of isolation by distance. *Annals of Human Genetics*, 40, 355–60.

Lalouel, J.M. & Morton, N.E. (1973). Bioassay of kinship in a South American Indian population. *American Journal of Human Genetics*, 25, 62–73.

Lamb, N.P. (1975). Papago Indian admixture and mating patterns in a mining town: A genetic cauldron. *American Journal of Physical Anthropology*, 42, 71–80.

Lame Deer, J. & Erdoes, R. (1972). *Lame Deer Seeker of Visions*. New York: Simon and Schuster, Inc.

Lampl, M. & Blumberg, B.S. (1979). Blood polymorphisms and the origin of New World populations. In *The First Americans: Origins, Affinities and Adaptations*, ed. W.S. Laughlin & A.B. Harper, pp. 107–23. New York: Gustav Fischer.

Lander, E.S. & Schork, N.J. (1994). Genetic dissection of complex traits. *Science*, 265, 2037–48.

Landsteiner, K. & Levine, P. (1927a). A new agglutinable factor differentiating individual human bloods. *Proceedings of the Society of Experimental Biology*, 24, 600–2.

Landsteiner, K. & Levine, P. (1927b). Further observations on individual difference of human blood. *Proceedings of the Society of Experimental Biology*, 24, 941–2.

Landsteiner, K. & Levine, P. (1930). On the inheritance and racial distribution of agglutinable properties of human blood. *Journal of Immunology*, 18, 87–94.

Landsteiner, K. & Weiner, A.S. (1940). An agglutinable factor in human blood recognized by immune sera for rhesus blood. *Proceedings of the Society of Experimental Biology*, 43, 223.

Lane, R.A. & Sublett, A. (1972). The osteology of social organization. *American Antiquities*, 37, 186–201.

Larsen, C.S. (1994). In the wake of Columbus: Native population biology in the postcontact Americas. *Yearbook of Physical Anthropology*, 37, 109–54.

Larsen, C.S., Schoeninger, M.J., Van Der Merwe, N.J., Moore, K.M. & Lee-Thorp, J. (1992). Carbon and nitrogen isotopic signatures of human dietary change in the Georgia Bight. *American Journal of Physical Anthropology*, 89, 197–214.

Lasker, G.W. (1952). Environmental growth factors and selective migration. *Human Biology*, 24, 262–89.

Lasker, G.W. (1954). Photoelectric measurement of skin color in a Mexican mestizo population. *American Journal of Physical Anthropology*, 12, 115–21.

Lasker, G.W. (1995). The study of migrants as a strategy for understanding human biological plasticity. In *Human Variability and Plasticity*, ed. C.G.N. Mascie-Taylor & B. Bogin, pp. 110–14. Cambridge: Cambridge University Press.

Lathrop, D. (1970). *The Upper Amazon*. New York: Praeger.

Laughlin, W.S. (1975). Aleuts: Ecosystem, holocene history and Siberian origin. *Science*, 189, 507–15.

Laughlin, W.S., Harper, A.B. & Thompson, D.D. (1979). New approaches to the pre- and postcontact history of Arctic peoples. *American Journal of Physical Anthropology*, 51, 579–88.

Layrisse, M. (1968). Biological subdivisions of the Indian on the basis of genetic traits. In *Biomedical Challenges Presented by the American Indian*, pp. 35–9. Washington: World American Health Organization.

Layrisse, M., Arends, T. & Dominguez Sisco, R. (1955). Nuevo grupo sanguineo encontrado en descendienes de Indios. *Acta Medica Venezolana*, 3, 132–8.

Layrisse, Z., Layrisse, M., Heined,. H.D. & Wilbert, J. (1976). The histocompatibility system in the Warao Indians of Venezuela. *Science*, 194, 1135–8.

Leaf, A. (1973). Getting old. *Scientific American*, 229, 44–7.

Leakey, L.S.B., Simpson, R.D.E. & Clements, T. (1968). Archeological excavation in the Calico Mountains, California: Preliminary report. *Science*, 160, 1022.

Leche, S.M. (1936a). Dermatoglyphics and functional lateral dominance in Mexican Indians: IV. Chamulas. In *Measures of Men*, Vol. 7, pp. 289–310. New Orleans: Tulane University Press.

Leche, S.M. (1936b). Dermatoglyphics and functional lateral dominance in Mexican Indians. In *Measures of Men*, Vol. 7, pp. 211–20. New Orleans: Tulane University Press.

Leche, S.M. (1936c). The dermatoglyphics of the Tarascan Indians of Mexico. In *Measures of Men*, Vol. 7, pp. 319–28. New Orleans: Tulane University Press.

Leche, S.M. (1936d). Dermatoglyphics and functional lateral dominance in Mexican Indians: III. Zapotecas and Mixtecas. In *Measures of Men*, Vol. 7, pp. 229–77. New Orleans: Tulane University Press.

Leche, S.M. (1934). Dermatoglyphics and functional lateral dominance in Mexican Indians (Mayas and Tarahumaras). In *Studies in Middle America*, Vol. 5, pp. 27–40. New Orleans: Tulane University Press.

Leche, S.M., Gould, H.N. & Tharp, D. (1944). Dermatoglyphics and functional dominance in Mexican Indians. IV. The Zinancantecs, Huixtecs, Amatenangos, and Finca Tzeltals. *Middle American Research Records*, 1, 21–84.

Lees, F.C. & Byard, P.J. (1978). Skin colorimetry in Belize: I. Conversion formulae. *American Journal of Physical Anthropology*, 48, 515–22.

Lees, F.C. & Crawford, M.H. (1976). Anthropometric variation in Tlaxcaltecan populations. In *The Tlaxcaltecans: Prehistory, Demography, Morphology and Genetics*, Publications in Anthropology 7, ed. M.H. Crawford, pp. 61–80. Lawrence: University of Kansas Press.

Leonard, W., Crawford, M.H., Comuzzie, A.G. & Sukernik, R.I. (1992). New light on nutrition and peopling of the New World. *Arctic Research of the United States*, 6, 13–16.

Leonard, W., Crawford, M.H., Comuzzie, A.G. & Sukernik, R.I. (1994). Correlates of low serum lipid levels among the Evenki herders of Siberia. *American Journal of Human Biology*, 6, 329–38.

Leonard, W., Katzmarzyk, P.T., Comuzzie, A.G., Crawford, M.H. & Sukernik, R.I. (1994). Growth and nutritional status of the Evenki reindeer herders of Siberia. *American Journal of Human Biology*, 6, 339–50.

Leslie, P.W. & Gage, T.B. (1989). Demography and human population biology: Problems and progress. In *Human Population Biology*, ed. M.A. Little & J.D. Haas, pp. 15–44. Oxford: Oxford University Press.

Levin, M.G. (1958). *Ethnic Anthropology and the Problem of Ethnogenesis of the Peoples of the Far East.* Moscow: Trudy. (In Russian.)

Levin, M.G. (1963). *Ethnic Origins of the Peoples of Northeastern Asia*, ed. H.N. Michael. Toronto: University of Toronto Press.

Levin, M.G. & Potapov, L.P. (1964). *The Peoples of Siberia.* Chicago: University of Chicago Press.

Lewin, R. (1987). The first Americans are getting younger. *Science*, 238, 1230–2.

Li, T.-K. (1983). The absorption, distribution and metabolism of ethanol and its effects on nutrition and hepatic function. In *Medical and Social Aspects of Alcohol Abuse*, ed. B. Tabakoff, P.B. Sutker & C.L. Randall, pp. 47–77. New York: Plenum Press.

Liberty, M., Hughey, D.V. & Scaglion, R. (1976). Rural and urban Omaha Indian fertility. *Human Biology*, 48, 59–71.

Liberty, M., Scaglion, R. & Hughey, D.V. (1976). Rural and urban Seminole Indian fertility. *Human Biology*, 48, 741–56.

Lin, P.M. (1984). Anthropometry of Black Caribs. In *Current Developments in Anthropological Genetics.* Vol. 3. *Black Caribs: A Case Study in Biocultural Adaptation*, ed. M.H. Crawford, pp. 189–214. New York: Plenum Press.

Lin, P.M., Crawford, M.H. & Oronzi, M. (1979). Universals in Dermatoglyphics. In *Dermatoglyphics – Fifty Years Later*, Birth Defects Original Article Series XV (6), ed. W. Wertelecki & C.C. Plato, pp. 63–84. New York: Alan Liss.

Lin, P.M., Encisco, V.B. & Crawford, M.H. (1983). Dermatoglyphic inter- and intrapopulation variation among indigenous New Guinea groups. *Journal of Human Evolution*, 12, 103–23.

Lin, P.M., Encisco, V.B., Crawford, M.H., Hutchison, J., Sank, D. & Firschein, B.S. (1984). Quantitative analyses of the dermatoglyphic patterns of the Black Carib populations of Central America. In *Current Developments in Anthropological Genetics.* Vol. 3. *Black Caribs: A Case Study in Biocultural Adaptation*, ed. M.H.

Crawford, pp. 241–68. New York: Plenum Press.

Lin, P.M. (1976). Factor analysis of the dentition of three Tlaxcaltecan populations. In *The Tlaxcaltecans: Prehistory, Demography, Morphology, and Genetics*, Publications in Anthropology 7, ed. M.H. Crawford, pp. 93–119. Lawrence: University of Kansas Press.

Lin, S.J., Tanaka, K., Leonard, W.R., Gerelsaikhan, T., Dashnyam, B., Nyamkhishig, S., Hida, A., Nakahori, Y., Omoto, K. Crawford, M.H. & Nakagome, Y. (1994). A Y-associated allele is shared among a few ethnic groups of Asia. *Japanese Journal of Human Genetics*, 39, 299–304.

Lisker, R. & Babinsky, V. (1986). Admixture estimates in nine Mexican Indian groups and five east coast localities. *La Revista Investigaciones Clinica (Mex.)*, 38, 145–9.

Lisker, R., Perez-Briceno, R., Granados, J. & Babinsky, V. (1988). Gene frequencies and admixture estimates in the State of Puebla, Mexico. *American Journal of Physical Anthropology*, 76, 331–5.

Lisker, R., Ramirez, E., Perez-Briceno, R., Granados, J. &â (1990). Gene frequencies and admixture estimates in four Mexican urban centers. *Human Biology*, 62, 791–801.

Lisker, R., Zarate, G. & Rodriguez, E. (1967). Studies on several hematological traits of the Mexican population. IV. Serum polymorphisms in several Indian tribes. *American Journal of Physical Anthropology*, 27, 27–32.

Livingstone, F.B. (1991). On the origin of syphilis: An alternative hypothesis. *Current Anthropology*, 32, 587–9.

Llop, E. & Rothhammer, F. (1988). A note on the presence of blood groups A and B in pre-Columbian South America. *American Journal of Physical Anthropology*, 75, 107–11.

Long, J.C., Chakravarti, A., Boehm, C.D., Antonarakis, S. & Kazazian, H.H. (1990). Phylogeny of human beta-globin haplotypes and its implications for human evolution. *American Journal of Physical Anthropology*, 81, 113–30.

Long, J.K. (1966). A test of multiple-discriminant analysis as a means of determining evolutionary changes and intergroup relationships in Physical Anthropology. *American Anthropologist*, 68, 444–64.

Long, J.C. & Smouse, P.E. (1983). Intertribal gene flow between the Ye'cuana and Yanomama: Genetic analysis of an admixed village.

Ameican Journal of Physical Anthropology, 61, 411–22.

Loomis, W.F. (1967). Skin pigment regulation of vitamin-D biosynthesis in man. *Science*, 157, 501–6.

Lorenz, J.G. & Smith, D.G. (1996). Distribution of four founding mtDNA haplotypes among native North Americans. *American Journal of Physical Anthropology*, 101, 307–23.

Lucciola, L., Kaita, H., Anderson, J. & Emery, S. (1974). The blood groups and red cell enzymes of a sample of Cree Indians. *Canadian Journal of Genetics and Cytology*, 16, 691–6.

Lynch, T.F. (1990). Glacial-age man in South America? A critical review. *American Antiquity*, 55, 12–36.

Lynch, T.F. & Kennedy, K.A.R. (1970). Early human cultural and skeletal remains from Guitarrero Cave, Northern Peru. *Science*, 169, 1307–9.

MacNeish, R. (1964). The origins of New World civilization. *Scientific American*, 211, 29–37.

MacNeish, R. (1976). Early man in the New World. *American Scientist*, 64, 316–27.

Maffei, M., Halaas, J., Ravussin, E., Prately, R.E., Lee, G.H., Zhang, Y., Fei, H., Kim, S., Lallone, R., Ranganathan, S., Kern, P.A. & Friedman, J.M. (1995). Leptin levels in human and rodent: Measurement of plasma leptin and ob RNA in obese and weight-reduced subjects. *Nature Medicine*, 11, 1155–61.

Majumder, P.A., Laughlin, W.S. & Ferrell, R.E. (1988). Genetic variation in the Aleuts of the Pribilof Islands and the Eskimos of Kodiak Island. *American Journal of Physical Anthropology*, 76, 481–8.

Malecot, G. (1948). *Les Mathematiques de l'Heredité*. Paris: Masson.

Malecot, G. (1950). Quelques schemas probabilistes sur la variabilite des populations naturelles. *Annales Universitatis de Lyon, Science Series A*, 13, 37–60.

Malecot, G. (1959). Les modeles stochastiques en genetique de population. *Publ. Inst. Statist. Univ. Paris*, 8, 173–210.

Malina, R.M. & Hines, J. (1977a). Seasonality of births in a rural Zapotec-speaking municipio. *Human Biology*, 49, 125–38.

Malina, R.M. & Hines, J. (1977b). Differential age effects in seasonal variation of mortality in a rural Zapotec-speaking municipio. *Human Biology*, 49, 415–28.

Mancilha-Carvalho, J.J. & Crews, D.E. (1990).

Lipid profiles of Yanomamo Indians. *Preventive Medicine*, 19, 66–75.

Mantel, N. (1967). The detection of disease clustering and generalized regression approach. *Cancer Research*, 27, 209–20.

Marcellino, A.J., Da Rocha, F.J. & Salzano, F.M. (1978). Size and shape differences among six South American Indian tribes. *Annals of Human Biology*, 5, 69–74.

Markow, T.A. & Martin, J.F. (1993). Inbreeding and developmental stability in a small human population. *Annals of Human Biology*, 20, 389–94.

Marshall, E. (1990). Clovis counterrevolution. *Science*, 249, 738–41.

Martin, P.S. (1973). The discovery of America. *Science*, 179, 969–72.

Matis, J.A. & Zwewer, T.J. (1971). Odontognathic discrimination of United States Indian and Eskimo groups. *Journal of Dental Research*, 50, 1245–8.

Matson, G.A. (1970). Distribution of blood groups in Mexico and Central America. In *Handbook of Middle American Indians*, Vol. 9, ed. T.D. Stewart, pp. 105–47. Austin: University of Texas Press.

Matson, G.A. & Schrader, H.F. (1933). Blood grouping among Blackfeet and Blood tribes of American Indians. *Journal of Immunology*, 25, 155–63.

Matsumoto, H. & Miyazaki, T. (1972). Gm and Inv allotypes of the Ainu in Hidaka area Hokkaido. *Japanese Journal of Human Genetics*, 17, 20–6.

Matsunaga, E. (1962). The dimorphism in human normal cerumen. *Annals of Human Genetics*, 25, 273–86.

Mayberry, R.H. & Lindemann, R.D. (1963). A survey of chronic disease and diet in Seminole Indians in Oklahoma. *American Journal of Clinical Nutrition*, 13, 127–34.

Mazess, R.B. & Forman, S.H. (1979). Longevity and age exaggeration in Vilcabamba, Ecuador. *Journal of Gerontology*, 34, 94–8.

Mazess, R.B. & Mathisen, R.W. (1982). Lack of unusual longevity in Vilcabamba, Ecuador. *Human Biology*, 54, 517–24.

McAlpine, P.J., Chen, S.H., Cox, D.W., Dossetor, J.B., Giblett, E., Steinberg, A.G. & Simpson, N.E. (1974). Genetic markers in blood in a Canadian Eskimo population with a comparison of allele frequencies in circumpolar populations. *Human Heredity*, 24, 114–42.

McAlpine, P.J. & Simpson, N.E. (1976). Fertility and other demographic aspects of the Canadian Eskimo communities of Igloolik and Hall Beach. *Human Biology*, 48, 113–38.

McComb, J. (1996). *The Effect of Unique Historic and Prehistoric Events on the Gene Pool of the Altai-Kizhi: A Study of Five Variable Number Tandem Repeat (VNTR) Loci*. Lawrence: University of Kansas. (Unpublished M.A. Thesis.)

McComb, J., Blagitko, N., Comuzzie, A.G., Schanfield, M.S., Sukernik, R.I., Leonard, W.R. & Crawford, M.H. (1995). VNTR DNA variation in Siberian indigenous populations. *Human Biology*, 67, 217–29.

McComb, J., Crawford, M.H., Leonard, W.R., Schanfield, M.S. & Osipova, L. (1996). Applications of DNA fingerprints for the study of genetic structure of human populations. In *Genomes of Plants and Animals: 21st Stadler Genetics Symposium*, ed. J.P. Gustafson & R.B. Flavell, pp. 31–46. New York: Plenum Press.

McComb, J., Crawford, M.H., Osipova, L., Karaphet, T., Posukh, O. & Schanfield, M.S. (1996). DNA inter-populational variation in Siberian indigenous populations: The Mountain Altai. *American Journal of Human Biology*, 8(5), 599–608.

McCullough, J.M. & Giles, E. (1970). Human cerumen types in Mexico and New Guinea: A humidity-related polymorphism in 'Mongoloid' peoples. *Nature*, 226, 460–2.

McGarvey, S.T. & Baker, P.T. (1979). The effects of modernization and migration on Samoan blood pressures. *Human Biology*, 51, 461–79.

Means, P.A. (1973). *Ancient Civilizations of the Andes*. New York: Charles Scribner's Sons.

Meier, R.J. (1966). Fingerprint patterns from Karluk Village, Kodiak Island. *Arctic Anthropology*, 111-1, 206–10.

Meier, R.J. (1978). Dermatoglyphic variation in five Eskimo groups from Northwestern Alaska. In *Dermatoglyphics: An International Perspective*, ed. J. Mavalwala, pp. 145–52. The Hague: Mouton Publications.

Melartin, L. & Blumberg, B. (1966). Albumin Naskapi: A new variant of serum albumin. *Science*, 153, 1664–6.

Melartin, L. & Blumberg, B. (1966). Inherited variants of human serum albumin. *Clinical Research*, 14, 482.

Meltzer, D.J. (1989). Why don't we know when the first people came to North America? *American Antiquity*, 54, 471–90.

Menozzi, P., Piazza, A. & Cavalli-Sforza, L.L. (1978). Synthetic maps of human gene frequencies in Europeans. *Science*, 201, 786–92.

Merbs, C.F. (1992). A New World of infectious disease. *Yearbook of Physical Anthropology*, 35, 3–42.

Merriwether, D.A., Hall, W.H., Vahlne, A. & Ferrell, R.E. (1996). mtDNA variation indicates Mongolia may have been the source for the founding population for the New World. *American Journal of Human Genetics*, 59, 204–212.

Merriwether, D.A., Rothhammer, F. & Ferrell, R.E. (1995). Distribution of the four founding lineage haplotypes in native Americans suggests a single wave of migration for the New World. *American Journal of Physical Anthropology*, 98, 411–30.

Mestriner, M.A., Simoes, A.L. & Salzano, F.M. (1980). New studies on the Esterase D polymorphism in South American Indians. *American Journal of Physical Anthropology*, 52, 95–101.

Miall, W.E. (1967). Age, sex, body habitus, family. In *Epidemiology of Hypertension*, ed. J. Stamler, R. Stamler & T.N. Pullman, pp. 60–9. New York: Grune and Stratton.

Michael, H.N. (1984). Absolute chronologies of late Pleistocene and early Holocene cultures of Northeastern Asia. *Arctic Anthropology*, 21, 1–68.

Milan, L.C. (1973). Ethnohistory of disease and medical care among the Aleut. *Anthropological Papers of the University of Alaska*, 16, 15–40.

Miller, V.T. (1994). Lipids, lipoproteins, women, and cardiovascular disease. *Atherosclerosis*, 108, S73–S82.

Millon, R. (1973). *Urbanization at Teotihuacan: The Teotihuacan Map*, Vol. 1. Austin: University of Texas Press.

Miura, T., Fukunaga, T., Igarishi, T., Yamashita, M., Ido, E., Funahashi, S.-I., Ishida, T., Washio, K., Ueda, S., Hashimoto, M., Yoshida, M., Osame, M., Singhal, B.S., Zaninovic, V., Cartier, L., Sonoda, S., Tajima, K., Ina, Y., Gojobori, T. & Hayami, M. (1994). Phylogenetic subtypes of human T-lymphotropic virus type I and their relations to the anthropological background. *Proceedings of the National Academy of Sciences, USA*, 91, 1124–7.

Mochanov, L.A. (1978a). The paleolithic of northeast Asia and the problem of the first peopling of America. In *Early Man in America from a Circum-Pacific Perspective*, ed. A.L. Bryan pp. 67–89. Edmonton, University of Alberta Press.

Mochanov, L.A. (1978b). Stratigraphy and absolute chronology of the Paleolithic of Northeast Asia. In *Early Man in America from a Circum-Pacific Perspective*, pp. 54–66. Edmonton, Canada, Archaeological Researches International.

Mooney, J. (1910). Population. In *Handbook of American Indians North of Mexico*, ed. F.W. Hodge, pp. 286–7. Washington, D.C.: Bureau of American Ethnology, Bulletin 30, Part 2.

Mooney, J. (1928). The aboriginal population of America north of Mexico. *Smithsonian Miscellaneous Collections*, 80, 1–40.

Mooree, J.G., Grundmann, A.W., Hall, H.J. & Fry, G.F. (1974). Human fluke infection in Glen Canyon at AD 1250. *American Journal of Physical Anthropology*, 41, 115–18.

Moorrees, C.F.A. (1957). *The Aleut Dentition*. Cambridge: Harvard University Press.

Morrell, V. (1990). Confusion in earliest America. *Science*, 248, 439–41.

Morrell, V. (1994). Did early humans reach Siberia 500,00 years ago? *Science*, 263, 611–12.

Morgan, K. (1968). *The Genetic Demography of a Small Navajo Community*. Ann Arbor: University of Michigan. (Ph.D. Dissertation.)

Morgan, K. (1973). Historical demography of a Navajo community. In *Methods and Theories of Anthropological Genetics*, ed. M.H. Crawford & P.L. Workman, pp. 263–314. Albuquerque: University of New Mexico Press.

Motulsky, A.G. & Morrow, A. (1968). Atypical cholinesterase gene E^a: 1. Rarity in Negroes and most orientals. *Science*, 159, 202–4.

Mourant, A.E., Kopec, A.C. & Domaniewska-Sobczak, K. (1976). *The Distibution of the Human Blood Groups*. London: Oxford University Press.

Mourant, A.E., Tills, D. & Domaniewska-Sobczak, K. (1976). Sunshine and the geographical distribution of the alleles of Gc system of plasma proteins. *Human Genetics*, 33, 307–14.

Mouro, I., Colin, Y., Cherif-Zahar, B., Cartron, J.-P. & Le Van Kim, C. (1993). Molecular genetic basis of the human Rhesus blood group system. *Nature Genetics*, 5, 62–5.

Murad, T.A. (1975). *The North Alaskan Eskimo Intra-population Variation for Palmar Dermatoglyphics*. Bloomington: Indiana University. (Unpublished Ph.D. Dissertation.)

Murillo, F., Rothhammer, F. & Llop, E. (1977). The Chipaya of Bolivia: Dermatoglyphics and ethnic relationships. *American Journal of Physical Anthropology*, 46, 45–50.

National Diabetes Data Group. (1986). Classification and diagnosis of diabetes mellitus and other categories of glucose intolerance. *Diabetes*, 28, 1039–57.

Neel, J.V. (1962). Diabetes mellitus: A "thrifty" genotype rendered detrimental by progress? *American Journal of Human Genetics*, 14, 353–62.

Neel, J.V. (1982) The thrifty genotype revisited. In *The Genetics of Diabetes Mellitus*, ed. J. Kobberling & R. Tattersall, pp. 283–93. New York: Academic Press.

Neel, J.V., Biggar, R.J. & Sukernik, R.I. (1994). Virologic and genetic studies relate Amerind origins to the indigenous people of the Mongolia/Manchuria/southeastern Siberia region. *Proceedings of the National Academy of Sciences, USA*, 91, 10737–41.

Neel, J.V., Centerwall, W.R., Chagnon, N.A. & Casey, H.L. (1970). Notes on the effects of measles and measles vaccine in a virgin-soil population of South American Indians. *American Journal of Epidemiology*, 91, 418–29.

Neel, J.V. & Chagnon, N. (1968). Demography of two tribes of primitive relatively unacculturated American Indians. *Proceedings of the National Academy of Science USA*, 59, 680–9.

Neel, J.V., Gershowitz, H., Spielman, R.S., Migliazzi, E.C., Salzano, F. & Oliver, W.J. (1977a). Genetic studies of the Macushi and Wapishana Indians. II. Data on 12 genetic polymorphism of the red cell and serum proteins: Gene flow between the tribes. *Human Genetics*, 37, 207–19.

Neel, J.V., Layrisse, M. & Salzano, F.M. (1977b). Man in the tropics: The Yanomama Indians. In *Population Structure and Human Variation*, ed. G.A. Harrison, pp. 109–42. Cambridge: Cambridge University Press.

Neel, J.V., Rothhammer, F. & Lingoes, J. (1974). The genetic structure of a tribal population, the Yanomama Indians. X. Agreement between representations of village distances based on a different set of characteristics. *American Journal of Human Genetics*, 26, 281–303.

Neel, J.V. & Salzano, F.M. (1967). Further studies on the Xavante Indians. X. Some hypotheses-generalizations resulting from the studies. *American Journal of Human Genetics*, 19, 554–74.

Neel, J.V. & Ward, R.H. (1972). Genetic structure of a tribal population, the Yanomama Indians. VI. Analysis by F-statistics, including a comparison with the Makiritare and Xavante. *Genetics*, 72, 639–66.

Neel, J.V. & Weiss, K.M. (1975). The genetic structure of a tribal population, the Yanomama Indians. *American Journal of Physical Anthropology*, 42, 25–52.

Nei, M. (1973). Analysis of gene diversity in subdivided populations. *Proceedings of the National Academy of Sciences USA*, 70, 3321–3.

Nei, M. (1977). F-statistics and analyses of genetic diversity in subdivided populations. *Annals of Human Genetics*, 41, 225–33.

Nei, M. (1985). *Molecular Evolutionary Genetics*. New York: Columbia University Press.

Nei, M., Stephens, J.C. & Saitou, N. (1985). Methods for computing the standard errors of branching points in an evolutionary tree and their application to molecular data from humans and apes. *Molecular Biology and Evolution*, 2, 66–85.

Neumann, G.K. (1952). Archeology and race in the American Indian. In *Archeology of Eastern United States*, ed. J.B. Griffin, pp. 13–34. Chicago: University of Chicago Press.

Newman, M.T. (1951). The sequence of Indian physical types in South America. In *Physical Anthropology of the American Indian*, ed. W.S. Laughlin, ppp. 69–97. New York: Viking Fund, Inc.

Newman, M.T. (1953). The application of ecological rules to the racial anthropology of the aboriginal New World. *American Anthropologist*, 55, 311–27.

Newman, M.T. (1960a). Adaptations in the physique of American aborigines to nutritional factors. *Human Biology*, 32, 288–313.

Newman, M.T. (1960b). Population analysis of finger and palm prints in highland and lowland Maya Indians. *American Journal of Physical Anthropology*, 18, 45–58.

Newman, M.T. (1970). Dermatoglyphics. In *Handbook of Middle American Indians (Physical Anthropology)*, Vol. 9, ed. R. Wauchope & T.D. Stewart, pp. 167–79. Austin: University of Texas Press.

Newman, M.T. (1976). Aboriginal New World epidemiology and medical care, and the impact of Old World disease imports. *American Journal of Physical Anthropology*, 45, 667–72.

Nikitin, Y.P. (1985). Some health problems of man in the Soviet Far North. In *Circumpolar Health 84*, ed. R. Fortuine, pp. 8–10. Seattle: University of Washington Press.

Niswander, J.D., Brown, K.S., Iba, B.Y., Leyshon, W.C. & Workman, P.L. (1970). Population studies on Southwestern Indian tribes. I. History, culture and genetics of the Papago. *American Journal of Human Genetics*, 23, 7–23.

Novoradovsky, A.G., Spitsyn, V.A., Duggirala, R. & Crawford, M.H. (1993). Population genetics and structure of Buryats from Lake Baikal region. *Human Biology*, 65, 689–710.

Novoradovsky, A., Tsai, S-J.L., Goldfarb, L., Peterson, R., Long, J.C. & Goldman, D. (1995). Mitochondrial aldehyde dehydrogenase polymorphism in Asian and American Indian population: Detection of new ALDH2 alleles. *Alcoholism: Clinical and Experimental Research*, 19(5), 1105–10.

Nutels, N. (1968). Medical problems of newly contacted Indian groups. In *Biomedical Challenges Presented by the American Indian*, pp. 68–76. Washington, D.C.: Pan American Health Organization.

Okladnikov, A.P. (1964). Ancient population of Siberia and its culture. In *The Peoples of Siberia*, ed. M.G. Levin & L.P. Potapov, pp. 13–98. Chicago: University of Chicago Press.

Okladnikov, A.P. (1965). The Soviet Far East in Antiquity: An Archaeological and Historical Study of the Maritime Region of the USSR, ed. H.N. Michael, Arctic Institute for North America, Toronto: University of Toronto Press.

Omoto, K. (1972). Polymorphic and genetic affinities of the Ainu of Hokkaido. *Human Biology of Oceania*, 1, 278–88.

O'Rourke, D.H. & Crawford, M.H. (1976). Odontometric analysis of four Tlaxcaltecan communities. In *The Tlaxcaltecans: Prehistory, Demography, Morphology and Genetics*, Publications in Anthropology 7, ed. M.H. Crawford, pp. 81–92. Lawrence: University of Kansas Press.

O'Rourke, D.H. & Crawford, M.H. (1980). Odontometric microdifferentiation of transplanted Mexican Indian populations: Cuanalan and Saltillo. *American Journal of Physical Anthropology*, 52, 421–34.

O'Rourke, D.H. & Lichty, A.S. (1989). Spatial analysis and gene frequency maps of Native

North American populations. *Collegium Antropologicum*, 13, 73–84.

O'Rourke, D.H. & Suarez, B.K. (1985). Patterns and correlates of genetic variation in South Amerindians. *Annals of Human Biology*, 13, 13–31.

O'Rourke, D.H., Suarez, B.K. & Crouse, J.D. (1985). Genetic variation in North Amerindian populations: Covariance with climate. *American Journal of Physical Anthropology*, 67, 241–50.

Ortner, D.J. (1972). Ecological factors in disease among North American archeological skeletal samples. *American Journal of Physical Anthropology*, 37, 447.

Ortner, D.J. & Putschar, W.G.J. (1985). *Identifications of Pathological Conditions in Human Skeletal Remains*. Smithsonian Contributions to Anthropology, No. 28. Washington D.C.: The Smithsonian.

Ortner, D.J., Tuross, N. & Stix, A.I. (1992). New approaches to the study of disease in archeological New World populations. *Human Biology*, 64, 337–600.

Ortner, D.J. & Utermohle, C.J. (1981). Polyarticular inflammatory arthritis in a pre-Columbian skeleton from Kodiak Island, Alaska, USA. *American Journal of Physical Anthropology*, 56, 23–32.

Osame, M., Usuku, K., Izumo, S., Ijichi, N., Amitani, H., Igata, A., Matsumoto, M. & Tara, M. (1986). HTLV-I associated myelopathy: A new clinical entity. *Lancet*, i, 8488, 1031–2.

Osborne, R.H. (1958). Serology in physical anthropology. *American Journal of Physical Anthropology*, 16, 187–95.

Ottensooser, F. (1944). Calculo do gran de mistura racial atranes dos grupos sanguineous. *Revista Brasileira de Biologia*, 4, 531–7.

Ousley, S.D. (1995). Relationship between Eskimos, Amerindians, and Aleuts: Old data, new perspectives. *Human Biology*, 67, 427–58.

Overfield, T.M. (1975). *Investigation of an Unusually High Frequency of the Silent Allele of Pseudocholinesterase in Southwestern Alaskan Eskimos*. Boulder: University of Colorado. (Unpublished Ph.D. Dissertation.)

Owsley, D.W. & Bass, W.M. (1979). A demographic analysis of skeletons from the Larson site (39ww2) Walworth County, South Dakota: Vital statistics. *American Journal of Physical Anthropology*, 51, 145–54.

Pääbo, S., Dew, K., Frazier, B.S. & Ward, R.H. (1990). Mitochondrial evolution and the peopling of the Americas. *American Journal of Physical Anthropology*, 81, 277.

Pääbo, S., Gifford, J.A. & Wilson, A.C. (1988). Mitochondrial DNA sequences from a 7000-year-old brain. *Nucleic Acids Research*, 16, 9775–87.

Pawson, S. & Milan, F.A. (1974). Cerumen types in two Eskimo communities. *American Journal of Physical Anthropology*, 41, 431–2.

Pena, H.F., Salzano, F.M. & Da Rocha, F.J. (1972). Dermatoglyphics of Brazilian Cayapo Indians. *Human Biology*, 44, 225–41.

Pena, S.D.J., Santos, F.R., Bianchi, N., Bravi, C.M., Carnese, F.R., Rothhammer, F., Gerelsaikhan, T. et al. (1995). Identification of a major founder Y-chromosome haplotype in Amerindians. *Nature Genetics*, 11, 15–6.

Petersen, W. (1975). A demographer's view of prehistoric demography. *Current Anthropology*, 16, 227–45.

Pielou, E.C. (1991). *After the Ice Age*. Chicago: University of Chicago Press.

Pinto-Cisternas, J. & Figueroa, H. (1968). Genetic structure of a population of Valparaiso. II. Distribution of two dental traits with anthropological importance. *American Journal of Physical Anthropology*, 29, 339–48.

Plaut, G., Ikin, E.W., Mourant, A.E., Sanger, R. & Race, R.R. (1953). A new blood-group antibody, anti-Jka. *Nature*, 171, 431.

Polednak, A.P. (1989). *Racial and Ethnic Differences in Disease*. Oxford: Oxford University Press.

Pollitzer, W.S., Phelps, D.S., Waggoner, R.E. & Leyshon, W.C. (1967). Catawba Indians: Morphology, genetics and history. *American Journal of Physical Anthropology*, 26, 5–14.

Polonovski, M. & Jayle, M.F. (1938). Existence dans le plasma sanguin d'une substance activant l'action peroxydasique de l'hemoglobine. *Compte Rendu Sociètè Biologie*, 129, 457.

Polonovski, M. & Jayle, M.F. (1940). Sur la preparation d'une nouvelle fraction des proteines plasmatiques, l'haptoglobine. *Compte Rendu Academie de Science*, 211, 517.

Popham, R.E. (1953). A comparative analysis of the digital patterns of Eskimo from Southampton Island. *American Journal of Physical Anthropology*, 11, 203–13.

Post, R.H, Neel, J.V. & Schull, W.J. (1968). Tabulations of phenotype and gene frequencies for 11 different genetic systems studied in American Indians. In *Biomedical Challenges Pre-*

sented by the American Indians, pp. 141–85. Washington, D.C.: Pan American Health Organization.

Posukh, O.L., Wiebe, V.P., Sukernik, R.I., Osipova, L.P., Karaphet, T.M. & Schanfield, M.S. (1990). Genetic study of the Evens, an ancient human population of Eastern Siberia. Human Biology, 62, 457–65.

Powell, M.L. (1986). Late prehistoric community health in the central deep south: Biological and social dimensions of the Mississippian chiefdom at Moundville, Alabama. In Skeletal Analysis in Southeastern Archeology, 24, North Carolina Council Publication.

Powell, M.L. (1996). Ancient diseases, modern perspectives: Tuberculosis and treponematosis in the age of agriculture. American Journal of Physical Anthropology, 22, 189.

Price, J.A. (1988). Indians of Canada. Cultural Dynamics. Salem: Sheffield Publishing.

Price, R.A., Charles, M.A., Petti, D.J. & Knowler, W.C. (1993). Obesity in Pima Indians: Large increases among post-World War II birth cohorts. American Journal of Physical Anthropology, 92, 473–9.

Ramenofsky, A.F. (1982). The Archaeology of Population Collapse: Native American Response to the Introduction of Infectious Disease, Seattle: University of Washington. (Ph.D. Dissertation.)

Ramenofsky, A.F. (1987). Vectors of Death: The Archaeology of European Contact. Albuquerque: University of New Mexico Press.

Ramenofsky, A.F. (1994). Book review of Disease and Death in Early Colonial Mexico: Simulating Amerindian Depopulation. American Antiquity, 59, 382–3.

Rasmussen, K. (1931). The Netsilik Eskimos. Report Fifth Thule Expedition 8. Copenhagen: Nordisk Forlag.

Ravussin, E., Bennett, P.H., Valencia, M.E., Schultz, L.O. & Esparza, J. (1994). Effects of a traditional lifestyle on obesity in Pima Indians. Diabetes Care, 17, 1067–74.

Reed, T.E. (1978). Racial comparison of alcohol metabolism: Background, problems, and results. Alcoholism: Experimental Research, 2, 83–7.

Reed, T.E., Kalant, H., Gibbons, R.J., Kapur, B.M. & Rankin, J.G. (1976). Alcohol and acetaldehyde metabolism in Caucasians, Chinese, and Amerinds. Canadian Medical Association Journal, 115, 851–5.

Reichs, K.J. (1989). Treponematosis: A possible case from the Late Prehistoric of North Carolina. American Journal of Physical Anthropology, 79, 289–304.

Reinhard, K.J. (1988). Cultural ecology of prehistoric parasitism on the Colorado Plateau as evidenced by coprology. American Journal of Physical Anthropology, 77, 355–66.

Relethford, J.H. (1980a). Bioassay of kinship from continuous traits. Human Biology, 52, 689–700.

Relethford, J.H. (1980b). Effects of English admixture and geographic distance on anthropometric variation and genetic structure in 19th century Ireland. American Journal of Physical Anthropology, 76, 111–24.

Relethford, J.H. (1991). Genetic drift and anthropometric variation in Ireland. Human Biology, 63, 155–65.

Relethford, J.H. & Lees, F.C. (1981). Admixture and skin color in the transplanted Tlaxcaltecan population of Saltillo, Mexico. American Journal of Physical Anthropology, 56, 259–67.

Relethford, J.H., Lees, F.C. & Crawford, M.H. (1981). Population structure and anthropometric variation in rural western Ireland: Isolation by distance and analysis of the residuals. American Journal of Physical Anthropology, 55, 233–45.

Rex, D.K., Bosron, W.F., Smialek, J.E. & Li, T.-K. (1985). Alcohol and aldehyde dehydrogenase isoenzymes in North American Indians. Alcohol Clinical Experimental Research, 9, 147–52.

Rife, D.C. (1968). Finger and palmar dermatoglyphics in Seminole Indians of Florida. American Journal of Physical Anthropology, 28, 119–26.

Rife, D.C. (1972). Dermatoglyphics of Cherokee and Mohawk Indians. Human Biology, 44, 81–5.

Rivet, P. (1924). Langues americaines. In Les Langues du Monde, pp. 597–712. Paris: Collection Linguistique.

Roberts, D.F. (1984). Anthropogenetics in a hybrid population. The Black Carib studies. In Current Developments in Anthropological Genetics. Vol. 3. A Case Study in Biocultural Adaptation, ed. M.H. Crawford, pp. 381–9. New York: Plenum Press.

Roberts, D.F. & Coope, E. (1975). Components of variation in a multifactorial character: A dermatoglyphic analysis. *Human Biology*, 47, 169–88.

Roberts, D.F. & Hiorns, R.W. (1962). The dynamics of racial intermixture. *American Journal of Human Genetics*, 14, 261–77.

Roberts, D.F. & Hiorns, R.W. (1965). Methods of analysis of the genetic composition of a hybrid population. *Human Biology*, 37, 38–43.

Rode, A., Shephard, R.J., Vloshinsky, P.E. & Kuksis, A. (1995). Plasma fatty acid profiles of Canadian Inuit and Siberian nGanasan. *Arctic Medical Research*, 54, 422–5.

Rodriguez, H., Rodriguez, E.D., Loria, A. & Lisker, R. (1963). Studies on several genetic hematological traits of the Mexican population. V. Distribution of blood group antigens in Nahuas, Yaquis, Tarahumaras, Tarascos and Mixtecos. *Human Biology*, 35, 350–60.

Rogan, P.K. & Lentz, S.E. (1994). Molecular genetic evidence suggesting treponematosis in pre-Columbian, Chilean mummies. *American Journal of Physical Anthropology*, S18, 171–2.

Rogers, R.A. (1985). Glacial geography and native North American languages. *Quaternary Research*, 23, 130–7.

Rogers, R.A. (1986). Language, human speciation, and ice age barriers in northern Siberia. *Canadian Journal of Anthropology*, 5, 11–22.

Rogers, R.A., Martin, L.D. & Nicklas, T.D. (1990). Ice-age geography and the distribution of native North American languages. *Journal of Biogeography*, 17, 131–43.

Rogers, R.A., Rogers, L.A., Hoffmann, R.S. & Martin, L.D. (1991). Native American biological diversity and the biogeographic influence of Ice Age refugia. *Journal of Biogeography*, 18, 623–30.

Rogers, R.A., Rogers, L.A. & Martin, L.D. (1992). How the door opened: The peopling of the New World. *Human Biology*, 64, 281–302.

Roisenberg, I. & Morton, N.E. (1970). Population structure of blood groups in Central and South American Indians. *American Journal of Physical Anthropology*, 32, 373–76.

Rokala, D.A., Polesky, H.F. & Matson, G.A. (1977). The genetic composition of reservation populations: The Blackfeet reservation, Montana, U.S.A. *Human Biology*, 49, 19–29.

Roosevelt, A.C., Lima da Costa, M., Lopes Machado, C., Michab, M., Mercier, N., Valladas, H., Feathers, J., Bernett, W., Imazio da Silveira, M., Henderson, A., Silva, J., Chernoff, B., Reese, D.S., Holman, J.A., Toth, N. & Schick, K. (1996). Paleoindian cave dwellers in the Amazon: The peopling of the Americas. *Science*, 272, 373–84.

Rosenblat, A. (1945). *La Poblacion Indigena de America desde 1492 Hasta La Actualidad*. Buenos Aires: Institution Cultural Espanola.

Rothhammer, F. (1990). Ethnogenesis and affinities to other South American aboriginal populations. In *The Aymara. Strategies in Human Adaptation to a Rigorous Environment*, ed. W.J. Schull, F. Rothhammer & S.A. Barton, pp. 203–10. Dordrecht: Kluwer Academic Publishers.

Rothhammer, F., Allison, M.J., Nunez, L., Standen, V. & Arriaza, B. (1985). Chagas' disease in pre-Columbian South America. *American Journal of Physical Anthropology*, 68, 495–8.

Rothhammer, F., Benado, M. & Pereira, G. (1971). Variability of two dental traits in Chilean Indian and mixed population. *Human biology*, 43, 309–17.

Rothhammer, F., Chakraborty, R. & Ferrell, R.E. (1990). Intratribal genetic differentiation as assessed through electrophoresis. In *The Aymara. Strategies in Human Adaptation to a Rigorous Environment*, ed. W.J. Schull, F. Rothhammer & S.A. Barton, pp. 193–202. Dordrecht: Kluwer Academic Publishers.

Rothhammer, F., Chakraborty, R. & Llop, E. (1977). A collation of gene and dermatoglyphic diversity at various levels of population differentiation. *American Journal of Physical Anthropology*, 46, 51–60.

Rothhammer, F., Chakraborty, R. & Llop, E. (1979). Dermatoglyphic variation among South American tribal populations and its association with marker gene, linguistic, and geographic distances. *Dermatoglyphics – Fifty Years Later. Birth Defects: Original Article Series*, ed. W. Wertelecki & C.C. Plato, pp. 269–76. New York: Alan R. Liss.

Rothhammer, F., Neel, J.V., da Rocha, F. & Sundling, G.Y. (1973). The genetic structure of a tribal population, the Yanomama Indians. VIII. Dermatoglyphic differences among villages. *American Journal of Human Genetics*, 25, 152–66.

Rothhammer, F. & Silva, C. (1989). Peopling of Andean South America. *American Journal of Physical Anthropology*, 78, 403–10.

Rothhammer, F. & Spielman, R.S. (1972). Anthropometric variation in the Aymara: Genetic, geographic, and topographic contributions. *American Journal of Human Genetics*, **24**, 371–80.

Rothschild, B.M., Turner, K.R. & Deluca, M.A. (1988). Symmetrical erosive peripheral polyarthritis in the Late Archaic Period of Alabama. *Science*, **241**, 1498–1501.

Rouse, I. (1949). The Caribs. In *Handbook of South American Indians*, 4, pp. 547–65. Washington, D.C.: Smithsonian Institution.

Roy, R.L. (1933). *The Book of Chilam of Chumayel*. Washington, D.C.: Carnegie Institute. (Translated and edited.)

Roychoudhury, A.K. (1977). Gene differentiation in three tribes of American Indians. *Human Heredity*, **27**, 389–92.

Roychoudhury, A.K. & Nei, M. (1988). *Human Polymorphic Genes – World Distribution*. Oxford: Oxford University Press.

Ruff, C.B. (1981). A reassessment of demographic estimates for Pecos Pueblo. *American Journal of Physical Anthropology*, **54**, 147–51.

Rychkov, Y.G., Perevozchykov, I., Sheremetyeva, V.A. & Volkova, T.V. (1969). On the population genetics of the native peoples of the Siberian Eastern Sayans. *Voprosi Antropologii*, **31**, 3–32. (In Russian.)

Rychkov, Y.G. & Sheremetyeva, V.A. (1980). The genetics of circumpolar populations of Eurasia related to the problem of human adaptation. *The Human Biology of Circumpolar Populations*, IBP 21, ed. F.A. Milan, pp. 37–80. Cambridge: Cambridge University Press.

Rychkov, Y.G. & Sheremetyeva, V.A. (1977). The genetic process in the system of ancient human isolates in North Asia. In *Population Structure and Human Variation*, ed. G.A. Harrison, pp. 47–108. London: Cambridge University Press.

Salo, W.L., Aufderheide, A.C., Buikstra, J. & Holcomb, T.A. (1994). Identification of *Mycobacterium tuberculosis* DNA in a pre-Columbian Peruvian mummy. *Proceedings of the National Academy of Sciences, USA*, **91**, 2091–4.

Salzano, F.M. (1961). Studies on the Caingang Indians. I. Demography. *Human Biology*, **33**, 110–30.

Salzano, F.M. (1961). Demographic studies on Indians from Santa Catarina, Brazil. *Acta Genetice Medicae et Gemellologiae*, **13**, 278–94.

Salzano, F.M. (1965). Selection intensity in Brazilian Caingang Indians. *Nature*, **199**, 514.

Salzano, F.M. (1968a). Survey of the unaccultured Indians of Central and South America. In *Biom-*

edical Challenges Presented by the American Indian, pp. 59–66. New York: Pan American Health Organization.

Salzano, F.M. (1968b). Intra- and inter-tribal genetic variability in South American Indians. *American Journal of Physical Anthropology*, **28**, 183–90.

Salzano, F.M. (1971). Demographic and genetic interrelationships among the Cayapo Indians of Brazil. *Social Biology*, **18**, 148–57.

Salzano, F.M. (1975). Interpopulation variability in polymorphic systems. In *The Role of Natural Selection in Human Evolution*, ed. F.M. Salzano, pp. 217–29. New York: Elsevier.

Salzano, F.M., Black F.L., Callegari-Jacques, S.M., Santos, S.E.B., Weimer, T.A., Mestringer, M.A., Kubo, R.R., Pandey, J.P. & Hutz, M.H. (1991). Blood genetic systems in four Amazonian tribes. *American Journal of Physical Anthropology*, **85**, 51–60.

Salzano, F.M. & Callegari-Jacques, S.M. (1979). Genetic demography of the Central Pano and Kanamari Indians of Brazil. *Human Biology*, **51**, 551–64.

Salzano, F.M. & Callegari-Jacques, S.M. (1988). *South American Indians. A Case Study in Evolution*. Oxford: Clarendon Press.

Salzano, F.M., Callegari-Jacques, S.M., Franco, M.H.L.P., Hutz, M.H., Weimer, T.A., Silva, R.S. & Da Rocha, F.J. (1980a). The Caingang revisted: Blood genetics and anthropometry. *American Journal of Physical Anthropology*, **53**, 513–24.

Salzano, F.M., Callegari-Jacques, S.M. & Neel, J.V. (1980b). Genetic demography of the Amazonian Ticuna Indians. *Journal of Human Evolution*, **9**, 179–191.

Salzano, F.M., Gershowitz, H., Mohrenweiser, H., Neel, J.V., Smouse, P.E., Mestriner, M.A., Weimer, T.A., Franco, M.H.L.P., Simoes, A.L., Constans, J., Oliveira, A.E. & Freitas, M.J.M. (1986). Gene flow across the tribal barriers and its effect among the Amazonian Icana river Indians. *American Journal of Physical Anthropology*, **69**, 3–14.

Salzano, F.M., Moreno, R., Palatnik, M. & Gershowitz, H. (1970). Demography and H-Le salivary secretion of the Maca Indians of Paraguay. *American Journal of Physical Anthropology*, **33**, 383–8.

Salzano, F.M., Neel, J.V., Gershowitz, H. & Migliazza, E.C. (1977). Intra and intertribal genetic variation within a linguistic group: The Ge-speaking Indians of Brazil. *American Journal of Physical Anthropology*, 47, 337–48.

Salzano, F.M., Neel, J.V. & Maybury-Lewis, D. (1967a). Further studies on the Xavante Indians. I. Demographic data on two additional villages: Genetic structure of the tribe. *American Journal of Human Genetics*, 19, 463–89.

Salzano, F.M. & De Olivera, R.C. (1970). Genetic aspects of the demography of Brazilian Terena Indians. *Social Biology*, 17, 217–23.

Salzano, F.M., Steinberg, A.G. & Tepfenhart, M.A. (1967b). Gm and Inv allotypes of Brazilian Cayapo Indians. *American Journal of Human Genetics*, 25, 167–77.

Sambuughin, N., Petrishchev, V.N. & Rychkov, Y.G. (1991). DNA polymorphism in Mongolian population – Analysis of restriction endonuclease polymorphism of mitochondrial DNA. *Genetika*, 27, 2143–51.

Sanger, R. & Race, R.R. (1951). The MNSs blood group system. *American Journal of Human Genetics*, 3, 332–43.

Sanger, R., Race, R.R. & Jack, J. (1955). The Duffy blood groups of New York Negroes: The phenotype Fy(a-b-). *British Journal of Haematology*, 1, 370–4.

Sanghvi, L.D. (1953). Comparison of genetical and morphological methods of a study of biological distance. *American Journal of Physical Anthropology*, 11, 385–404.

Santos, F.R., Hutz, M., Coimbra, C.E.A., Santos, R.V., Salzano, F.M. & Peena, S.D.J. (1995). Further evidence for the existence of a major founder Y chromosome haplotype in Amerindicans. *Brazilian Journal of Genetics*, 18, 669–72.

Santos, F.R., Rodriguez-Delfin, L., Pena, S.D.J., Moore, J. & Weiss, K.M. (1996). North and South Amerindians may have the same major founder Y chromosome haplotype. *American Journal of Human Genetics*, 58, 1369–70.

Santos, S.E.B., Salzano, F.M. & Franco, M.H.L.P. (1983). Mobility, genetic markers, susceptibility to malaria and race mixture in Manaus, Brazil. *Journal of Human Evolution*, 12, 373–81.

Sapper, K. (1924). Die Zahl und die Volksdichte der Indianischen Bevolkerung der Gegebwart. In *Proceedings of the 21st International Congress of the Americanists*. pp. 95–104. Leiden: E.J. Brill.

Schanfield, M.S. (1976). Immunoglobulin haplotypes in Tlaxcaltecan and other populations. In *The Tlaxcaltecans: Prehistory, Demography, Morphology and Genetics*, Publications in Anthropology 7, ed. M.H. Crawford, pp. 150–4. Lawrence: University of Kansas Press.

Schanfield, M.S. (1980). The anthropological usefulness of highly polymorphic systems. HLA and Immunoglobulin allotypes. In *Current Developments in Anthropological Genetics. Vol. 1*. Theory and Methods, ed. J.H. Mielke and M.H. Crawford, pp. 65–85. New York: Plenum Press.

Schanfield, M.S. (1992). Immunoglobulin allotypes (GM and KM) indicate multiple founding populations of Native Americans: Evidence of at least four migrations to the New World. *Human Biology*, 64, 381–402.

Schanfield, M.S., Alexeyeva, T.E. & M.H. Crawford. (1980). Studies on the immunoglobulin allotypes of Asiatic populations. VIII. Immunoglobulin allotypes among the Touvinians of the U.S.S.R. *Human Heredity*, 30, 343–9.

Schanfield, M.S., Brown, R. & Crawford, M.H. (1984). Immunoglobulin allotypes in the Black Caribs and Creoles of Belize and St. Vincent. In *Current Developments in Anthropological Genetics. Vol. 3. Black Caribs. A Case Study in Biocultural Adaptation*, ed. M.H. Crawford, pp. 345–63. New York: Plenum Press.

Schanfield, M.S., Crawford, M.H., Dossetor, J.B. & Gershowitz, H. (1990). Immunoglobulin allotypes in several North American Eskimo populations. *Human Biology*, 62, 773–89.

Schanfield, M.S., Fudenberg, H.H., Crawford, M.H. & Turner, K.R. (1978). The distribution of immunoglobulin allotypes in two Tlaxcaltecan populations. *Annals of Human Biology*, 5, 577–90.

Scheidegger, J.J., Martin, E. & Riotton, G. (1956). L'apparition des diverses composantes antigeniques du serum au cours du developpement foetal. *Schwizerische Medizinische Wochenschrift*, 86, 224.

Schell, L.M. & Blumberg, B.S. (1977). The genetics of human serum albumin. In *Albumin Structure, Function and Uses*, ed. V. Rosenoer, M. Oratz & M.A. Rothschild, pp. 113–41. New York: Pergamon Press.

Scheurlen, P.G. (1955). Uber serumeiweissveranderungen beim Diabetes Mellitus. *Klinische Wochenschrift*, 33, 198–205.

Schraer, C.D., Lanier, A.P., Boyko, E.J., Gohndes, D. & Murphy, N.J. (1988). Prevalence of diabetes mellitus in Alaskan Eskimos,

Indians and Aleuts. *Diabetes Care*, 11, 693–700.

Schrire, C. & Stiger, W.L. (1974). A matter of life and death: An investigation into the practice of female infanticide in the Arctic. *Man*, 9, 161–84.

Schull, W.J. & MacCluer, J.W. (1968). Human genetics: Structure of population. *Annual Review of Genetics*, 2, 279–304.

Schurr, T.G., Ballinger, S.W., Gan, Y.-Y., Hodge, J.A., Merriwether, D.A., Lawrence, D.N., Knowler, W.C., Weiss, K.M. & Wallace, D.C. (1990). Amerindian mitochondrial DNAs have rare Asian mutations at high frequencies indicating they derived from four primary maternal lineages. *American Journal of Human Genetics*, 46, 613–23.

Scotch, N.A. (1963). Sociocultural factors in the epidemiology of Zulu hypertension. *American Journal of Public Health*, 53, 1205–13.

Scott, E.C. (1979). Increase of tooth size in prehistoric coastal Peru, 10,000 B.P. – 1,000 B.P. *American Journal of Physical Anthropology*, 50, 251–8.

Scott, E.M. (1973). Inheritance of two types of deficiencey of human serum cholinesterase. *Annals of Human Genetics*, 37, 139–43.

Scott, E.M. & Wright, R.C. (1978). Polymorphism of red cell enzymes in Alaskan ethnic groups. *Annals of Human Genetics*, 41, 341–6.

Scott, E.M. & Wright, R.C. (1983). Genetic diversity of Central Yupik Eskimos. *Human Biology*, 55, 409–415.

Scott, G.R., Yap Potter, R.H., Noss, J.F., Dahlberg, A.A. & Dahlberg, T. (1983). The dental morphology of Pima Indians. *American Journal of Physical Anthropology*, 61, 13–31.

Sever, P.S., Gordon, D., Peart, W.S. & Beighton, P. (1980). Blood-pressure and its correlates in urban and tribal Africa. *Lancet*, i, 60–4.

Shapiro, H.L. (1939). *Migration and Environment*. Oxford: Oxford University Press.

Sharlin, K., Heath, G.W., Ford, E.S. & Welty, T.K. (1994). Hypertension and blood pressure awareness among American Indians of the Northern Plains. *Ethnicity and Disease*, 3, 337–43.

Shephard, R.J. & Rode, A. (1996). *The Health Consequences of 'Modernization': Evidence from Circumpolar Peoples*. Cambridge: Cambridge University Press.

Short, G.B. (1972). *Mating Propinquity, Inbreeding and Biological Fitness as Measured by Differential Fertility and Offspring Vitality from Papago Breeding Unions*. Boulder, Colorado: University of Colorado. (Unpublished Ph.D. Dissertation.)

Shows, T.B., McAlpine, P.J., Boucheix, C., Collins, F.S. et al. (1987). Guidelines for human gene nomenclature. An international system for human gene nomenclature (ISGN, 1987). *Cytogenetics and Cellular Genetics*, 46, 11–28.

Shutler, R. (1983). *Early Man in the New World*. Beverly Hills, California: Sage Publications.

da Silva, E.M. (1949). Blood groups of Indians, Whites and White-Indian mixtures in southern Mato Grosso, Brazil. *American Journal of Physical Anthropology*, 7, 575–86.

Simoes, A.L., Kompf, J., Ritter, H., Luckenbach, C., Zischler, Z. & Salzano, F.M. (1989). Electrophoretic and isoelectric focusing studies in Brazilian Indians: Data on four systems. *Human Biology*, 61, 427–38.

Simpson, N.E. (1968). Genetics of esterase in man. *Annals of the New York Academy of Sciences*, 151, 699–709.

Smith, D.G. (1980). Fertility differentials within a subdivided population: A controlled comparison of Four Sells Papago isolates. *Human Biology*, 52, 325–42.

Smithies, O. (1955). Zone electrophoresis in starch gels: Group variations in the serum proteins of normal human adults. *Biochemistry Journal*, 61, 629.

Smithies, O. (1957). Variations in human serum beta-globulins. *Nature*, 180, 1482.

Smithies, O. & Walker, N.F. (1956). Notation for serum protein groups and the genes controlling their inheritance. *Nature*, 178, 694.

Smouse, P.E. (1982). Genetic architecture of swidden agricultural tribes from the lowland rain forests of South America. In *Current Developments in Anthropological Genetics*. Vol. 2. *Ecology and Population Structure*, ed. M.H. Crawford and J.H. Mielke, pp. 139–204. New York: Plenum Press.

Snow, C.E. (1948). Indian Knoll Skeletons of Site Oh 2. Ohio County, Kentucky. *University of Kentucky Reports in Anthropology*, 4, No. 3.

Snyder, R.G., Dahlberg, A.A., Snow, C.C. & Dahlbeg, T. (1969). Trait analysis of the dentition of the Tarahumara Indians and Mestizos of the Sierra Madre Occidental, Mexico. *American Journal of Physical Anthropology*, 31, 65–76.

Soafer, J., Niswander, J., MacLean, C. & Workman, P.L. (1972). Population studies on Southwestern Indian tribes. V. Tooth morphology as an indicator of biological distance. *American Journal of Physical Anthropology*, 37, 357–66.

Sokal, R.R. (1988). Genetic, geographic and linguistic distances in Europe. *Proceedings of the National Academy of Sciences, USA*, 85, 1722–6.

Sokal, R.R. & Friedlaender, J. (1982). Spatial autocorrelation analysis of biological variation on Bougainville Island. *Current Developments in Anthropological Genetics*. Vol. 2. *Ecology and Population Structure*, ed. M.H. Crawford & J.H. Mielke, pp. 205–27. New York: Plenum Press.

Sokal, R.R. & Oden, N.L. (1978). Spatial autocorrelation in biology. 1. Methodology. *Biology Journal of the Linnaean Society*, 10, 199–228.

Sokal, R.R., Smouse, P.E. & Neel, J.V. (1986). The genetic structure of a tribal population, the Yanomama Indians. XV. Patterns inferred by autocorrelation analysis. *Genetics*, 114, 259–87.

Soodyall, H., Vigilant, L., Hill, A.V., Stoneking, M. & Jenkins, T. (1996). mtDNA control-region sequence variation suggests multiple independent origins of an "Asian-specific" 9-bp deletion in Sub-Saharan Africans. *American Journal of Human Genetics*, 58, 595–608.

Spencer, N., Hopkinson, D.A. & Harris, H. (1964). Phosphoglucomutase polymorphism in man. *Nature*, 204, 742–5.

Spielman, R.S. (1973a). Differences among Yanomana Indian villages: Do patterns of allele frequencies, anthropometrics and map locations correspond? *American Journal of Physical Anthropology*, 39, 461–80.

Spielman, R.S. (1973b). Do the natives all look alike? Size and shape components of anthropometric differences among Yanomana Indian villages. *American Naturalist*, 107, 694–708.

Spielman, R.S., Migliazza, E.C. & Neel, J.V. (1974). Regional linguistic and genetic differences among Yanomama Indians. *Science*, 184, 634–7.

Spielman, R.S. & Smouse, P.E. (1976). Multivariate classification of human populations. I. Allocation of Yanomama Indians to villages. *American Journal of Human Genetics*, 28, 317–31.

Spinden, H.J. (1928). The population of ancient America. *Geographic Review*, 18, 641–60.

Spitsyn, V.A. (1985). *Human Biochemical Polymorphism. Anthropological Aspects*. Moscow: Moscow University. (In Russian.)

Spitsyn, V.A. (1972). Transferrins: A study of serum factors in Siberia. *Voprosi Antropologii*, 41, 36–47.

Spuhler, J.N. (1962). Empirical studies on quantitative human genetics. In *The Use of Vital Statistics for Genetic and Radiation Studies*. New York: United Nations – World Health Organization.

Spuhler, J.N. (1963). The scope for natural selection in man. In *Selection in Man*, ed. W.J. Schull, pp. 1–99. Ann Arbor: University of Michigan Press.

Spuhler, J.N. (1972). Genetic, linguistic and geographical distances in Native North America. In *The Assessment of Population Affinities in Man*, ed. J.S. Weimer & J. Huizinga, pp. 72–95. Oxford: Clarendon Press.

Spuhler, J.N. (1979). Genetic distance, trees, and maps of North American Indians. In *The First Americans: Origins, Affinities, and Adaptations*, ed. W.S. Laughlin & A.B. Harper, pp. 135–83. New York: Gustav Fischer.

Spuhler, J.N. (1989). Update to Spuhler and Kluckhohn's 'Inbreeding coefficients of the Ramah Navajo population'. *Human Biology*, 61, 726–30.

Spuhler, J.N. & Kluckhohn, C. (1953). The inbreeding coefficients of the Ramah Navajo population. *Human Biology*, 25, 295–317.

St. Hoyme, L.E. (1969). On the origins of New World paleopathology. *American Journal of Physical Anthropology*, 31, 295–302.

Starr, F. (1902). The physical characters of the Indians of southern Mexico. *University of Chicago Decennial Publications*, 4, 53–109.

Steggerda, M. (1950). Anthropometry of South American Indians. In *Handbook of South American Indians*. Vol. 6. *Physical Anthropology, Linguistics, and Cultural Geography of South American Indians*, Bureau of American Ethnology Bulletins 143, ed. J.H. Steward, pp. 57–69. Washington: Smithsonian Institution.

Steinberg, A.G., Cordova, M.S. & Lisker, R. (1967). Studies on several hematologic traits of Mexicans. XV. The Gm allotypes of some Indian tribes. *American Journal of Human Genetics*, 19, 747–56.

Steinberg, A.G., Stauffer, R., Blumberg, B.S. & Fudenberg, H. (1961). Gm phenotypes and genotypes in U.S. Whites and Negroes; in

American Indians and Eskimos; in Africans; and in Micronesians. *American Journal of Human Genetics*, 13, 205–13.

Steinberg, A.G., Tillikainen, A., Eskola, M.-R. & Eriksson, A.W. (1974). Gamma globulin allotypes in Finnish Lapps, Finns, Aland Islanders, Maris (Cheremis), and Greenland Eskimos. *American Journal of Human Genetics*, 26, 223–43.

Stepanova, E.G. & Shubnikov, E.W. (1991). Diabetes, glucose tolerance and some risk factors of diabetes mellitus in natives and newcomers of Chukotka. In *Circumpolar Health 90*, ed. B.D. Postl et al., pp. 413–4. Winnipeg: Canadian Society for Circumpolar Health.

Stern, C. (1953). Model estimates of the frequency of white and near-white segregants in the American Negroe. *Acta Genetica (Basel)*, 4, 281–98.

Steward, J.H. & Faron, L.C. (1959). *Native Peoples of South America*. New York: McGraw-Hill.

Stewart, T.D. (1960). A physical anthropologist's view of the peopling of the New World. *Southwest Journal of Anthropology*, 16, 259–73.

Stewart, T.D. (1973). *The People of America*, Scribners, New York, 1973.

Stewart, T.D. & Newman, M.T. (1951). An historical resume of the concept of differences in Indian types. *American Anthropologist*, 53, 19–36.

Stinson, S. (1982). The interrelationship of mortality and fertility in rural Bolivia. *Human Biology*, 54, 299–313.

Stinson, S. (1990). Variation in body size and shape among South American Indians. *American Journal of Human Biology*, 2, 37–51.

Storey, R. (1986). Perinatal mortality at pre-Columbian Teotihuacan. *American Journal of Physical Anthropology*, 69, 541–8.

Strotz, C.R. & Shorr, G.I. (1973). Hypertension in the Papago Indians. *Circulation*, 48, 1299–1302.

Sturtevant, W.C. (1978–1986). *Handbook of North American Indians*. Twenty volume series (seven volumes published). Washington, D.C.: Smithsonian Institution.

Suarez, B.K., Crouse, J.D. & O'Rourke, D.H. (1985). Genetic variation in North Amerindian populations: The geography of gene frequencies. *American Journal of Physical Anthropology*, 67, 217–32.

Sukernik, R.I., Karaphet, T.M., and Osipova, L.P. (1978). Distribution of blood groups, serum markers and red cell enzymes in two human populations from northern Siberia. *Human Heredity*, 28, 321–7.

Sukernik, R.I., Lemza, S.V., Karaphet, T.M. & Osipova, L.P. (1981). Reindeer Chukchi and Siberian Eskimos: Studies on blood groups, serum proteins, and red cell enzymes with regard to genetic heterogeneity. *American Journal of Physical Anthropology*, 55, 121–9.

Sukernik, R.I. & Osipova, L.P. (1982). Gm and Km immunoglobulin allotypes in Reindeer Chukchi and Siberian Eskimos. *Human Genetics*, 61, 148–53.

Sukernik, R.I., Osipova, L.P., Karaphet, T.M. & Abanina, T.A. (1980). Studies on blood groups and other genetic markers in Forest Nentzi: Variation among subpopulations. *Human Genetics*, 55, 397–404.

Sukernik, R.I., Osipova, L.P., Karaphet, T.M., Vibe, V. & Kirpichnikov, G.A. (1986a). Genetic and ecological studies of aboriginal inhabitants of northeastern Siberia. I. Gm haplotype and their frequencies in ten Chukchi populations. Genetic structure of reindeer Chukchi. *Genetika*, 22, 2361–8.

Sukernik, R.I., Vibe, V., Karaphet, T.M., Osipova, L.P. & Kirpichnikov, G.A. (1986b). Genetic and ecological studies of aboriginal inhabitants of North-Eastern Siberia. II. Polymorphic blood systems in Asiatic Eskimos. *Genetika*, 22, 2369–80.

Swanton, J.R. (1952). *The Indian Tribes of North America*. Smithsonian Institution, Bureau of American Ethnology Bulletin No. 145. Washington, D.C.: G.P.O.

Szathmary, E.J.E. (1981). Genetic markers in Siberian and North American Populations. *Yearbook of Physical Anthropology*, 24, 37–73.

Szathmary, E.J.E. (1983). Dogreb Indian of the Northwest Territories, Canada: Genetic diversity and genetic relationship among sub-arctic Indians. *Annals of Human Biology*, 10, 147–62.

Szathmary, E.J.E. (1985). The search for genetic factors controlling plasma glucose levels in Dogreb Indians. In *Diseases of Complex Etiology in Small Populations*, ed. R. Chakraborty & E. Szathmary, pp. 199–26. New York: Liss Inc.

Szathmary, E.J.E. (1986a). Comments on an article, The Settlement of the Americas. *Current Anthropology*, 27, 490-1.

Szathmary, E.J.E. (1986b). Diabetes in Arctic and Subarctic populations undergoing acculturation. *Collegium Antropologicum*, 10, 145-58.

Szathmary, E.J.E. (1987). Genetic and environmental risk factors. In *Diabetes in the Canadian Native Population: Biocultural Perspectives*, ed. T.K. Young, pp. 27-66. Toronto: Canadian Diabetes Association.

Szathmary, E.J.E. (1990). Diabetes in Amerindian populations: The Dogreb studies. In *Health and Diseases of Populations in Transition*, ed. G. Armelagos & A. Swedlund, pp. 75-103. New York: Bergin and Garvey.

Szathmary, E.J.E. (1993). Genetics of aboriginal North Americans. *Evolutionary Anthropology*, 1, 202-20.

Szathmary, E.J.E. (1994). Non-insulin dependent diabetes among aboriginal North Americans. In *Annual Review of Anthropology*, pp. 457-82. Palo Alto: Stanford University Press, Palo Alto.

Szathmary, E.J.E. & Auger, F. (1983). Biological distances and genetic relationships within Algonkians. In *Boreal Forest Adaptations*, ed. A.T. Steegman, pp. 289-315. New York: Plenum Press.

Szathmary, E.J.E., Cox, D.W., Gershowitz, H., Ruchnagel, D.L. & Schanfield, M.S. (1974). The Northern and Southeastern Ojibwa: Serum proteins and red cell enzyme systems. *American Journal of Physical Anthropology*, 40, 49-66.

Szathmary, E.J.E., Ferrell, R.E. & Gershowitz, H. (1983). Genetic differentiation in Dogrib Indians: Serum protein and erythrocyte enzyme variation. *American Journal of Physical Anthropology*, 62, 249-54.

Szathmary, E.J.E., Mohn, J.F., Gershowitz, H., Lambert, R.M. & Reed, T.E. (1975). The Northern and Southeastern Ojibwa: Blood group systems and the cause of genetic divergence. *Human Biology*, 47, 351-68.

Szathmary, E.J.E. & Ossenberg, N.S. (1978). Are the biological differences between North American Indians and Eskimos truly profound? *Current Anthropology*, 19, 673-701.

Szathmary, E.J.E. & Reed, T.E. (1972). Caucasian admixture in two Ojibwa Indian communities in Ontario. *Human Biology*, 44, 655-71.

Szathmary, E.J.E. & Reed, T.E. (1978). Calculation of the maximum amount of genetic admixture in a hybrid population. *American Journal of Physical Anthropology*, 48, 29-34.

Thieme, F.P. & Otten, C.M. (1957). The unreliability of blood typing aged bone. *American Journal of Physical Anthropology*, 15, 387-97.

Thomasson, H.R., Zeng, D., Ching, C. & Crawford, M.H. (1997). Alcohol and aldehyde dehydrogenase genotypes among ethnic groups of Russia, China and North America. (Unpublished manuscript).

Thomasson, H.R., Edenberg, H.J., Crabb, D.W., Mai, X.-L., Jerome, R.E., Li, T.-K., Wang, S.-P., Lin, Y.-T., Lu, R.-B. & Lin, S.-J. (1991). Alcohol and aldehyde dehydrogenase genotypes and alcoholism in Chinese men. *American Journal of Human Genetics*, 48, 677-81.

Thompson, P.R., Childers, D.M. & Hatcher, D.E. (1967). Anti-Di – first and second examples. *Vox Sanguinus*, 13, 314-8.

Thornton, R. & Marsh-Thornton, J. (1981). Estimating prehistoric American Indian population size for United States area: Implications of the nineteenth century population decline and nadir. *American Journal of Physical Anthropology*, 55, 47-53.

Tobler, W.R. (1970). A computer movie simulating urban growth in the Detroit region. *Economic Geography*, 46, 234.

Torroni, A., Schurr, T.G., Cabell, M.F., Brown, M.D., Neel, J.V., Larsen, M., Smith, D.G. et al. (1993a). Asian affinities and continental radiation of the four founding Native American mtDNAs. *American Journal of Human Genetics*, 53, 563-90.

Torroni, A., Sukernik, R.I., Schurr, T.G., Starikovskaya, Y.B., Cabell, M.F., Crawford, M.H., Comuzzie, A.G. & Wallace, D.C. (1993b). Mitochondrial DNA variation of aboriginal Siberians reveal distinct genetic affinities with native Americans. *American Journal of Human Genetics*, 53, 591-608.

Torroni, A. & Wallace, D.C. (1995). mtDNA haplotypes in Native Americans. *American Journal of Human Genetics*, 56, 1234-6.

Turner, C.G. (1971). Three-rooted mandibular first permanent molars and the question of American Indian origins. *American Journal of Physical Anthropology*, 34, 229-42.

Turner, C.G. (1985). The dental search for Native American origins. In *Out of Asia. Peopling the Americas and the Pacific*, pp. 31-78. Canberra: Australian National University.

Turner, C.G. (1987). Telltale teeth. *Natural History*, 96, 6-10.

Turner, C.G. (1990). The major features of Sundadonty and Sinodonty, including suggestions about East Asian microevolution, population history, and late Pleistocene relationships with Australian aboriginals. *American Journal of Physical Anthropology*, 82, 295–317.

Turner, J.H., Crawford, M.H. & Leyshon, W.C. (1975). Phenotypic karyotypic localization of the human Rh-locus on chromosome 1. *Journal of Heredity*, 66, 97–9.

Turner, K.R. (1976). Computer simulation of transplanted Tlaxcaltecan populations. In *The Tlaxcaltecans. Prehistory, Demography, Morphology and Genetics*, Publications in Anthropology 7, ed. M.H. Crawford, pp. 48–60. Lawrence: University of Kansas Press.

Ubelaker, D.H. (1974). *Reconstruction of Demographic Profiles from Ossuary Skeletal Samples: A Case Study from the Tidewater Potomac, Smithsonian Contributions to Anthropology*, No. 18. Washington, D.C.: The Smithsonian.

Ubelaker, D.H. (1976a). The sources and methodology for Mooney's estimates of North American Indian populations. In *The Native Population of the Americas in 1492*, ed. W.M. Denevan, pp. 243–88. Madison: Univeristy of Wisconsin Press.

Ubelaker, D.H. (1976b). Prehistoric New World population size: Historical review and current appraisal of North American estimates. *American Journal of Physical Anthropology*, 45, 661–6.

Ubelaker, D.H. (1988). North American Indian population size, A.D. 1500 to 1985. *American Journal of Physical Anthropology*, 77, 289–94.

Ubelaker, D.H. & Jantz, R.L. (1986). Biological history of the aboriginal populations of North America. In *Rassengeschichte der Menschheit. Amerika I: Nordamerika*, ed. I. Schwidetzky, pp. 7–79. Munich: Oldenbourg Verlag.

Underhill, R. (1957). Religion among American Indians. *Annals of the American Academy of Political and Social Sciences*, 311, 127–36.

Underhill, P., Jin, L., Zemans, R., Oefner, P.J. & Cavalli-Sforza, L.L. (1996). A pre-Columbian Y chromosome-specific transition and its implications for human evolutionary history. *Proceedings of the National Academy of Sciences, USA*, 93, 196–200.

U.S. Department of Health and Human Services. (1994). *National Heart, Lung, and Blood Institute Report of the Task Force on Research in Epidemiology and the Prevention of Cardiovas-*

cular Diseases. Bethesda: Public Health Services, National Institutes of Health.

Vague, J. (1956). The degree of masculine differentiation of obesities: A factor determining predisposition to diabetes, atherosclerosis, gout and uric calculous disease. *American Journal of Clinical Nutrition*, 4, 20–34.

Van Loghem, E. (1971). Stability of Gm polymorphism. In *Human, Antihuman Gammaglobulins*, ed. R. Grubb and G. Samuelson, pp. 29–37. Oxford: Pergamon Press.

Van Stone, J.W. (1958). Commercial whaling in the Arctic Ocean. *Pacific Northwest Quarterly*, 49, 1–10.

Vasilevich, G.M. (1946). Drevneyshiye etnonimy Azii nazvaniya evenkiyskikh rodov. *Sovetskaya Etnografiya*, 4, Moscow.

Von Endt, D.W. & Ortner, D.J. (1982). Amino acid analysis of bone from a possible case of prehistoric iron deficiency anemia from the American Southwest. *American Journal of Physical Anthropology*, 59, 377–85.

Wahlund, S. (1928). Zusammensetzung von Populationen und Korrelationsersheinungen vom Standpunkt den Vererbungslehre ausbetrachtet. *Hereditas*, 11, 65–106.

Wainscoat, J.S., Hill, A.V.S., Boyce, A.L., Flint, J., Hernandez, M., Thein, S.L., Old, J.M., Lynch, J.R., Falusi, A.G., Weatherall, D. & Clegg, J.B. (1986). Evolutionary relationships of human populations from an analysis of nuclear DNA polymorphisms. *Nature*, 319, 491–3.

Wallace, D.C., Garrison, K. & Knowler, W.C. (1985). Dramatic founder effects in Amerindian mitochondrial DNAs. *American Journal of Physical Anthropology*, 68, 149–55.

Wallace, D.C., Ye, J., Neckelmann, S.N., Singh, G., Webster, K.A. & Greenberg, B.D. (1987). Sequence analysis of cDNAs for the human and bovine ATP synthase beta subunit: Mitochondrial genes sustain seventeen times more mutations. *Current Genetics*, 12, 81–90.

Walsh, R.J. & Montgomery, C. (1947). A new human isoagglutinin subdividing the MN blood groups. *Nature*, 160, 504.

Ward, R.H. (1973). Some aspects of genetic structure in the Yanomama and Wakiritare: Two tribes of Southern Venezuela. In *Methods and Theories of Anthropological Genetics*, ed. M.H. Crawford & P.L. Workman, pp. 367–88. Albuquerque: University of New Mexico Press.

Ward, R.H., Chin, P.G. & Prior, I.A.M. (1979). Genetic epidemiology of blood pressure in a migrating isolate: Prospectus. In *Genetic Analysis of Common Diseases. Predictive Factors in Coronary Heart Disease*, ed. C.F. Sing & M. Skolnick, pp. 695–709. New York: A.R. Liss, Inc.

Ward, R.H., Frazier, B.S., Dew, K. & Pääbo, S. (1991). Extensive mitochondrial diversity within a single Amerindian tribe. *Proceedings of the National Academy of Sciences, USA*, **88**, 8720–9.

Weiner, J.S., Sebag-Montefiore, N.C. & Peterson, J.N. (1963). A note on the skin color of Aguarana Indians of Peru. *Human Biology*, **35**, 470–3.

Weiss, K.M. (1973). Demographic models for anthropology. *Society for American Archaeology, Memoir 27*.

Weiss, K.M. (1985). Phenotype amplification, as illustrated by cancer of the gallbladder in New World peoples. In *Disease of Complex Etiology in Small Populations: Ethnic Differences and Research Approaches*, ed. R. Chakraborty & E. Szathmary, pp. 179–98. New York: Alan Liss Press.

Weiss, K.M., Ferrell, R.E. & Hanis, C.L. (1984). New World Syndrome of metabolic diseases with a genetic and evolutionary basis. *Yearbook of Physical Anthropology*, **27**, 153–78.

Weitkamp, L.R., Arends, T., Gallango, M.L., Neel, J.V., Schultz, J. & Schreffler, D.C. (1972). The genetic structure of a tribal population, the Yanomama Indians. III. Seven serum protein systems. *Annals of Human Genetics*, **35**, 271–9.

Welch, S. (1974). Red cell esterase D polymorphism in Gambia. *Humangenetik*, **21**, 365–7.

Whitmore, T.M. (1992). *Disease and Death in Early Colonial Mexico: Simulating Amerindian Depopulation*. Boulder: Westview Press.

Wieme, R.J. (1959). Albumindoppelzacken als verebbare Blutenweissanomalie. *Schweizerische Medizinische Wochenschrift*, **89**, 150–2.

Wieme, R.J. (1962). On the presence of two albumins in certain normal human sera and its genetic determination. In *Protides of the Biological Fluids. Proceedings of the Ninth Colloquium, Bruges, 1961*, ed. H. Peeters, pp. 221–74. New York: Elsevier.

Williams, R.C, Steinberg, A.G., Gershowitz, H., Bennett, P.H., Knowler, W.C., Pettitt, D.J., Butler, W.J., Baird, R., Dowda-Rea, L., Burch, T.A., Morse, H.G. & Smith, D.G. (1985). GM allotypes in native Americans: Evidence for three distinct migrations across the Bering land bridge. *American Journal of Physical Anthropology*, **66**, 1–19.

Williams-Blangero, S., Blangero, J. & Towne, B. (eds) (1990). Special Issue on Quantitative Traits and Population Structure. *Human Biology*, **62**, 1–162.

Wolff, P.H. (1973). Vasomotor sensitivity to alcohol in diverse Mongoloid populations. *American Journal of Human Genetics*, **25**, 193–9.

Womble, W.H. (1951). Differential systematics. *Science*, **114**, 315–22.

Woolf, C.M. & Dukepoo, F.C. (1969). Hopi Indians, inbreeding, and albinism. *Science*, **164**, 30–7.

World Health Organization Expert Committee on Diabetes Mellitus. (1980). *Second Report. Technical Report Series 646*, Geneva: WHO.

Workman, P.L., Harpending, H.C., Lalouel, J.M., Lynch. C., Niswander, J.D. & Singleton, R. (1973). Population studies on Southwestern Indian tribes. VI. Papago population structure: A comparison of genetic and migration analyses. In *Genetic Structure of Populations*, ed. N.E. Morton, pp. 166–94. Honolulu: University of Hawaii Press.

Workman, P.L. & Jorde, L.B. (1980). The genetic structure of the Aland Islands. In *Population Structure and Genetic Disease*, ed. A.W. Eriksson, H. Forsius, H.R. Nevanlinna & P.L. Workman, pp. 487–508. New York: Academic Press.

Workman, P.L. & Niswander, J.D. (1970). Population studies on Southwestern Indian tribes. II. Local genetic differentiation in the Papago. *American Journal of Human Genetics*, **22**, 24–49.

Workman, P.L., Niswander, J.D., Brown, K.S. & Leyshon, W.C. (1974). Population studies on southwestern Indian tribes. IV. The Zuni. *American Journal of Physical Anthropology*, **41**, 119–32.

World Health Organization Expert Committee on Diabetes Mellitus. (1980). *Second Report. Technical Report Series 646*, Geneva: WHO.

Wright, S. (1921). Systems of mating. *Genetics*, **6**, 111–78.

Wright, S. (1943). Isolation by distance. *Genetics*, **28**, 114–38.

Wright, S. (1951). The genetical structure of populations. *Annals of Eugenics*, **15**, 323–54.

Wright, S. (1965). The interpretation of population structure by F-statistics with special regard to systems of mating. *Evolution*, 19, 395–420.

Wright, S. (1978). *Evolution and the Genetics of Populations*. Vol. 4. Variability Within and Among Natural Populations. Chicago: University of Chicago Press.

Yarho, A.I. (1947). *Turkic Tribes of the Altay-Sayan Region: A Physical Anthropological Sketch*. Abakan. (In Russian.)

Young, T. (1842). *Narrative of a Residence on the Mosquito Shore*. London: Smith, Elder, and Co.

Young, T.K. (1993). Diabetes mellitus among native Americans in Canada and the United States: An epidemiological review. *American Journal of Human Biology*, 5, 399–413.

Zavala, C., Alatorre, S. & Lisker, R. (1982). Kinship coefficients and genetic distance among Mexican Indian tribes. *Acta Anthropogenetica*, 6, 265–74.

Zavala, C., Cobo, A. & Lisker, R. (1971). Dermatoglyphic patterns in Mexican Indian groups. *Human Heredity*, 21, 394–401.

Zegura, S. (1985). The initial peopling of the Americas: An overview. *Out of Asia. Peopling the Americas and the Pacific*, ed. R. Kirk & E. Szathmary, pp. 1–18. Canberra: Australian National University.

Zhang, Y., Proenca, R., Maffei, M., Barone, M., Leopold, M. & Friedman, J.M. (1994). Positional cloning of the mouse obese gene and its human homologue. *Nature*, 372, 425–32.

Author index

Page numbers in *italics* refer to figures. Page numbers in **bold** refer to tables.

Subject index

Page numbers in *italics* refer to figures. Page numbers in **bold** refer to tables.

blood-group systems (*cont.*)
  Lutheran 107
  Miltenberger 107
  MNSs 101–3
  P 107
  Penny 107, 108
  Rhesus (RH) 98–101
    geographical distribution *102*
  Scianna 107
  South American Indians and 31
  Stoltzfus 107
  Sutter 107, 108
  Vel 107
  XG 107, 108
Book of Chilam Balam 53

cardiovascular system
  blood pressure 251–5, *253, 254*
  coronary heart disease 247
  hypertension 253
  lipids and 248–51
Central America
  analyses of anthropometric data 199–202
  census 72
  dentition studies 226–8
  dermatoglyphic studies 214–19, *215, 217, 218, 219*
  population size at Contact 36–8, *37*
cold screen theory 51–2
Contact
  see also epidemics; religion; warfare
  depopulation, regional 41–9
  description of 261
  population sizes at **34, 36, 37, 39, 40**
Cordilleran Ice Sheet 12–13, *13*
craniometry 14, 30, 31, 195, 196

De Genetica Humani Varietate Nativa 3
demography
  age and sex distributions in skeletal populations 64–5, **65**
  causes of death 82–3
  census sizes 70–3
  definition 63
  dynamics of colonizing populations 15–16
  estimating problems 63–4
  fertility studies
    Black Caribs 73
    Tlaxcala 75–7, *76*
    urban/rural 77
    Yanomama Indians 74–5
  index of opportunity for selection 80–1, *81*

life-expectancy evidence 53, **68**, *78, 79*, 82
migrations (contemporary) 83–6, **85**
migrations evidence 69–70
mortality (contemporary) 78–83, **84**
mortality evidence 67–9
population density calculations 66–7
dendrograms
  description 166–7
  examples *5, 168, 169*
dentition
  see also anthropometry; dermatoglyphics; skin color
  age determination 64
  evidence for Asian origins 7, 220, 222–3
  geographic variation 220–9
dermatoglyphics
  description 205, *206, 207*
  'field theory' in 206–8
  population comparisons 208–20
  traits within populations 205–8, *206, 207*
diseases
  see also epidemics
  causes of death (contemporary) 82–3, **84**
  cold screen theory 51–2
  contemporary
    alcoholism 255–9
    coronary heart disease 247–51
    diabetes mellitus 243–5
    gall-bladder disease 245–6
    hypertension 251–5
    New World syndrome 243
    prevalences 241, *242*
  New World 53–4
    American mucocutaneous leishmaniasis 57
    arthritis 59–60
    Carrion's disease 57
    Chagas' disease 57
    food poisoning 58–9
    genetic syndromes 60–1
    hookworm infection 58
    human fluke infection 57–8
    nutritional deficiency diseases 60
    otitis media 6–7
    roundworm infection 58
    syphilis 54–5
    tuberculosis 49–50, 241
    viral diseases 59
  Old World
    impact 50–3
    imported 61
  population densities and 52
  zoonoses 52

Printed in the United States
By Bookmasters